£19-95

The Molecular
Fabric of Cells

BOOKS IN THE BIOTOL SERIES

BIOTECHNOLOGY BY OPEN LEARNING

The Molecular
Fabric of Cells

PUBLISHED ON BEHALF OF :

Open universiteit   and   **Thames Polytechnic**

Valkenburgerweg 167   Avery Hill Road
6401 DL Heerlen   Eltham, London SE9 2HB
Nederland   United Kingdom

Butterworth–Heinemann Ltd
Linacre House, Jordan Hill, Oxford OX2 8DP

 PART OF REED INTERNATIONAL BOOKS

OXFORD LONDON BOSTON
MUNICH NEW DELHI SINGAPORE SYDNEY
TOKYO TORONTO WELLINGTON

First published 1991

British Library Cataloguing in Publication Data
A catalogue record for this book is available from the British
Library

**Library of Congress Cataloguing in Publication
Data**
A catalogue record for this book is available from the Library
of Congress

ISBN 0 7506 1499 4

Composition by Thames Polytechnic
Printed and bound in Great Britain by
Thomson Litho, East Kilbride, Scotland

life saine
cells
biology
amino acids
lipids
enzymes
Carbohydrates
proteins

The Biotol Project

The BIOTOL team

**OPEN UNIVERSITEIT,
THE NETHERLANDS**
Dr M. C. E. van Dam-Mieras
Professor W. H. de Jeu
Professor J. de Vries

**THAMES POLYTECHNIC,
UK**
Professor B. R. Currell
Dr J. W. James
Dr C. K. Leach
Mr R. A. Patmore

This series of books has been developed through a collaboration between the Open universiteit of the Netherlands and Thames Polytechnic to provide a whole library of advanced level flexible learning materials including books, computer and video programmes. The series will be of particular value to those working in the chemical, pharmaceutical, health care, food and drinks, agriculture, and environmental, manufacturing and service industries. These industries will be increasingly faced with training problems as the use of biologically based techniques replaces or enhances chemical ones or indeed allows the development of products previously impossible.

The BIOTOL books may be studied privately, but specifically they provide a cost-effective major resource for in-house company training and are the basis for a wider range of courses (open, distance or traditional) from universities which, with practical and tutorial support, lead to recognised qualifications. There is a developing network of institutions throughout Europe to offer tutorial and practical support and courses based on BIOTOL both for those newly entering the field of biotechnology and for graduates looking for more advanced training. BIOTOL is for any one wishing to know about and use the principles and techniques of modern biotechnology whether they are technicians needing further education, new graduates wishing to extend their knowledge, mature staff faced with changing work or a new career, managers unfamiliar with the new technology or those returning to work after a career break.

Our learning texts, written in an informal and friendly style, embody the best characteristics of both open and distance learning to provide a flexible resource for individuals, training organisations, polytechnics and universities, and professional bodies. The content of each book has been carefully worked out between teachers and industry to lead students through a programme of work so that they may achieve clearly stated learning objectives. There are activities and exercises throughout the books, and self assessment questions that allow students to check their own progress and receive any necessary remedial help.

The books, within the series, are modular allowing students to select their own entry point depending on their knowledge and previous experience. These texts therefore remove the necessity for students to attend institution based lectures at specific times and places, bringing a new freedom to study their chosen subject at the time they need and a pace and place to suit them. This same freedom is highly beneficial to industry since staff can receive training without spending significant periods away from the workplace attending lectures and courses, and without altering work patterns.

Contributors

AUTHORS

Dr R.D.J. Barker, Leicester Polytechnic, Leicester, UK

Dr C. K. Leach, Leicester Polytechnic, Leicester, UK

EDITOR

Dr C. K. Leach, Leicester Polytechnic, Leicester, UK

SCIENTIFIC AND COURSE ADVISORS

Dr M. C. E. van Dam-Mieras, Open universiteit, Heerlen, The Netherlands

Dr C. K. Leach, Leicester Polytechnic, Leicester, UK

ACKNOWLEDGEMENTS

Grateful thanks are extended, not only to the authors, editors and course advisors, but to all those who have contributed to the development and production of this book. They include Mrs A. Allwright, Miss J. Skelton and Professor R. Spier. The development of BIOTOL has been funded by COMETT, The European Community Action programme for Education and Training for Technology, by the Open universiteit of The Netherlands and by Thames Polytechnic.

Project Manager: Dr J.W. James

Contents

How to use an open learning text

An open learning text presents to you a very carefully thought out programme of study to achieve stated learning objectives, just as a lecturer does. Rather than just listening to a lecture once, and trying to make notes at the same time, you can with a BIOTOL text study it at your own pace, go back over bits you are unsure about and study wherever you choose. Of great importance are the self assessment questions (SAQs) which challenge your understanding and progress and the responses which provide some help if you have had difficulty. These SAQs are carefully thought out to check that you are indeed achieving the set objectives and therefore are a very important part of your study. Every so often in the text you will find the symbol Π, our open door to learning, which indicates an activity for you to do. You will probably find that this participation is a great help to learning so it is important not to skip it.

Whilst you can, as a open learner, study where and when you want, do try to find a place where you can work without disturbance. Most students aim to study a certain number of hours each day or each weekend. If you decide to study for several hours at once, take short breaks of five to ten minutes regularly as it helps to maintain a higher level of overall concentration.

Before you begin a detailed reading of the text, familiarise yourself with the general layout of the material. Have a look at the contents of the various chapters and flip through the pages to get a general impression of the way the subject is dealt with. Forget the old taboo of not writing in books. There is room for your comments, notes and answers; use it and make the book your own personal study record for future revision and reference.

At intervals you will find a summary and list of objectives. The summary will emphasise the important points covered by the material that you have read and the objectives will give you a check list of the things you should then be able to achieve. There are notes in the left hand margin, to help orientate you and emphasise new and important messages.

BIOTOL will be used by universities, polytechnics and colleges as well as industrial training organisations and professional bodies. The texts will form a basis for flexible courses of all types leading to certificates, diplomas and degrees often through credit accumulation and transfer arrangements. In future there will be additional resources available including videos and computer based training programmes.

Preface

This is the first of two BIOTOL texts which focus on cells as the basic operational units of biological systems. The central theme of both of these texts is to consider cells as biological 'factories'. Cells are, indeed, outstanding 'factories'. Each cell type takes in its own set of chemicals and making its own collection of products. The range of products is quite remarkable and encompass chemically simple compounds such as ethanol and carbon dioxide as well as the extremely complex proteins, carbohydrates, lipids, nucleic acids and secondary products.

To understand how a factory operates requires knowledge of the tools and equipment available within the factory and how these tools are organised. We might anticipate that our biological factories will be comprised of structural and functional elements. This text examines the nature and properties of the molecular components of these elements. The knowledge it provides underpins our understanding of the properties and activities of cells. The arrangement, and organisation of these activities, are the topics of the second cell-orientated BIOTOL text 'Infrastructure and Activities of Cells'. The two texts together therefore provide a broad and integrated understanding of the structure and properties of cells.

In many ways this text fills a unique position in the matrix of BIOTOL texts. Clearly, if we are to understand and contribute to biotechnology and applied biology we need to understand how biomolecules behave and how they fulfil their biological roles. These self-same molecules are themselves often the desired products of biotechnology. Thus knowledge of the molecules which make up cells not only provide the basis upon which an understanding of cell function may be developed, but also provides the knowledge needed to manipulate and purify biochemicals. The text therefore, holds an important and key position in both applied biology and biotechnology.

Finally, this text is written on the assumption that the reader is familiar with chemistry at high school level. Although this assumption has been made, extensive use of chemical 'reminders' have been given within this text and readers are not, therefore, completely abandoned to their own memory of chemistry. We recommend readers to attempt the in-text activities and self-assessed questions built within the text, as these will greatly enhance learning.

Scientific and Course Advisors: Dr M.C.E. van Dam-Mieras
Dr C.K. Leach

Cells - the basic units of living systems

Cells - the basic units of living systems

1.1 Introduction

macroscopic
and
microscopic
diversity

Even the most casual observer of nature cannot help but be impressed by the great diversity of living systems. Plants and animals both large and small show a bewildering array of forms and modes of life. Those who have had opportunity to observe both micro-organisms and thin sections of plants and animals will also know that the incredible diversity of the macroscopic forms is also observed at the microscopic level. The prospective biologist should not, however, be intimidated by the apparent complexity of biological systems. Developments in our knowledge of these systems has enabled the emergence of relatively straightforward principles and we are now able to offer rational explanations for many biological phenomena.

cell theory

One of the earliest and most fundamental of the unifying principles of biology was the development of the 'cell theory'. It is now more-or-less universally accepted that all living systems, irrespective of size, are made up of cells and that each type of cell exhibit many common features and carry out analogous, if not identical, processes. Cells are, in fact, the fundamental unit of function within living systems. An understanding of the structure and activities of cells is, therefore, vital to all who work within the sphere of biology. The study of cells is often called cell biology. This book and its partner BIOTOL text 'The Infrastructure and Activities of Cells' provide access to the key facts and concepts of cell biology.

The purpose of this chapter is to generate an overview of living systems and cell biology in order to provide a context within which the remainder of the book may be studied. This is achieved by first exploring the special properties of living systems. Then historical development of cell biology is examined and a brief review of cell structure is given. This overview serves as a starting point for the indepth consideration of the composition of cells dealt with in subsequent chapters and for the detailed analysis of sub-cellular organisation discussed in the partner BIOTOL text. This chapter will also explain the rationale for the division of cell biology into two BIOTOL texts.

1.2 Living systems as agents of chemical change

∏   Think for a moment and write a list of special properties of living systems.

You may well have been able to list a number such as: the ability to nourish themselves; the ability to bring about chemical reactions to release energy for their own purposes; the ability to eliminate waste products; the ability to grow and to multiply. We would like to focus on three special attributes of all living systems. Firstly, all living systems are capable of bringing about an enormous number of chemical changes. For example, consider a tree: it can use carbon dioxide from the atmosphere and a few mineral salts from the soil to make the chemically complex structures of leaves, branches, stems and roots. Likewise animals can take in one set of chemicals (their food) and make a completely new set, the chemicals which make up their own bodies. Even the humble yeast can use simple sugars and mineral salts to make the many thousands of complex chemicals which make up the cells of yeast. We can therefore regard living systems as superb chemical factories, taking in one set of chemicals (nutrients) and converting them into new products. The range of chemicals made is, as you will learn in later chapters, far greater than all of the chemicals produced by man-made factories.

living systems as chemical factories

∏ Can you list other differences between biological systems and chemical factories?

The chances are that you can list many. The two we wish to focus on are that:

- biological systems carry out their chemical reactions in an aqueous environment at quite low temperatures whereas reactions in chemical factories are often conducted at very high temperatures.

- biological systems are capable of multiplying themselves, chemical factories do not.

The three special attributes of living systems that we have focused on are therefore:

- the diversity and complexity of the chemical reactions they carry out;

- the ability to carry out these reactions at quite low temperatures;

- the ability to multiply.

These three special attributes are also attributable to the cells which make up living systems. It is not surprising, therefore, that we have adopted the underpinning theme of cells as very special chemical factories for this and the partner BIOTOL text.

To understand how these factories operate requires knowledge of the structure and properties of the components which make up these factories and 'tools' which are available within them. This aspect of cell biology is the topic of this text. Once this knowledge has been gained it is possible to examine how these 'tools' are used and managed in a co-ordinated manner to provide an effective and efficient unit. These facets of cell biology are the focus of the partner BIOTOL text 'The Infrastructure and Activities of Cells'.

Before we plunge into the composition of cells it is important that we gain a general idea of the layout of these 'factories'. In the next sections, we first provide a brief description of the stages in the development of cell biology before moving onto a general description of cell structure.

∏   Before moving on, make a list of all the useful products that we derive from living
    systems.

We suspect that you will have generated an enormous list including items of food, leather, timber, cotton, wool, alcohol, medicines and so on. In making this list you should have convinced yourself that living systems are indeed remarkable chemical factories making an enormous array of products. You should have also realised that biological systems, as chemical factories, make an enormous contribution to human society by providing an almost endless range of useful products.

1.3 The development of the cell theory

The theory that all living systems are made up of cells was proposed in ignorance of the chemical reactions that such systems carried out. The term 'cell' was first applied, by Robert Hooke in 1665, to the box-like structures he found in plant material. The contents of these cells he described as 'nourishing juices'. Subsequently, through the work of Turpin (1826) and the French student Dujardin (1835), Schleiden and Schwann (1838) were able to produce the theory that all living systems were composed of cells. The following few decades saw greater elaboration of this 'cell theory'. First Naegeli (1854) and then Virchow (1858) were able to claim that all living things were constructed of units (cells) each of which owed its origin to the pre-existence of other units (cells). These workers also argued that the contents of the cells of plants were of greater importance than the outer coat (cell wall).

protoplasm

The term protoplasm, first coined by Purkinje in 1859, became widely accepted for the jelly-like contents of cells from all sources especially after von Mohl and Schulze (1861) described similarities between the jelly-like material in both plant and animal system.

The term 'cell' thus became generally used to describe a protoplasmic unit (protoplast) whether plant or animal, together with any material it may produce in or around itself.

eukaryotic and prokaryotic cells

The intervening years since the mid 1800s witnessed further development of light microscopy and staining procedures which enabled microscopists to describe some of the major sub-cellular components. Nucleus, plasmids and cytoplasm became part of the biologists' vocabulary. The development of the electron microscope during the 1950's led to much greater resolution of the fine structure of cells. These studies clearly demonstrated that cells can be divided into two quite distinct types described as prokaryotic and eukaryotic. Eukaryotic cells are structurally the more complex of the two and are found in all types of plants and animals including the microscopic forms (eg algae, fungi, protozoans). Prokaryotic cell organisation is confined to the bacteria.

eubacteria

archaebacteria

These developments in microscopy were paralleled by developments in biochemical analysis and this allowed assignation of particular chemicals to particular cell structures. At the same time, greater knowledge was gained of the function of sub-cellular particles. The chemical analysis of cells also revealed that prokaryotic cells can be divided into two sub-groups. Those which contain chemicals similar to those of eukaryotic cells were called the true bacteria or, more properly, the eubacteria. Those which were chemically quite distinct (especially in the structure of the fats they contained) from eukaryotes were called archaebacteria. Currently we accept that there are three basic cell groups; the eukaryotic, the eubacterial and the archaebacterial types. A summary of these groups is provided in Table 1.1.

Cell Structure	Group	Properties	Examples
Prokaryotic	Eubacteria	Simple structure, chemically similar to eukaryotes	Most bacteria, including disease organisms, green photosynthetic bacteria, cyanobacteria (blue-green algae), purple photosynthetic organisms
Prokaryotic	Archaebacteria	Simple structure, chemically quite different from eukaryotes	Thermoacidophiles, halophiles, methanogens
Eukaryotic	Eukaryotes	Can be unicellular or multicellular, basic cell architecture is much more complex than with prokaryotes	Microbes (algae, fungi, protozoa), plants (mosses, ferns, seed plants), animals (invertebrates, vertebrates)

Table 1.1   The major divisions of cell types.

SAQ 1.1

1) Which one of the following terms would we use to describe the 'nourishing juices' of Hooke's (1665) thin sections of plants?

    nucleus, plasmids, protoplasm, cells

2) Which of the following organisms do not show eukaryotic organisation of their cells?

    an oak tree; the protozoan which causes dysentery; an elephant; the bacterium *Salmonella typhi*; the large seaweeds that grow on rocky shores; the house fly.

Let us now turn our attention to the general morphology of these cell types. Since both eubacteria and archaebacteria are morphologically similar, we will examine them both together under the heading of prokaryotic cells.

1.4 Prokaryotic cells

size and shape of prokaryotes

The prokaryotic cell form is confined to the bacteria. They are simple, usually unicellular, organisms that are often only a few (2-5) µm long. (Note 1 µm is equivalent to 10^{-6}m or 10^{-3}mm). They exist in several different shapes (see Figure 1.1) and often possess a tough, protective coat called the cell wall.

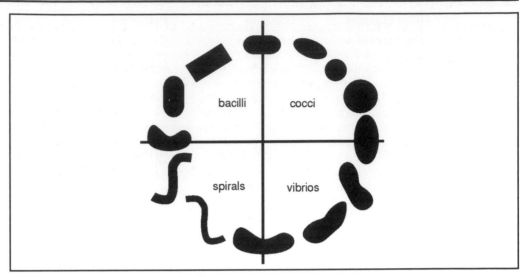

Figure 1.1 The shapes of prokaryotic cells.

plasma
membrane

cytoplasm

genetic material

Within the cell wall is a membrane called a plasma membrane (or plasmalemma) which encloses a jelly-like substance called the cytoplasm (see Figure 1.2). Note that the cytoplasm and the plasma membrane together make up the protoplasm. Under the electron microscope, the cytoplasm may appear more-or-less homogenous although in some cases some particulate material (often storage materials) may be present. The cytoplasm is, however, not homogenous. Embedded in the cytoplasm is the genetic material (DNA). We will learn more of the structure and function of this material in later sections, for now we confine ourselves to saying that this material stores the information needed for the cell to carry out its functions. Copies of this information are passed onto daughter cells when the cell multiplies.

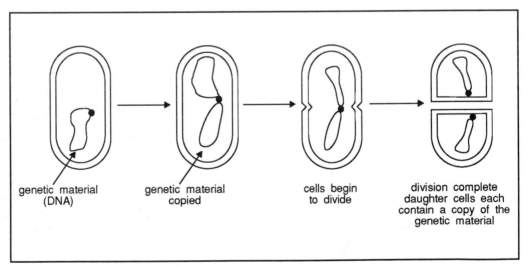

Figure 1.2 Transmission of genetic material to daughter cells.

cytoplasm the
site of many
chemical
reactions

fuelling
reactions

anabolism

catabolism

The cytoplasm is the site where many of the chemical changes take place. These chemical changes include those processes which lead to production of a usable form of energy, reducing power and a series of simple organic molecules. These processes we may describe as the fuelling reactions of the cell. If the starting materials used in these processes are themselves organic molecules, they are collectively referred to as catabolism. The products of the fuelling reactions are used to drive the synthesis of new chemicals (biosynthesis) which may be assembled into new cell structures. Biosynthesis is also known as anabolism. Thus we may view the cytoplasm and its surrounding plasma membrane as being the workshop of the chemical factory. The fuelling reactions and biosynthesis of new cell material are together referred to as metabolism. Thus the chemical changes brought about by cells can be described as metabolism which can be sub-divided in the following way.

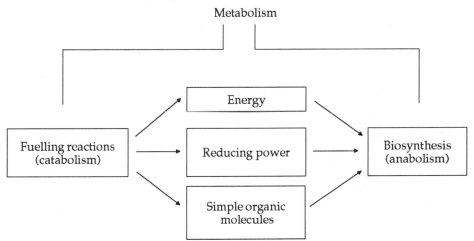

Thus in our analogy of cells as chemical factories we see the genetic material as the management of the factory specifying and controlling which processes (chemical changes) take place, whilst the remainder of the cytoplasm and plasma membrane act as the shop floor - this is where the processes take place.

binary fission

Prokaryotes multiply by dividing into two by a process of binary fission, shown below.

Although apparently simple, careful thought will have revealed to you that this process must, at least at a molecular level, be quite complex. For example the genetic material has to be replicated (copied) and a mechanism established to ensure that each daughter cell receives a copy. There must also be a mechanism to ensure that the cross wall is made in the right place.

rapid growth

An amazing feature of these cells is that, in the presence of a plentiful supply of food and under optimal conditions, they may grow and divide every twenty minutes or so. In principle, a single cell could give rise to 4×10^9 cells within about 11 hours!! You will learn in later chapters that these seemingly simple cells are composed of many, very complex chemicals. The rapid multiplication of cells, therefore, represent an enormous capability to synthesise chemicals. Prokaryotic cells are amazing chemical factories.

Before we leave the prokaryotic cell form, you should be made aware that, although confined to the bacteria, cells displaying the prokaryotic arrangement are more or less ubiquitous in nature. The eubacteria are found in large numbers and varieties in soil, water and on the surface of other, larger, living systems. Archaebacteria are more frequently encountered in more 'severe' environments including hot springs and salt lakes.

1.5 Eukaryotic cells

differentiation

division of labour

tissues and organs

Cells displaying eukaryotic features are characteristic of all plants and animals. They may exist singly as in the unicellular algae and protozoans or, more commonly, in larger groups in the macroscopic plants and animals. In these multicellular forms, different cells within the same organisms display different features. They are said to be differentiated. Differentiation is therefore a process by which cells develop specialised features to carry out specific functions. Some for example may be involved with the transport of nutrients, others with defence against infection while others are involved with excretion. There is therefore a division of labour amongst the cells in multicellular systems. Cells of related function are often grouped together into tissues. In turn, tissues may be grouped together into organs such as the lungs and hearts of animals, and the leaves and flowers of plants, each of which perform specific tasks in maintaining and propagating the system.

Despite this specialisation, certain features of the cells are common to all. The description of eukaryotic cells given here is a generalised one. The reader should bear in mind that each cell type within a multicellular organism will have its own special characteristics. For convenience it is appropriate to consider the cells of plants and animals separately. Nevertheless, the reader will be made aware of the many similarities between the two. We begin by describing a typical animal cell.

1.5.1 Animal cells

nucleus

Animals cells vary greatly in terms of size and shape (Figure 1.3). Typically, however, they are much larger than prokaryotic cells having linear dimensions of about 10-100 μm. Under the light microscope they appear transparent except for a highly refractile nucleus.

The cytoplasm is surrounded by the plasma membrane (plasmalemma).

∏ Compare Figures 1.2 and 1.3. Apart from their size, what is the main difference between the cytoplasm of prokaryotic and eukaryotic cells?

The conclusion you should have reached is that whereas the cytoplasm of prokaryotic cells appears to be more-or-less homogenous, many sub-cellular structures are embedded in the cytoplasm of eukaryotic cells.

organelles

The sub-cellular structures of eukaryotic cells are surrounded by membranes and each carries out particular functions. They have analogies with the specialised functions of organs and are often referred to as organelles. Here we will briefly review the functions of the organelles described in Figure 1.3.

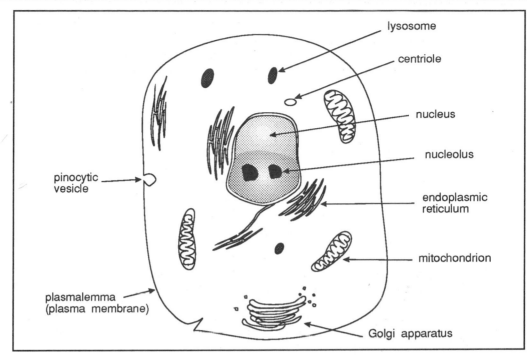

Figure 1.3 Generalised animal cell.

histones

chromosomes

The nucleus contains the genetic material (DNA). Unlike the DNA in prokaryotic systems, the genetic material is wrapped up around proteins called histones to form thread-like structures called chromosomes. The number of chromosomes in each nucleus is characteristic of each organism. In prokaryotic cells there is no such packaging of DNA which remains more-or-less naked within the 'jelly' of the cytoplasm. Also embedded in the nucleus may be one-or-more dark staining bodies called nucleoli (singular nucleolus). These are the sites in which ribosomes are assembled. Ribosomes are small particles which are the sites for making proteins. The ribosomes of prokaryotic and eukaryotic cells show some quite fundamental differences although both act as protein synthesis factories. You will learn much more about them later.

nucleolus

ribosomes

Outside of the nucleus are many small, often discoid, organelles called mitochondria. These are the sites where the final oxidation of the nutrients used for energy takes place. In other words these act as the power house of the cell, providing the energy for the cell to carry out its particular functions.

mitochondria

The endoplasmic reticulum is a membrane system which ramifies through the cytoplasm. Ribosomes attach to this membrane system to give the endoplasmic reticulum a rather granular appearance (the so called rough endoplasmic reticulum). The endoplasmic reticulum is responsible for 'processing' the protein products made by the ribosomes.

endoplasmic
reticulum

Areas of the endoplasmic reticulum which do not have ribosomes attached are much smoother in appearance and are usually referred to as smooth endoplasmic reticulum.

Golgi apparatus

The Golgi apparatus, another membranous structure embedded in the cytoplasm, is also involved in the processing of macromolecules made within the cell. Its special properties are for modifying cell products so that they can be exported from the cell. In our chemical factory analogy, they are the packaging and exporting department.

lysosomes   The lysosomes are small membrane bound vesicles which contain enzymes (catalysts) for breaking down macromolecules. They are involved in the turnover of macromolecules within the cell.

Storage materials (fats and glycogen) may also be deposited in the cytoplasm.

centriole   In addition to these organelles, we also find a granular structure called the centriole. This is often surrounded by a region of cytoplasm sometimes known as a centrosome. The centriole functions in the process of cell division.

centrosome

As with the prokaryotes, eukaryotic cells grow and divide by binary fission. Here the process is even more complex since the cells must replicate their genetic material and arrange for the daughter cells to receive a copy of each chromosome. This is achieved

mitosis   by a process called mitosis.

The structure, function and roles of these sub-cellular organelles and the process of mitosis are described in fuller detail in the BIOTOL text 'The Infra-structure and Activities of Cells'.

In the brief description of animal cells given above, you should have realised that cells are much more than an amorphous jelly in a membranous sack. There is considerable organisation and division of labour within cells. You should also anticipate that cells of different function display some variations on this general design. You might expect that cells responsible for secreting large amounts of proteins (eg the cells which produce and secrete digestive enzymes) or are rapidly growing will contain large amounts of rough endoplasmic reticulum and possess a dominant Golgi apparatus. Cells which are not rapidly growing or are on the way towards cell death (eg the cells on the outer layers of the skin) may lose many of the characteristics of actively growing and metabolising cells. Thus they progressively lose or breakdown mitochondria, ribosomes and endoplasmic reticulum. We would make one final point about cell specialisation. Unspecialised cells can grow and multiply by binary fission. Often, as cells become more and more specialised, they become less able to multiply in this way. Red blood cells in mammals, for example, ultimately break down their mitochondria, ribosomes and nuclei and thus do not have the machinery for further cell growth and division. Such differentiated cells can only be replaced by cells derived from relatively

stem cells   undifferentiated cells called stem cells. We can represent the growth and development of a multicellular system in the following way:

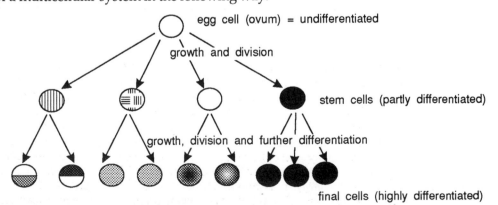

Thus some of the daughter cells of the stem cells become more differentiated to carry out new, specialised functions. Some of the progeny of the stem cells remain as stem cells.

1.5.2 Plant cells

Much of what we have learnt about animal cells is also true of plant cells. Thus the cytoplasm is surrounded by a plasma membrane (plasmalemma) (Figure 1.4). Embedded in the cytoplasm is a nucleus containing nucleoli, mitochondria and endoplasmic reticulum; lysosomes and Golgi apparatus are also present and carry out analogous functions to those described in animal cells.

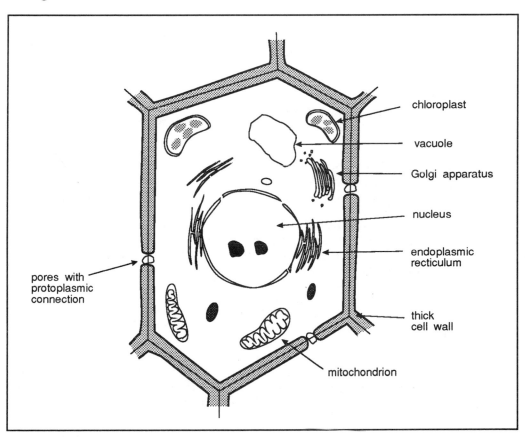

Figure 1.4 Generalised plant cell (young).

Plant cells, however, differ from animal cells in some important respects. The difference between the firmness of plant material (for example the branch of a tree) and the relative softness of animal tissues is well known. This difference can be attributed to the production of a thick cell wall by plant cells. This cell wall gives rigidity to the cell and provides it with a distinctive shape. Many plant cells can be readily identified from their shape. Some, like the outer layer of cells from leaves (epidermis), are often box-like. Others, such as the cells responsible for transporting water and nutrients (xylem and phloem) are long and cylindrical. A second major difference between plant and animal cells is that plant cells often contain vacuoles. When plant cells are young, they do not contain vacuoles but, as they mature, they often elongate and become vacuolated. In fully grown plant cells, which may reach 100μm or more in linear measurement, these

cell walls

epidermis

xylem and phloem

vacuoles

vacuoles may occupy more than 90% of the cell (see Figure 1.5). The vacuoles are surrounded by membrane and are filled with an aqueous solution.

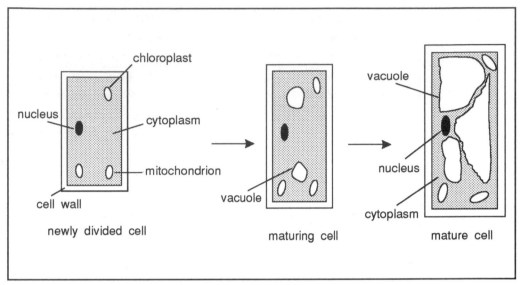

Figure 1.5 The maturation of a plant cell (stylised).

∏ What happens to a plant if it is deprived of water? Explain what is happening within its cells.

We would anticipate that you will have explained that the plant would at first wilt (droop) and ultimately die. The deprivation of water would mean that at first the amount of water in the cells' vacuoles would be diminished and the cytoplasm and cell walls would not be stretched. In other words the cell would loose its turgor and this would result in the plant becoming flaccid and, therefore, wilt.

chloroplasts

photosynthesis

A further difference between plants and animal cells is that plant cells, especially those from leaves, contain membrane bounded, highly pigmented organelles called chloroplasts. These organelles harvest the energy of light and convert it to a form that can be used by the cell. They also collect carbon dioxide from the atmosphere and convert it into a form suitable for the cell to make new cell constituents. These processes together are called photosynthesis. Plant cells may also contain deposits of stored compounds especially starch and oils. These may be deposited in the cytoplasm or in the chloroplasts.

plasmodesmata

Finally, another difference between plant and animals cells is that plant cells are in direct contact with each other through pores in the cell wall. The protoplasm of one cell is in direct communication with the protoplasm of neighbouring cells through these pores. The pores, which are called plasmodesmata, are thought to be involved in the transport of metabolites from one part of the plant to another.

SAQ 1.2

Pair each of the organelles listed with its primary function.

Organelle	Function
nucleolus	houses the genetic material
Golgi apparatus	harvests the energy of light
rough endoplasmic reticulum	break down macromolecules
chloroplast	site of terminal oxidation of organic nutrients
mitochondria	processing of macromolecules for export
lysosomes	helps in organising the process of cell division
nucleus	processing the products of protein synthesis
centriole	site of ribosome construction
ribosomes	carries out protein synthesis

SAQ 1.3

Examine the following drawing of a cell carefully.

Is this cell from a plant or from an animal? Give 3 reasons for your choice.

SAQ 1.4

Indicate whether the following statements are true or false.

1) Prokaryotic cells do not contain mitochondria.

2) Prokaryotic cells do not contain ribosomes.

3) Mammalian red blood cells are capable of growing and multiplying.

4) Histones, the proteins associated with the genetic material, are found in all living systems.

5) The archaebacteria are chemically more closely related to the eukaryotes than are the eubacteria.

6) Prokaryotic cells are much smaller than eukaryotic cells.

Summary and objectives

We have thus far established that cells are superb chemical factories and that the activities of cells are carried out in an orderly and controlled manner. We have also learnt that there are two basic levels of cell organisation, namely the prokaryotic and eukaryotic organisation. In eukaryotic cells, we can distinguish many different structures or organelles which enable the cells to act effectively. Thus in eukaryotes, the information needed to carry out and control cellular processes is stored within the nucleus. This information is translated by the ribosomes into a series of proteins made on the endoplasmic reticulum. The endoplasmic reticulum may modify these proteins to produce the final products. Products which are to be exported from the cell are processed by the Golgi apparatus.

Now that you have completed this chapter you should be able to:

- distinguish between eukaryotic and prokaryotic cell organisation;

- identify the major organelles of eukaryotic cells;

- assign functions to the major organelles of eukaryotic cells;

- distinguish between cells from animals and plants;

- describe in general terms why living systems in general, and cells in particular, may be regarded as excellent chemical factories.

Amino acids

Amino acids

2.1 Introduction and roles

Although this chapter is primarily about amino acids, we shall begin by considering proteins. The reason is that proteins are polymers constructed from amino acids. Thus by establishing the importance of proteins we can more readily appreciate why amino acids should be studied.

Of all the types of compounds found in cells, proteins (from the Greek, proteios = 'first') are arguably the most important. Whilst it is true that deoxyribonucleic acid (DNA) holds the genetic information for the cell, without proteins nothing else in the cell (including DNA) would be made. In this and the next two chapters, we are going to explore proteins and nucleic acids, in order to establish some fundamental properties of these molecules. We will see how the structure of proteins is responsible for their shapes and properties: these in turn determine their uses or functions. Thus understanding the structure of biological molecules is vital for a thorough understanding of what they can do. By studying the functional properties of proteins and the roles they perform in cells, as well as by obtaining insights into how cells carry out and regulate chemical processes ('metabolism'), we may also identify how we could use proteins. This will introduce you to the idea of using cells or parts of cells to accomplish processes and show you how technology is now applicable to biology.

metabolism

technology

Proteins are present in large quantities in cells. They typically constitute about 50% of the dry mass of cells. Their importance is in part a result of the enormous variety of structures (and hence properties) which is possible through the manner of their construction. Before considering details of structure, let us review their functions. These are summarised in Table 2.1. Several comments should be made about the roles of proteins. It would be difficult to overstate the importance of proteins - just think of the significance of any one of the categories identified in Table 2.1, let alone all of them.

importance of
proteins

Note also the diversity of roles. Proteins should be perceived as being sufficiently 'adaptable' to fulfil a huge range of requirements. The ability of proteins to fulfil these roles is a consequence of the enormous range of three-dimensional structures which proteins can take - each protein has a precise structure, from which its properties derive. Underpinning an understanding of proteins is the need to understand the components from which proteins are composed, amino acids. In this chapter we will examine the structure and properties of these building blocks before examining proteins in the next chapter. We will particularly focus on their ionisation behaviour as this is vitally important if we are to understand the properties of both amino acids and proteins.

diversity of
roles of proteins

Enzymes	The catalysts of biochemical reactions
Immune/protective proteins	Antibodies: recognise and bind to foreign substances Complement: complexes with some antibody - antigen complexes and causes destruction of pathogens Fibrinogen and thrombin: involved in blood clotting
Transport proteins	Serum albumin - transport of fatty acids Haemoglobin - transport of oxygen Ceruloplasmin or transferrin - transport of iron
Storage proteins	Ovalbumin - in egg white Casein - in milk Ferritin - storage of iron Storage proteins in seeds eg beans
Structural proteins	Collagen - in skin Elastin - in elastic tissues Keratin - in hair and nails Viral coat proteins Membrane structural proteins
Contractile/motile proteins	Myosin, actin - involved in movement
Hormones and their receptors	Insulin, growth hormone. Receptors for signal reception and for transport of material into cells
Regulating proteins	Selective stimulation or inhibition of expression of DNA

Table 2.1 The roles of proteins.

2.2 Amino acids

amino groups

carboxyl groups

α carbon atom

All proteins are composed of compounds called amino acids, which are frequently thought of as the building blocks of proteins. Amino acids consist of at least one amino group (-NH$_2$), and at least one carboxyl group (-COOH). Whilst an enormous number of amino acids probably exist, those which are incorporated into proteins are all of a type called α-amino acids. Routinely, only 20 different α-amino acids are found in proteins. An α-amino acid is one in which both an amino group and a carboxyl group are joined to the same carbon atom, which is known as the α-carbon atom (Figure 2.1). A hydrogen atom is also attached to the α-carbon atom, as well as a fourth group, known collectively as a sidechain or R-group. It is through the structure of this sidechain that differences in amino acids arise. The simplest sidechain consists of a hydrogen atom (R=H), giving the amino acid glycine. The types of sidechains will be discussed shortly.

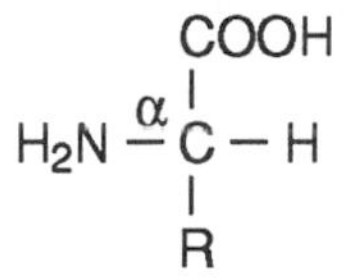

$$H_2N \overset{\alpha}{-} \underset{R}{\overset{COOH}{\underset{|}{\overset{|}{C}}}} - H$$

Figure 2.1 The general structure of α- amino acids. The sidechain, which is different in different amino acids, is shown by the letter R.

So far, we have depicted an α-amino acid as 2-dimensional. In reality, it is of course 3-dimensional. Carbon has a valency of 4, meaning that it can form 4 single covalent bonds with other atoms. When a carbon atom does have 4 substituent groups, the bonds will be evenly distributed in space: the bonds will point to the four corners of a **tetrahedron**, with the α-carbon atom at the centre of tetrahedron. A widely used way of describing chemical structures is by ball-and-stick models. If we do this for α-amino acids, and we symbolise each substituent group of the α-carbon atom by a different coloured ball (-NH₂, -COOH, -H and -R), we find that we can draw two structures. You will find this easier to understand if you build your own models using a ball-and-stick molecular modelling kit; plasticine and matchsticks will also work! Take as an example the amino acid alanine, in which the sidechain is a methyl group (R=CH₃). The two forms shown in Figure 2.2 both contain the same numbers of each substituent atoms.

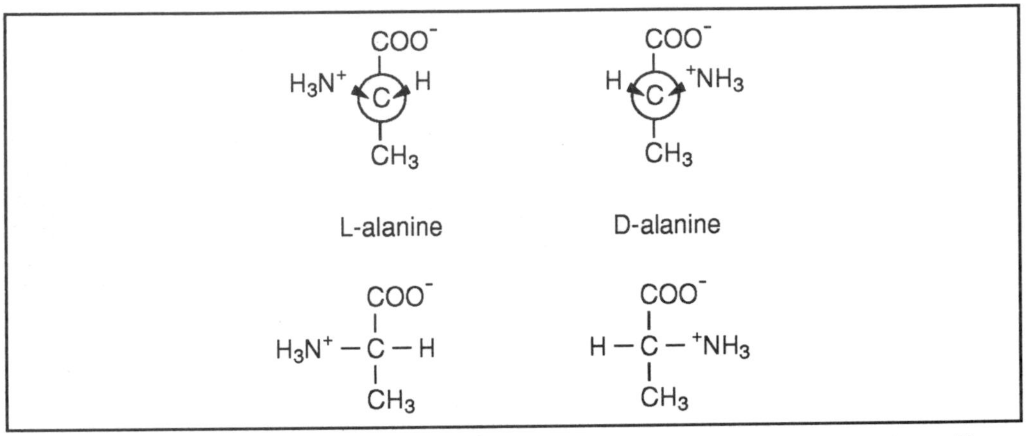

Figure 2.2 Three-dimensional representation of α- amino acids. Ball-and-stick models permit the positions of groups in space to be visualised. Note that two alternative forms of alanine, designated D- and L-, exist.

They also contain the same groups of atoms (amino, carboxyl, methyl), yet they are **isomers** different structures. Such pairs of compounds are known as isomers. When, as in this case, they are mirror images of each other, they are known as stereoisomers. Any compound containing a carbon atom with 4 different substituent groups can be depicted as two stereoisomers: the central carbon atom (the α-carbon atom of α-amino **asymmetric carbon atom** acids) is then known as an asymmetric carbon atom or asymmetric centre, and it confers particular properties on the molecule (for example, ability to rotate plane-polarised light). The important aspect here is that all α-amino acids except for glycine (where the sidechain group is a hydrogen atom) possess an asymmetric carbon atom.

absolute configuration Stereoisomers are assigned an absolute configuration by comparison with the structure of glyceraldehyde. Glyceraldehyde is used as a reference compound. To determine the absolute configuration of an amino acid, the molecule is written with the carbon atoms in a vertical line, with the α-carboxyl group (ie the carboxyl group attached to the α-carbon atom) behind the plane of the paper. The α-carbon atom is taken as being in the plane of the paper, whilst the sidechain group will be behind the plane of the paper. The hydrogen atom attached to the α-carbon atom and the amino group will be in front of the plane of the paper. If the amino group is to the right of the α-carbon atom, as you look at it, then it is a D-amino acid; if the amino group is to the left, then it is a L-amino acid. Whilst isomerism will be further developed when we discuss carbohydrates, the importance of isomerism at this stage is as follows: of the two forms of amino acids which can be made (D- and L-forms) only the L-form is incorporated into proteins. Whilst D-amino acids do occur in living organisms and are sometimes incorporated into macromolecular structures (eg the peptidoglycan cell wall of bacteria), for most purposes D-amino acids can be ignored.

∏   What is the absolute configuration of these three amino acids? Hint: make 3-dimensional models of these molecules, based on the convention given in the text. Try this before reading on.

COOH
|
H₂N — C—H
|
CH₂OH

CH₃
|
H — C —NH₂
|
COOH

H
|
HOOC — CH₂ — C — COOH
|
NH₂

1) This amino acid is written according to the description in the text. Since the amino group is to the left of the carbon skeleton, it is an L-amino acid (it is L-serine).

2) Not quite so simple! Although this one is written with the carbon atoms in a vertical line, the α-carboxyl group is at the bottom. If the molecule is rotated by 180° degrees in the plane of the paper, the α-carboxyl group will now be at the top. The methyl and α-carboxyl groups will still be behind the plane of the paper (as in the convention); the amino group and hydrogen atom will be in front of the plane of the paper, with the amino group to the left. This is an L-amino acid (this one is L-alanine). In summary, what we have done is:

(CH₃
|
H—C—NH₂
|
COOH)

rotate as
shown, giving

COOH
|
H₂N — C—H
|
CH₃

3) If this amino acid is turned 90° to the left, in the plane of the paper, the α-carboxyl and sidechain groups (whilst now 'vertical') will still be in front of the plane of the paper ie (representations 1 and 2). To get them into the conventional display, we must now twist along the vertical axis, giving representation 3 above. Thus this is L-aspartic acid.

1) starting position

2) after 90° twist

3) after rotation about
the vertical axis

H
|
HOOC — CH₂ ⟳ COOH
|
NH₂

COOH
H ⟳ NH₂
CH₂ COOH

COOH
H₂N ⟳ H
CH₂ COOH

2.2.1 Amino acid sidechains

The sidechains of amino acids make a major contribution to the properties of both amino acids and of proteins constructed from them. It is important to gain a 'feel' for the general structures of the sidechains, although you do not need to know precise structures.

polarity of
sidechains

Amino acids are routinely classified according to the general properties of the sidechains. Whilst they could be classified according to chemical nature (sulphydryl, aliphatic, aromatic, etc), a more useful classification is on the basis of polarity of the sidechain. This is related to the extent to which charge separation occurs, whether complete (resulting in ionisation), partial, or minimal. This approach is useful because it most satisfactorily reflects the way the sidechains interact (with each other or with other compounds) and the possible functional role of each amino acid.

Four groups are identified:

a) Non-polar or hydrophobic sidechains

alanine, valine,
leucine,
isoleucine,
methionine,
phenylalanine,
tryptophan,
proline

Hydrophobic means 'water-hating' and is a general term applied to chemical groups or molecules, particularly hydrocarbons (whether aliphatic or aromatic), which display only very limited solubility in water. This poor solubility in water (which is a polar solvent) is a consequence of their non-polar character. This group contains aliphatic amino acids: alanine, valine, leucine, isoleucine and methionine; aromatic amino acids: phenylalanine and tryptophan. It also includes proline, in which the side chain is linked to the α-amino group; strictly, this makes it an imino acid, since the nitrogen is now part of a secondary rather than a primary amine. The structures of these amino acids are shown in Figure 2.3. Amino acids are frequently identified by means of 3-letter abbreviations; these are also shown in Figure 2.3.

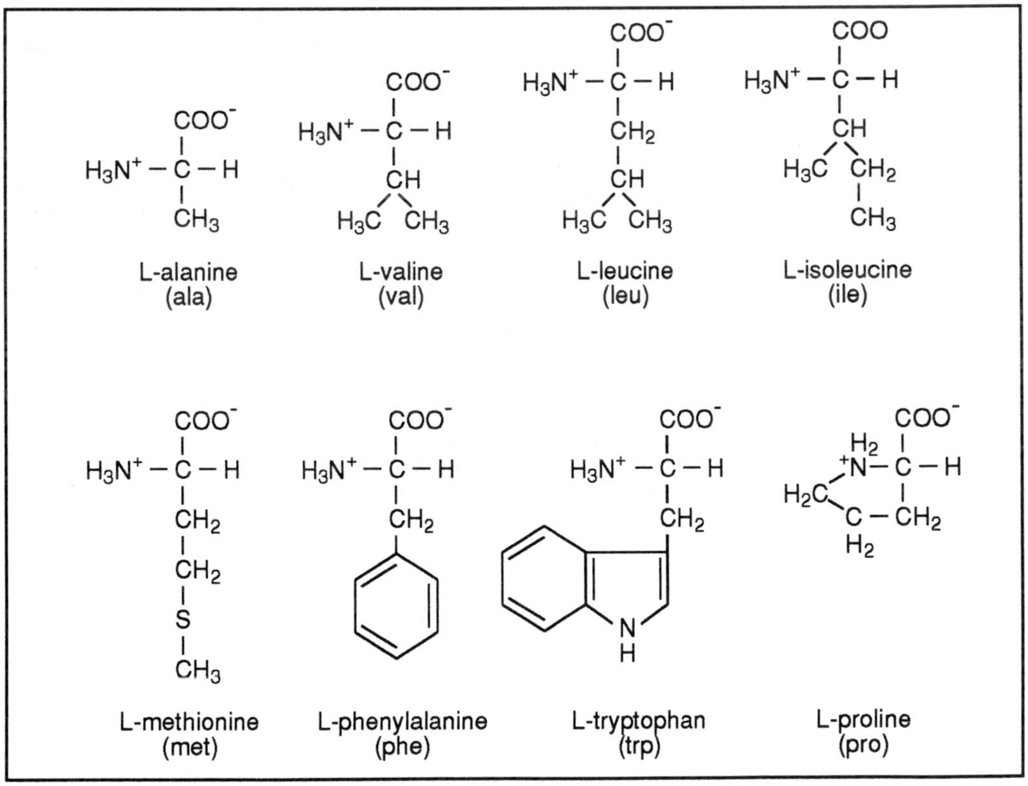

Figure 2.3 Amino acids with non-polar, hydrophobic sidechains. Also shown are the 3-letter abbreviations for the amino acids, which are widely used.

∏ It might be helpful here to make yourself a revision table on a separate sheet of paper. For example:-

Group	Amino Acids
Non-polar (hydrophobic)	ala, val, leu, ile, met, phe, trp, pro

b) Polar but uncharged sidechains

These sidechains show partial charge separation, which arises through electrons not being evenly distributed around atoms (as they are with carbon - hydrogen bonds), but being unevenly distributed. In the example shown in Figure 2.4, electrons associated with the hydroxyl group are attracted to the oxygen atom at the expense of the hydrogen atom.

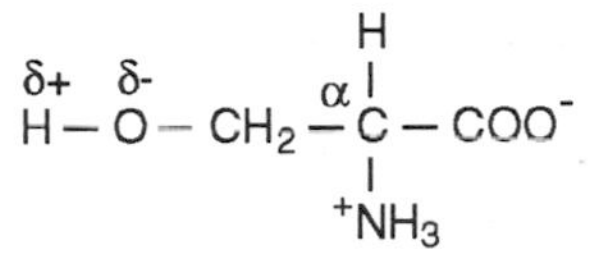

$$\overset{\delta+}{H} - \overset{\delta-}{O} - CH_2 - \overset{\overset{H}{|}}{\underset{\underset{^+NH_3}{|}}{\overset{\alpha}{C}}} - COO^-$$

Figure 2.4 Partial charge separations within a neutral sidechain. The electronegative oxygen atom attracts bonding electrons towards it, resulting in the hydrogen atom which is attached to it becoming slightly deficient in electrons. This leads to the slight charges shown, which creates a polar group; these tend to form hydrogen bonds.

partial charge separation

electronegativity

This creates partial charges, shown as $\delta+$ and $\delta-$. Atoms which tend to attract electrons in this way are said to be electronegative. Groups in which partial charge separation occurs (as in Figure 2.4) tend to form hydrogen bonds (see Section 3.3.2). The amino acids in this group are shown in Figure 2.5, together with their 3-letter abbreviations.

glutamine, asparagine, serine, threonine, tyrosine, glycine and cysteine

Amino acids in this group contain amides (glutamine, asparagine), hydroxyl groups (serine, threonine and tyrosine) and sulphydryl groups (cysteine). All these sidechains are polar but not ionised at neutral pH. Glycine is also included in this group, primarily because the sidechain is small and, although a hydrogen atom, lacks hydrophobic character.

c) Amino acids with negatively charged sidechains

This group includes aspartic acid and glutamic acid, which each contain a carboxyl group in their sidechain (Figure 2.6).

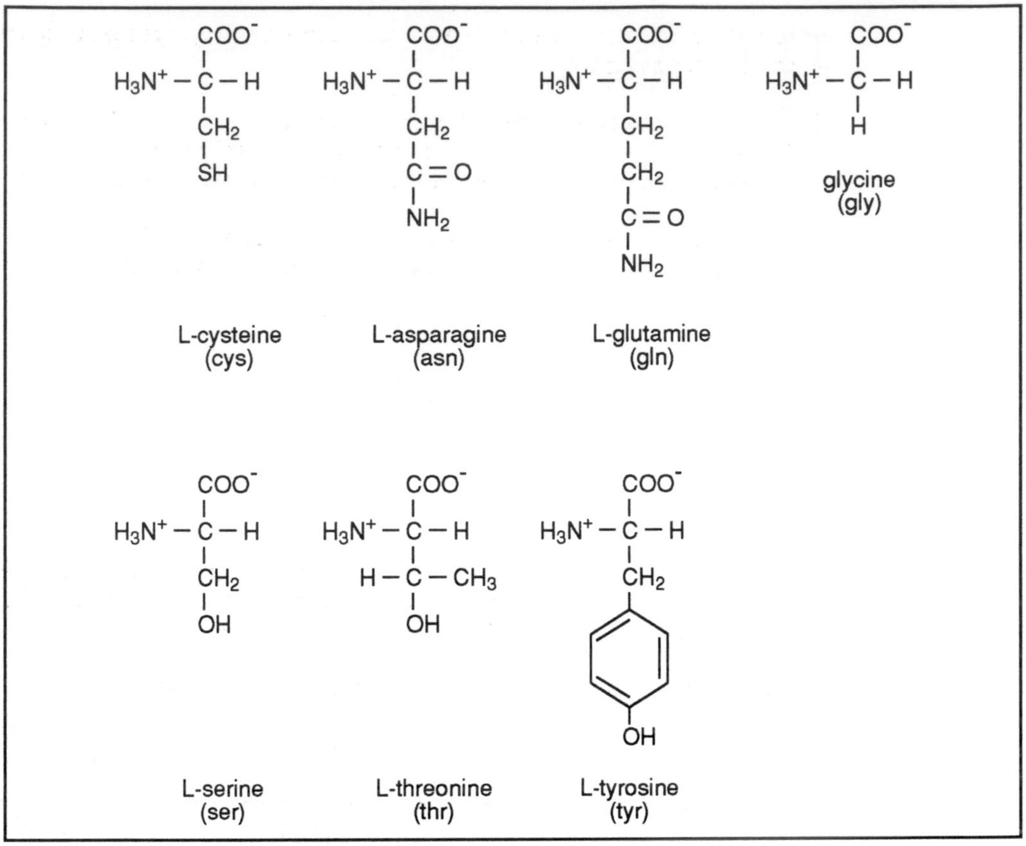

Figure 2.5 Amino acids with polar but uncharged sidechains.

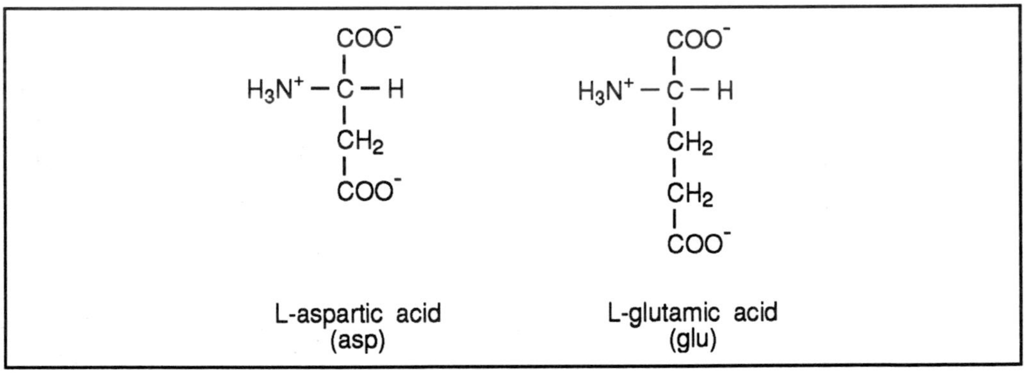

Figure 2.6 Amino acids with negatively charged sidechains. Strictly, when these sidechain groups are deprotonated, they should be described as aspartate and glutamate, respectively.

aspartic acid

glutamic acid

At neutral pH these groups will be dissociated (ie ionised) and will possess a net charge of -1. In dissociated form these amino acids should, strictly, be referred to as aspartate and glutamate. When discussing the carboxyl groups, in order to distinguish the sidechain carboxyl from the α-carboxyl group, the sidechain carboxyl is named after the carbon atom to which it is attached. In aspartic acid, this is the β-carbon atom (ie that next to the α-carbon atom); hence the sidechain carboxyl group in aspartic acid is known as the β-carboxyl group.

∏ If the sidechain carboxyl group in aspartic acid is called the β-carboxyl group, what is the sidechain carboxyl group of glutamic acid called? (You will need to know the first few letters of the Greek alphabet to be able to answer this).

The answer is the γ-carboxyl group, since it is attached to the third carbon atom (where the α-carbon atom is the first carbon atom). These identifications are shown here:

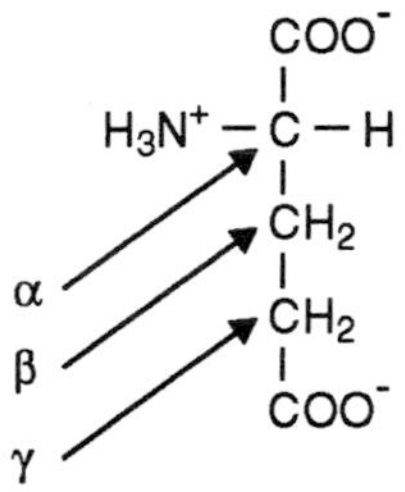

$$
\begin{array}{c}
COO^- \\
| \\
H_3N^+ - C - H \\
| \\
CH_2 \\
| \\
CH_2 \\
| \\
COO^-
\end{array}
$$

α
β
γ

A similar analysis applied to the sidechain amino group of lysine (see amino acids of group d) means that the lysine sidechain amino group is known as the ε-amino group. Check it! (NB. The Greek alphabet begins α, β, γ, δ, ε).

d) Amino acids with positively charged sidechains

arginine
lysine

There are three amino acids which possess a positively charged sidechain. The sidechain of arginine and lysine (Figure 2.7) will be fully charged at neutral pH.

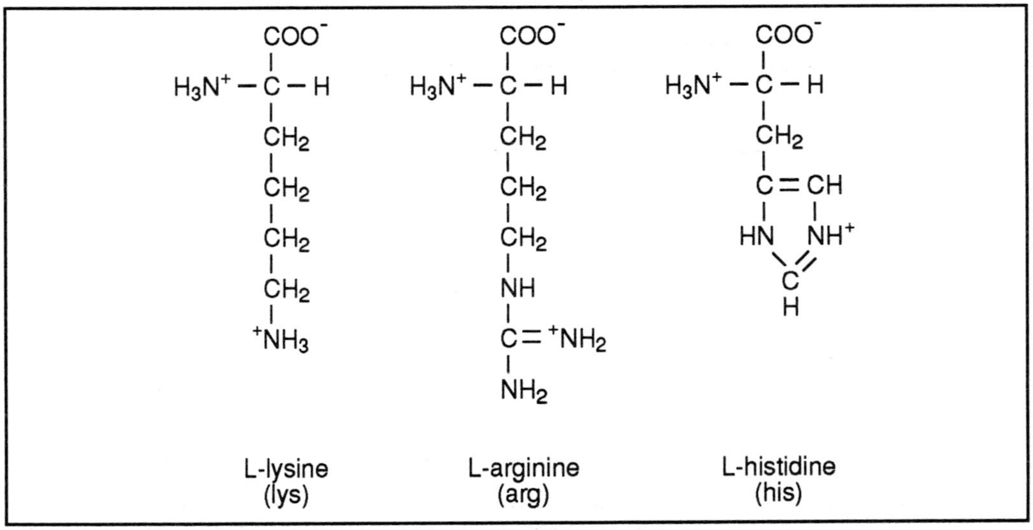

Figure 2.7 Amino acids with positively charged sidechains.

histidine

That of histidine (also shown in Figure 2.7) will depend on the precise pH, since it is a weak acid with a pKa of 6.0. The meaning and significance of pKa values will be discussed in the next section. Dissociation or association of groups which can ionise has important implications for the properties of amino acids and proteins, as we shall see.

| **SAQ 2.1** | One of the α- amino acids found in proteins does not exist in D- and L- forms. Which is it and briefly explain why? (Examine Figures 2.3-2.7 for a clue). |

| **SAQ 2.2** | Inspect the five amino acids shown below. Write down, giving brief reasons: |

1) which of them is a non-polar hydrophobic amino acid;

2) which of them have polar sidechains;

3) which of them is not an α- amino acid.

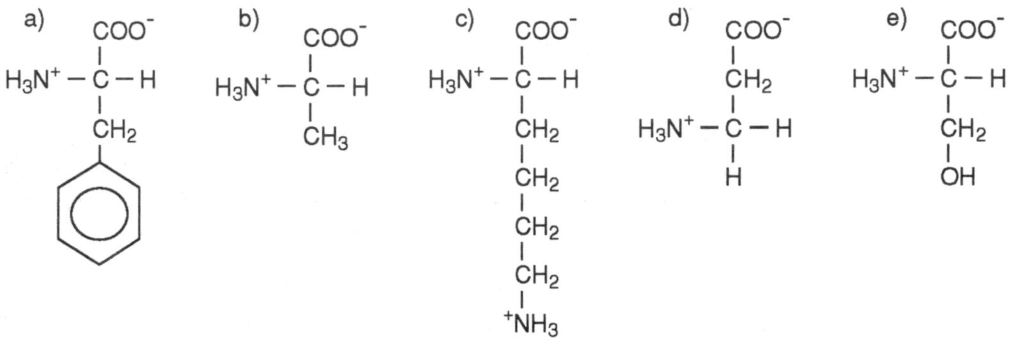

a)
$$COO^-$$
$$H_3N^+ - C - H$$
$$CH_2$$
(phenyl ring)

b)
$$COO^-$$
$$H_3N^+ - C - H$$
$$CH_3$$

c)
$$COO^-$$
$$H_3N^+ - C - H$$
$$CH_2$$
$$CH_2$$
$$CH_2$$
$$CH_2$$
$$^+NH_3$$

d)
$$COO^-$$
$$CH_2$$
$$H_3N^+ - C - H$$
$$H$$

e)
$$COO^-$$
$$H_3N^+ - C - H$$
$$CH_2$$
$$OH$$

2.3 Ionisation of amino acids

You will have noticed that in the last section amino acids were depicted in ionised form: carboxyl groups had lost a proton to form -COO⁻ groups and amino groups were shown protonated as $-NH_3^+$. The reason is that this more correctly represents their structure at neutral pH. For a number of reasons (titration behaviour, solubility in water and insolubility in non-polar organic solvents such as ether or chloroform, melting points of crystals of amino acids) it is strongly believed that amino acids exist in the forms shown in Figures 3.3 to 3.7. These ionised structures are given the name 'zwitterion' (from the German zwitter, meaning 'hybrid') in that the amino acid is a hybrid of negative and positive charges.

zwitterion

pKa

Whilst this accurately depicts the ionic status of amino acids at neutral pH, the situation may differ as the pH is altered. Fortunately, for each group which can ionise there is a pKa value, which provides valuable information on the extent of ionisation under given conditions. pKa is the pH at which each group exists half in an ionised form and half in an unionised form. Thus for every carboxyl, amino and ionising sidechain group there is an associated pKa value, which can be used to establish or predict what ionic form the group is in (protonated or deprotonated). These are important concepts. It is vital that you understand the ionisation behaviour or amino acids (and other weak acids). You may already be familiar with the concepts of ionisation of weak acids, pKa values and the relationship between pH and pKa. If not, then careful study of Appendix 1 should enable you to understand these concepts. In some ways this section is quite long reflecting the importance we attach to the concepts and the care with which they are explained. You will learn in later sections that the ionising properties of amino acids are important in understanding the principles behind the purification of amino acids and proteins as well as predicting the effects of pH on the properties and activities of proteins. Understanding the ionisation behaviour of weak acids is also important in understanding how buffers work and how to prepare them.

2.3.1 Ionisable groups of amino acids

The groups of amino acids which are able to ionise can all be treated as weak acids. This is equally applicable to carboxyl groups:

$$-COOH \rightleftharpoons -COO^- + H^+$$

to amino groups:

$$-{}^+NH_3 \rightleftharpoons -NH_2 + H^+$$

and to the sidechain groups which ionise, for example:

$$-SH \rightleftharpoons -S^- + H^+$$

Each of them has a particular pKa value. pKa values for α-carboxyl groups are usually around pH 2.0-2.5 (there are minor variations, as a consequence of the different sidechains) whilst those for α-amino groups are usually in the pH range 9.5-10.0. Before considering the amino acids with sidechains which can ionise, we will first analyse how charge changes in amino acids with non-ionising sidechains (groups 1 and 2).

If we take a solution of L-alanine which has been adjusted to pH 1.0 with hydrochloric acid and then titrate it with sodium hydroxide (NaOH), the result would be as shown in Figure 2.8. As NaOH is added, pH rises but not in an even way.

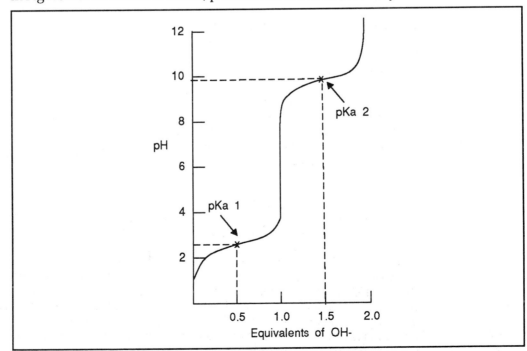

Figure 2.8 Titration of L-alanine. The amino acid was initially adjusted to pH 1.0 with hydrochloric acid. The graph shows the effect of addition of sodium hydroxide on the pH of the solution. The positions of the pKa values for the α-carboxyl group (pKa$_1$) and the α-amino group (pKa $_2$) are shown.

Instead there are two plateau regions, where increase in pH is minimal for addition of standard amounts of NaOH. In other regions (pH 3.5-8.0) change in pH is rapid. The plateau regions represent buffering by the amino acid. In the region of rapid pH change no buffering is occurring. The behaviour of alanine is precisely analogous to that of the weak acids (discussed in Appendix 1). The resistance to pH change at low pH (centred around pH 2.3) results from ionisation of the α-carboxyl group. The mid-point is at pH 2.3, which is the pKa of the α-carboxyl group. Appendix 1 contained an in text question which dealt with the pH ranges over which alanine would act as a buffer. If you apply the Henderson-Hasselbalch equation to the ionisation of the α-carboxyl group, you will find that protonated α-COOH represents 91% of the carboxyl groups at pH 1.3, 50% at pH 2.3 and 9% at pH 3.3. This buffering is a consequence of the ability to change ionic form and is confined to a pH range of pKa ± 1.

A similar region of buffering occurs around pH 9.7. This is a result of the α-amino group changing its ionic state.

∏ Predict the % of the -$^+NH_3$ form of the α-amino group at pH 8.7, 9.7 and 10.7 if the pKa is 9.7.

The percentage of -$^+NH_3$ is as follows: pH 8.7 = 91%, pH 9.7 = 50%, pH 10.7 = 9%.

You should have used the Henderson-Hasselbalch equation, $pH = pKa + \log \frac{[A^-]}{[HA]}$

Thus for pH 8.7, $8.7 = 9.7 + \log \frac{[A^-]}{[HA]}$

∴ $\log \frac{[A^-]}{[HA]} = 8.7 - 9.7 = -1$

∴ $\frac{[A^-]}{[HA]} = $ Antilog -1   ∴ $\frac{[A^-]}{[HA]} = 0.1$ or $\frac{1}{10}$

therefore HA represents 10/11 of the total or 91%.

For pH 9.7, $9.7 = 9.7 + \log \frac{[A^-]}{[HA]}$

∴ $\log \frac{[A^-]}{[HA]} = 0$ ∴ $\frac{[A^-]}{[HA]} = $ Antilog 0 = 1.0

ie the ratio is 1:1 and 50% of molecules are in the form HA.

For pH 10.7, $10.7 = 9.7 + \log \frac{[A^-]}{[HA]}$

∴ $\log \frac{[A^-]}{[HA]} = 1.0$ ∴ $\frac{[A^-]}{[HA]} = \frac{10}{1}$

ie 10/11 or 91% of molecules are present as A⁻.

Hence HA represents $(100-91) = 9\%$.

If you had difficulties with these calculations, we would again suggest you read Appendix 1.

So what has happened to the structure and charge of alanine as a result of changing the pH? At low pH (pH 1.0), the amino acid is fully protonated, with a net charge of +1 (Figure 2.9). As deprotonation of the carboxyl group occurs (pH range 1-4) the carboxyl group progressively acquires a negative charge. At pH 2.3, half the carboxyl groups will be negative - this means that, at any instant in time, half of the carboxyl groups will have a negative charge. Individual groups will fluctuate between being protonated and deprotonated, thus the population as a whole behaves as though the carboxyl group is half protonated (ie net charge on the carboxyl group is -0.5). The overall net charge on alanine will be +0.5. By pH 4 or so, the carboxyl group is fully deprotonated and the net charge is zero. In reaching pH 6.0 we find that an amount of NaOH equal to that of the amino acid has been added (ie if 1 mmol amino acid was present, then 1 mmol NaOH has been used). You may have noticed that in reaching pH 2.3 (the pKa of the carboxyl group) half an equivalent amount of NaOH was needed - this corresponds to the half titration of the carboxyl group (ie [COOH] = [COO']).

∏   Predict the net charge on L- alanine at pH 9.7 and at pH 12.0.

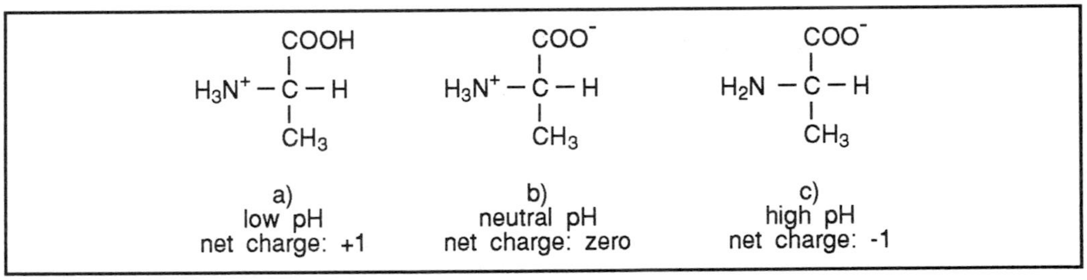

Figure 2.9 Effect of pH on the ionic state of L- alanine a) At low pH, L- alanine is fully protonated, with a net charge of +1. At neutral pH b), the α- carboxyl group is deprotonated, giving a net zero charge. c) At high pH, deprotonation of the α- amino group results in a net charge of -1.

The net charge on L- alanine at pH 9.7 is -0.5 and at pH 12.0 it will be -1. The detailed reasoning is as follows. Above pH 5 or so the carboxyl group is fully deprotonated and carries a charge of -1. The amino group is fully protonated (charge +1) at pH 5-7, but deprotonates in the region of its pKa (9.7). At pH 9.7 the amino group is half-deprotonated.

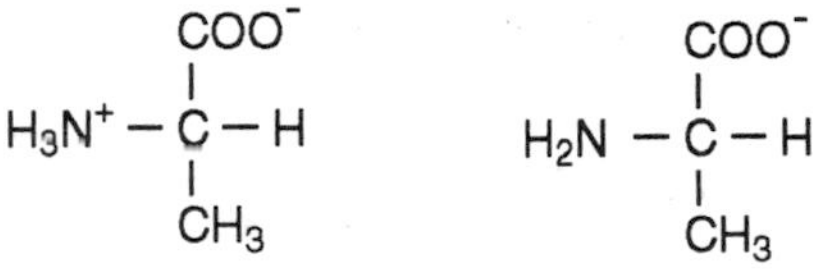

Thus, at any instant, half the molecules are in ionic state a) and half are in ionic state b). The overall effect is of a charge of +0.5 on the amino group, giving a net charge on L-alanine at pH 9.7 of -0.5. By pH 12.0, the α-amino group has completely dissociated. Thus the net charge is -1.

The titration pattern given by L- alanine is characteristic of amino acids whose sidechains do not ionise. Thus all amino acids of groups a and b will show two regions of buffering, and their charges will vary from +1 (low pH) through zero (neutral pH) to -1 at high pH. This enables us to appreciate the behaviour of amino acids during separation procedures.

2.3.2 Electrophoresis of amino acids

If amino acids are placed in a direct current electric field, they will migrate according to their overall charge. If they are positively charged, they migrate towards the cathode; if negatively charged, they migrate towards the anode. Those with no net charge do not move. The speed of movement depends mainly on the size of the charge and the applied voltage. This is utilised in practice in paper electrophoresis, where chromatography paper serves as a supporting medium. Samples of amino acids are applied as spots on an origin line on dry paper. The paper is wetted up with an appropriate buffer (giving whatever pH is required); electrodes are connected and a voltage applied. After separation has been accomplished (typically 1-2 hours), the paper is dried. Amino acids are visualised (and hence their movement determined) by treatment with ninhydrin (see section 2.4).

∏ Assume that L- alanine was subjected to paper electrophoresis at 3 different pH values. Below is shown the result after electrophoresis at pH 1.0. Predict, and write into this table, the result for electrophoresis at the other 2 pH values given (assume that all other conditions are the same).

pH	Direction and distance migrated
1.0	2.8cm, towards cathode
6.0	
12.0	

Your reasoning in answering this should have been as described below. At pH 1.0, both ionising groups are essentially fully protonated, giving a net charge of +1 (there may be very slight dissociation of the α-carboxyl group). Hence migration towards the cathode.

At pH 6.0, the α-carboxyl group will be fully deprotonated, since pH is well above the pKa (ie charge -1). Since the pH is well below its pKa, the α-amino group will still be fully protonated (charge +1). Hence overall charge is zero and the amino acid remains where it was applied.

At pH 12.0 both groups will be deprotonated, since the pH is well above both pKa values. Hence a net charge of -1 and migration 2.8-3 cm towards the anode. The distance migrated is about the same or slightly more than at pH 1.0, reflecting the fact that the pH is 2.3 units away from the α-amino pKa, giving more complete ionisation than at pH 1.0 (which is only 1.3 units below the α-carboxyl pKa).

2.3.3 Ionisation of amino acid sidechains

With a sidechain which can ionise, the titration curve is changed. Sidechain pKa values depend entirely on the particular group. These are listed in Table 2.2. So that you do not forget that the sidechain ionisation is in addition to α-carboxyl and α-amino groups, these are listed as well. You need to know which amino acids can ionise, how their charge changes (is it -XH$^+$ ⇌ -X + H$^+$ or -YH ⇌ -Y+ H$^+$?) and at roughly what pH this occurs. Precise details of structure are much less important.

Amino acid	Sidechain	pKa α-COOH	pKa α-NH$_2$	pKa sidechain
Aspartic acid	$- CH_2 - COOH$ (carboxyl)	2.0	9.9	3.9
Glutamic acid	$-CH_2 - CH_2 - COOH$ (carboxyl)	2.1	9.5	4.1
Histidine	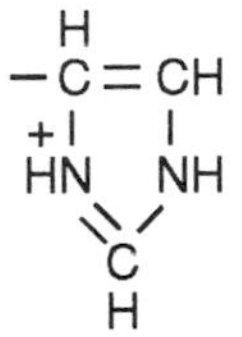 (imidazole)	1.8	9.3	6.0
Cysteine	$- CH_2 - SH$ (sulphydryl)	1.9	10.7	8.4
Lysine	$- (CH_2)_4 - {}^+NH_3$ (amino)	2.2	9.1	10.5
Tyrosine	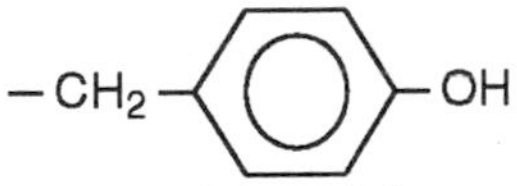 $- CH_2 -$ ⬡ $- OH$ (phenolic)	2.2	9.2	10.5
Arginine	$- (CH_2)_3 - N(H) - C(={}^+NH_2) - NH_2$	1.8	9.0	12.5

Table 2.2 pKa values of amino acids with ionising sidechains.

Figure 2.10 contains the results of titration of aspartic acid and of lysine. Both are found to have an additional region of buffering and require 3 equivalents of NaOH per equivalent of amino acid. This is because there are 3 groups ionising. Aspartic acid has another ionisation at low pH and twice as much NaOH is required to reach pH 6.0 as for alanine. This is because of the sidechain carboxyl group (the β-carboxyl). After pH

6.0, the titration is like that of alanine. For lysine, the low pH region is like that of alanine (hence a single ionisation in this region), but there is now an additional ionisation at high pH, corresponding to the ionisation of the sidechain amino group (the ε-amino).

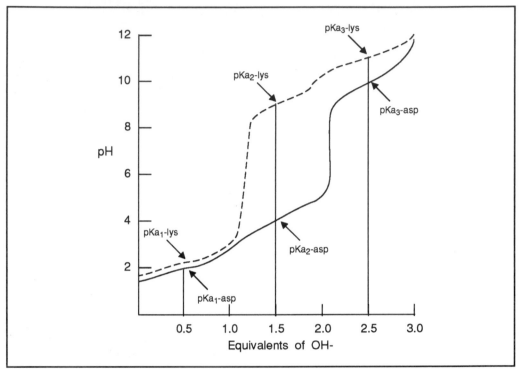

Figure 2.10 Titration curves for L- aspartic acid (continuous line) and L- lysine (dashed line). L- aspartic acid has an additional ionisation at low pH (β- carboxyl group; pKa = 3.9). L- lysine has an additional ionisation at high pH (ε- amino group; pKa = 10.5).

How does the charge change as the pH is raised? Ionic forms of aspartic acid at various pHs are shown in Figure 2.11. At low pH, net charge is +1; as the α-carboxyl group deprotonates the net charge drops to zero.

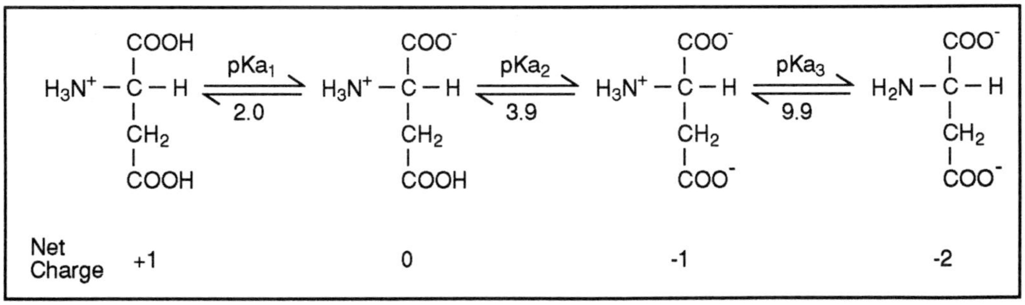

Figure 2.11 The ionic forms of L- aspartic acid which occur at various pH values. Four ionic forms occur. pKa values show the pH of the midpoint of the transition from one form to the next.

As the sidechain (-COOH) ionisation occurs, the overall charge becomes -1, so that aspartic acid is negatively charged at pH 6.0. Deprotonation of the α-amino group around pH 9.9 leads to an ionic species of net charge -2.

∏ Use a similar analysis to establish the net charge on lysine at pH 1.0, 6.0 and 12.0. (Do this before looking at the next figure).

Ionic forms of lysine, and the pKa values which 'control' their interconversions, are as follows:

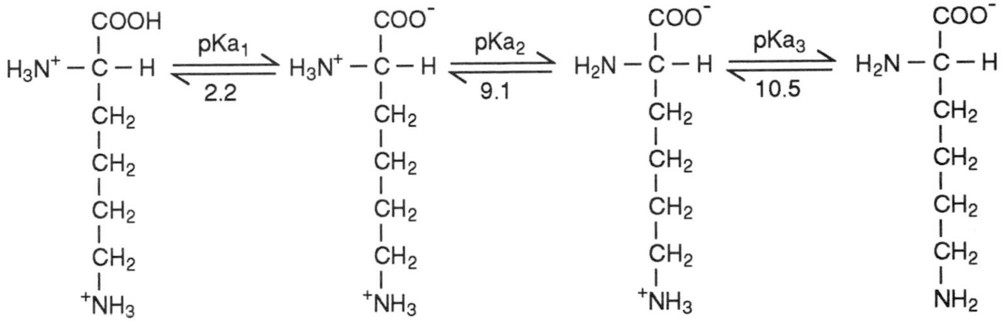

| Net Charge | +2 | +1 | 0 | -1 |

Hence, at low pH (1.0) the net charge is +2. At pH 6.0, the α-carboxyl will have deprotonated but both amino groups will still be protonated, giving a net charge of +1. At pH 12.0, all groups will be deprotonated, giving a net charge of -1.

Amino acids with ionising sidechains thus show a greater range of charges than those without. Furthermore, they may have a net charge at neutral pH.

SAQ 2.3

Which of the following representations of L glutamic acid will predominate at pH 6.0? The pKa values given in Table 2.2 may help you decide.

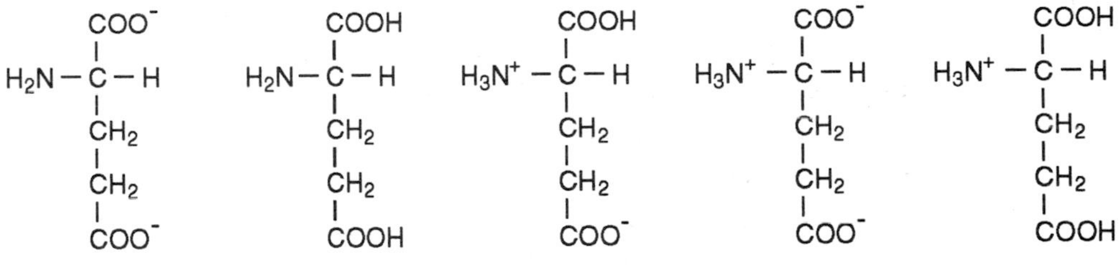

SAQ 2.4

We introduced paper electrophoresis of amino acids in an earlier in text activity. Assume that a similar experiment was conducted on:

1) aspartic acid, 2) leucine, 3) lysine.

Sketch the expected outcome for electrophoresis at pH 1.0 and 6.0, given the result shown below as a start. This shows that, at pH 6.0, aspartic acid migrated towards the anode. Draw in where the other amino acids should be:

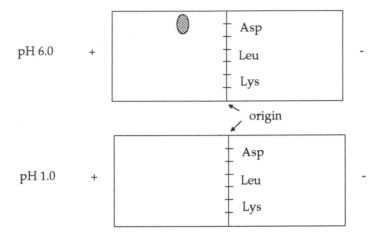

SAQ 2.5

We have displayed titration curves for some amino acids. Sketch similar graphs for the following amino acids. Hint: use the axes of Figure 2.10.

1) glycine, 2) histidine, 3) arginine.

You may have to look up what the sidechain groups are and how they ionise.

2.3.4 Isoelectric point of amino acids

definition of isoelectric point

We have just seen that the charge on an amino acid varies with pH and changes from positive at low pH to negative at high pH. At some point there must be zero net charge. We stress that the **net** charge is zero because the amino acid still has charges. It is in its 'zwitterion' form, and the charges neutralise each other. The term isoelectric point or isoelectric pH is used to describe the pH at which an amino acid has no net charge and hence will not migrate in an electric field. This pH value is sometimes written as pI.

pI

How may the isoelectric point be determined? It can be measured experimentally (at what pH does no movement occur during paper electrophoresis? Some cases have already been seen in previous SAQs and in in text activities). It can also be calculated. A simple way is as follows. If we start with the zwitterion form of alanine and we change the pH, the overall charge will change as the pH approaches the pKa of either ionising group (Figure 2.12.).

You have already calculated the overall charge at various pHs. We can thus think of the pKa values as indicators or 'controllers' of the change in the charge state. In reaching pH 2.3, the pKa of the α-carboxyl group, the net charge will be +0.5; in reaching pH 9.7 (the pKa of the α-amino group) the charge will be -0.5. Since ionisation is symmetrical

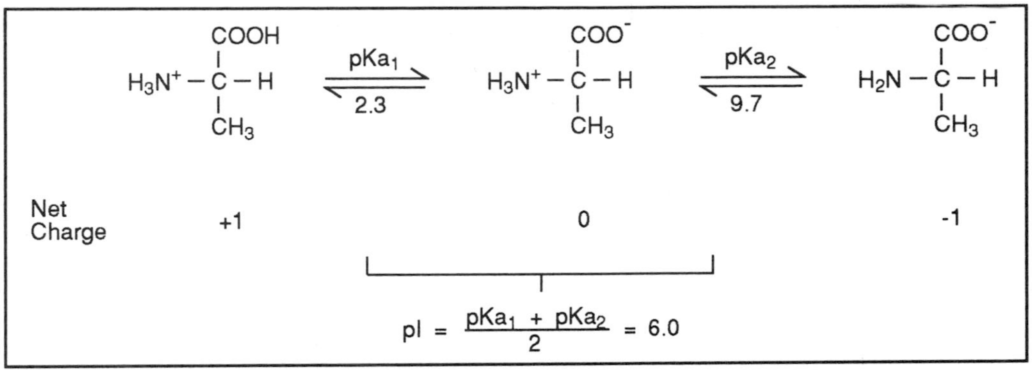

Figure 2.12 The isoelectric point for L- alanine is mid-way between the ionisations which 'border' the zwitterion, ie the ionic species with no net charge.

around the pKa, it follows that precisely zero net charge must be equidistant between the pKa values which border the zero charge form. For alanine these are pKa_1 (α-carboxyl group) and pKa_2 (α-amino group) and pI (the isoelectric point) is given by

$$pI = \frac{pKa_1 + pKa_2}{2}$$

ie for alanine $pI = \dfrac{2.3 + 9.7}{2} = 6.0$

What happens with an ionising sidechain? Display the various ionic species, linked by the various pKa values (as in Figure 2.12) and identify the charge for each form. The pI is given by the mean of the pKa values on either side of the ionic form with zero charge. This was shown in Figure 2.11 for aspartic acid. From this we can see that pKa_1 and pKa_2 are the pKa values on either side of (ie which border) the form with no net charge. From this the isoelectric point is obtained as the mean of these 2 pKa values, ie 2.95. Note that this is a simplified strategy which works for amino acids found in proteins but would not invariably be accurate.

∏   Determine the pI of histidine, for which pKa values are as follows: α-carboxyl = 1.8; α-amino = 9.3; sidechain imidazole = 6.0. Hint: you will need to know (or look up) how the sidechain ionises!

The various forms of histidine must be displayed. When this is done, you can easily see which ionic species has no net charge.

The isoelectric point is given by the mean of pKa_2 + pKa_3, since these are the pKa values of ionisations leading to development of a net charge (ie they are on either side of the ionic species with zero charge). In this case $pI = \dfrac{6.0 + 9.3}{2} = 7.65$.

Structure 1 (charge +2):
$$H_3N^+ - \overset{\displaystyle COOH}{\underset{\displaystyle CH_2}{C}} - H$$
with side chain $HC = CH$, HN^+ NH, ring carbon C, H.

$\xrightarrow[1.8]{pKa_1}$

Structure 2 (charge +1):
$$H_3N^+ - \overset{\displaystyle COO^-}{\underset{\displaystyle CH_2}{C}} - H$$
with side chain $C = CH_2$, HN^+ NH, ring carbon C, H.

$\xrightarrow[6.0]{pKa_2}$

Structure 3 (charge 0):
$$H_3N^+ - \overset{\displaystyle COO^-}{\underset{\displaystyle CH_2}{C}} - H$$
with side chain $C = CH_2$, N NH, ring carbon C, H.

$\xrightarrow[9.3]{pKa_3}$

Structure 4 (charge -1):
$$H_2N - \overset{\displaystyle COO^-}{\underset{\displaystyle CH_2}{C}} - H$$
with side chain $C = CH$, N NH, ring carbon C, H.

| +2 | +1 | 0 | -1 |

In cases where there are 3 pKa values (as here), a common error is to take the mean of all 3 pKa values. Look again at the various ionic forms of histidine and their charges. The first ionisation (α-carboxyl) changes the charge from +2 to +1. This change (and hence this pKa) clearly has nothing to do with the pH of no net charge - it simply determines whether the charge is positive or more positive! Thus it must be ignored

and, in this case, $pI = \dfrac{pKa_2 + pKa_3}{2}$

2.3.5 Significance of ionisation in amino acids

What is the significance of the changes in ionic state which we have discussed. Several comments need to be made.

separation by electrophoresis

From the electrophoretic experiments, it is quite obvious that charge influences the movement of the amino acids. Their charge is controlled by pH. Thus separation of amino acids is strongly influenced by pH. Careful choice of pH allows particular amino acids to be separated from each other by electrophoresis. Other ways to separate amino acids are based on binding to insoluble resins which carry charged groups. This is the basis of ion-exchange chromatography; this is capable of separating all the amino acids which occur in proteins, which paper electrophoresis would not so easily accomplish. As a simple illustration of ion-exchange chromatography, examine Figure 2.13, where a negatively charged resin is being used to separate 3 amino acids. At the pH used (pH 6.0), lysine binds, whereas glycine and aspartic acid do not and can be removed. The adsorbed lysine can be removed by using a salt solution or by changing the pH.

separation by ion-exchange chromatography

∏ 1) What pH should the solution be adjusted to in order to desorb (ie remove) lysine from the ion-exchange resin? Hint: you have got to eliminate the net positive charge on the lysine!
2) How could the glycine and aspartic acid be separated? Hint: think what would happen if a resin with positive charges was now used.

The answer to 1) is to raise the pH to about 10.5. Lysine has been bound by the resin because at pH 6.0 it has a net positive charge, whilst the resin has a negative charge. Look at the response to the earlier in text activity on the charge on lysine. If you raise the pH to about 10.5, then the net charge on lysine will be slightly negative and lysine will be released.

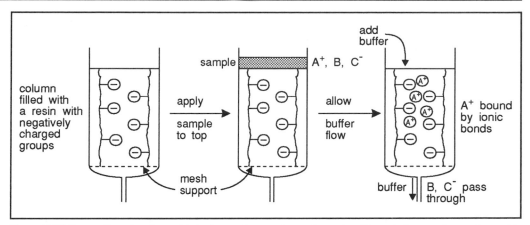

Figure 2.13 Separation of amino acids by ion-exchange chromatography. The net charge of an amino acid determines whether, and how tightly, it will bind to charged groups on a resin. The resin is packed into a column, so that the sample and buffer may be pumped through it. At pH 6.0 and with fixed negatively charged groups on the resin, only amino acids with no net charge (eg glycine) or a net negative charge (eg aspartate) pass through.

The answer to 2) is if the mixture of glycine and aspartic acid is applied, at pH 6.0, to a column packed with a resin with positive charges, the aspartic acid will bind. At pH 6.0 aspartic acid will be negatively charged. Glycine will have no net charge at pH 6.0 and will not be adsorbed. Removal of bound aspartic acid would subsequently require lowering of pH to abolish the net negative charge on it.

We have now seen that making use of charge, and the way it is influenced by pH, has considerable potential for separating amino acids. Note that, as proteins also possess charged groups (the sidechains of appropriate amino acids), they are also amenable to separation by these sorts of methods.

importance of amino acids

At this stage, note particularly that virtually all of the α-amino and α-carboxyl groups of amino acids in proteins are condensed together in forming peptide bonds, and can no longer ionise. The only free α-amino group in a polypeptide is that attached to the first amino acid. The only free α-carboxyl group is that at the other end of the polypeptide chain. Thus the vast majority of the charges present in a protein are those in sidechains of amino acid residues. This is why it is important to have some idea of the behaviour of these groups.

sidechain ionisation in proteins

Having stressed the importance of charge for separation of amino acids, and the relevance of sidechain groups to the charge of proteins, how important is this for the behaviour and properties of proteins? The charge of a protein is a function of which amino acid residues are present and the pH. Proteins, like amino acids, have an isoelectric point (pI) at which they possess no net charge. For proteins it cannot be deduced by simple calculation but has to be determined experimentally. At pH values below their isoelectric point proteins will have a net positive charge. If the pH is above the isoelectric point, they have a net negative charge. Since proteins differ in their isoelectric points, the charges of different proteins at a particular pH will differ. This allows them to be separated and illustrates how the behaviour of proteins (eg migration in an electric field or binding to ion-exchange resins) is pH-dependent.

influence of pH
on the
activities of
proteins

pH can have dramatic effects on protein function. This is also, of course, a consequence of change in ionic state of amino acid sidechains. We will learn in the next chapter that the ionic state of the sidechains of the amino acids greatly influences the structure of a protein which in turn greatly affects their activities. Let us examine an example to illustrate thus. Enzymes are proteins which speed up (catalyse) chemical reactions. For enzymes to work it is usually necessary for particular amino acid sidechains to be in a particular ionic form. This may be to enable substrate binding to occur, or for catalysis. For example, the enzyme lysozyme displays a bell-shaped curve for activity against pH (Figure 2.14).

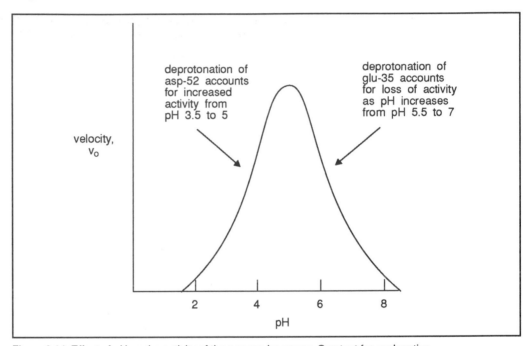

Figure 2.14 Effect of pH on the activity of the enzyme lysozyme. See text for explanation.

This is because 2 active site groups must be in a particular form for catalysis to occur. At the pH optimum they are both in the 'correct' form. As pH is lowered, the carboxyl group of an aspartate residue protonates, with loss of activity which precisely mirrors the degree of protonation. As pH is raised above pH 5.0, activity also declines, this time because of deprotonation of a glutamic acid sidechain. This example is typical of the requirement for a precise charge distribution for biological activity (whether enzymatic or not). It also justifies the importance we have given to the effect of pH on ionisation of amino acids and their sidechains.

SAQ 2.6

Assume that binding of L-alanine to an enzyme relies solely on an ionic bond forming between the carboxyl group of alanine and a histidine residue in the active site of the enzyme. At which of the following pH values will binding be strongest? Give reasons for your answer.

1) pH 1.0, 2) pH 4.0, 3) pH 7.0.

Assume that the active site histidine is in the middle of the polypeptide chain of the enzyme.

2.4 Identification and reactions of amino acids

Amino acids may be separated by electrophoresis and by chromatography. When the complete amino acid composition of a sample is required, chromatography is necessary. This separation is usually performed by ion-exchange chromatography. Here we consider how the separated amino acids can be detected, identified and their quantity determined.

ninhydrin
reaction

Amino acids give the reactions expected of amino groups. The one which is widely used for visualisation and quantitation is the ninhydrin reaction (Figure 2.15).

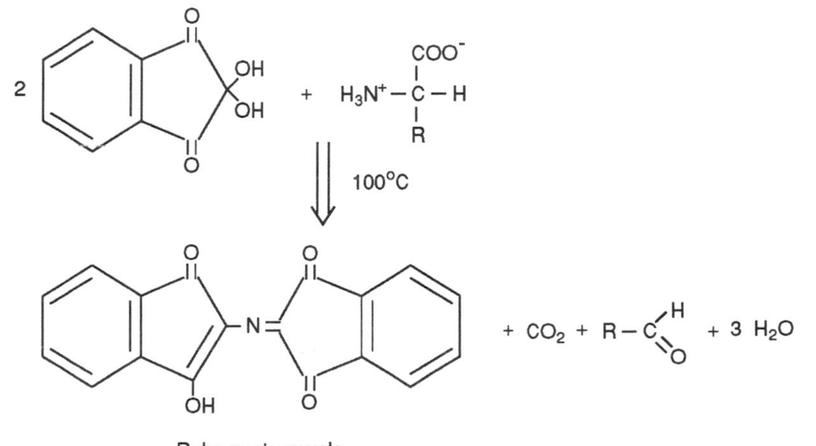

Figure 2.15 The ninhydrin reaction. The amino group of the amino acid reacts (at elevated temperature) to form an intense purple-coloured compound (Ruheman's purple).

Ninhydrin (dissolved in acetone) may be sprayed onto the paper used as a support in electrophoresis or paper chromatography. Alternatively, when amino acids are in solution, a solution of ninhydrin may be mixed with them. Heating (at around 100°C) produces a characteristic purple colour (Ruheman's purple) whose intensity is proportional to the amount of the amino acid present. Samples in solution can therefore, be quantified by measuring the amount of purple colour produced (ie measurement of absorbance).

chromogenic
reactions

Whilst amino acids also give reactions expected of carboxylic acids, many other biological materials do so as well. Note that other amino-containing compounds will also react with ninhydrin. Some amino acids can be identified and quantified by more specific reactions of their sidechains. There are various colour-forming reactions (often known as chromogenic reactions), such as those for cysteine (reaction of DTNB with the sulphydryl group) and for arginine (the Sakaguchi reaction). DTNB is 5,5'-dithio-bis (2 nitrobenzoic acid). We deliberately will not discuss these in detail; just note that reactions selective for particular chemical groups exist.

Summary and objectives

In this chapter we have examined the range of amino acids which make up proteins. We have examined aspects of their stereochemistry and dwelt upon their ionisation behaviour. Using knowledge of these properties we have learnt how amino acids may be separated by ion exchange chromatography and by electrophoresis. We have also learnt how amino acids can act as buffers.

Now that you have completed this chapter (and Appendix 1) you should be able to:

* identify amino acids from their sidechains;

* group amino acids according to the nature of their side groups;

* from three dimensional drawings, work out if an amino acid is in an L- or D- configuration;

* calculate the ratio of ionised to unionised forms of a weak acid from given pH and pKa values;

* explain why weak acids act as buffers around their pKa values;

* select from a list of compounds with known pKa values, suitable candidates to use as buffers at particular pH values;

* predict the ionic forms of amino acids at different pH values;

* predict the behaviour of amino acids in electric fields at given pH values.

Proteins

Proteins

At the beginning of Chapter 2, we established that proteins are very important in biological systems. Now that we have learnt much about the properties of the building blocks of proteins, the amino acids, we are in a better position to examine and understand the structure and properties of proteins.

In this chapter, we will begin by examining the structure of proteins and why they take up particular conformations. This will, of course, enable us to make certain predictions about their properties. We will then go on to briefly consider how proteins may be purified from complex mixtures.

3.1 The peptide bond and primary structure of proteins

Proteins consist of an unbranched chain of amino acids linked by peptide bonds. A peptide bond is formed between the α-carboxyl group of one amino acid and the α-amino group of a second by a condensation reaction in which a molecule of water is eliminated. Whilst the overall reaction is shown in Figure 3.1, you should note that, in cells, the reaction does not proceed directly in this way.

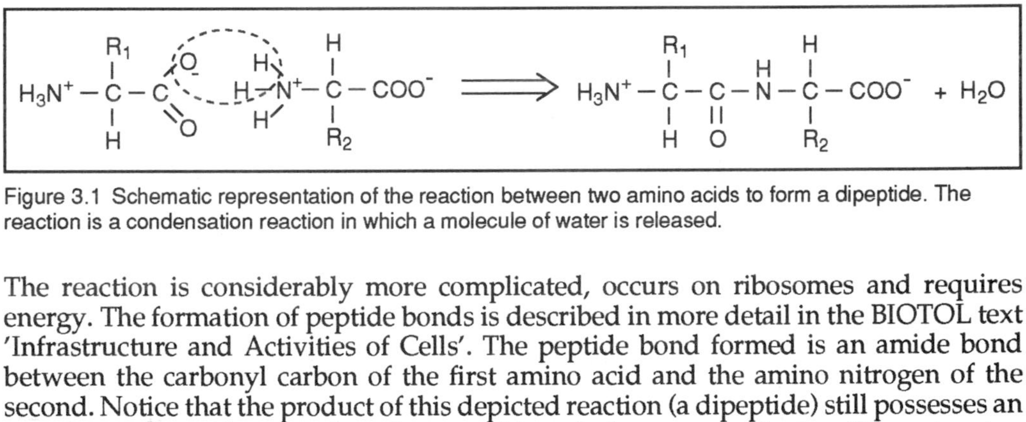

Figure 3.1 Schematic representation of the reaction between two amino acids to form a dipeptide. The reaction is a condensation reaction in which a molecule of water is released.

The reaction is considerably more complicated, occurs on ribosomes and requires energy. The formation of peptide bonds is described in more detail in the BIOTOL text 'Infrastructure and Activities of Cells'. The peptide bond formed is an amide bond between the carbonyl carbon of the first amino acid and the amino nitrogen of the second. Notice that the product of this depicted reaction (a dipeptide) still possesses an α-amino group and an α-carboxyl group, although these are attached to different amino acid residues. Note that after incorporation into proteins, amino acids are referred to as amino acid residues, since part of each amino acid is lost in the formation of the peptide bond; what remains is a 'residue'. Since α-amino and α-carboxyl groups still occur, more peptide bonds can be made. Proteins vary in the number of the amino acid residues from a few (an oligopeptide) to many hundreds (a polypeptide). The precise number of amino acid residues, which amino acids they are and their order are determined by the DNA of the gene which codes for them. Each protein has a precise sequence of amino acid residues, giving a macromolecule of a precise size. The sequence of a protein ie the precise order and identity of amino acid residues within it, is called the primary structure of the protein. There is considerable evidence, some of which is presented in Section 3.7, that the primary structure determines the final shape of the protein and, through its shape, its biological properties. Thus the amino acid sequence is vital to produce the correct conformation and biological activity.

amino acid residues

dipeptide, oligopeptide, polypeptide

amino acid sequence

primary structure

The peptide bond as it is usually depicted is shown in Figure 3.2a. This implies that the carbon-oxygen bond of the carbonyl group is a double bond, whilst the carbonyl-carbon to amino-nitrogen bond is a single bond. In fact, the extra pair of electrons which would be expected to be associated with the carbon-oxygen bond are partially shared with the peptide bond (Figure 3.2b). The electrons of the double bond are said to be delocalised; the outcome of this sharing is that both the carbon-oxygen and the carbon-nitrogen bonds have partial double bond character. This has several consequences, of which those for bond lengths and bond rotation are notable.

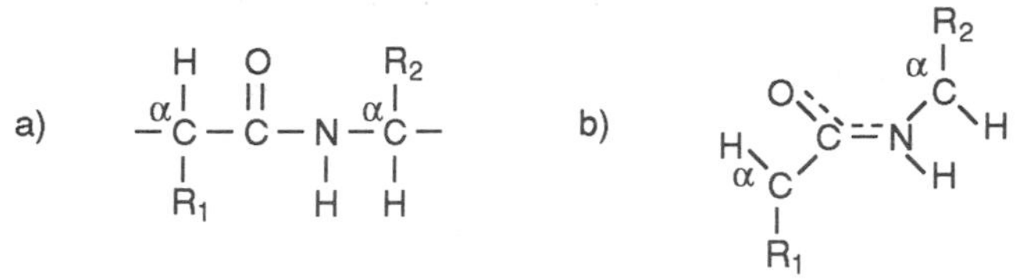

delocalised electrons in the peptide bond

Figure 3.2 The peptide bond a) Conventional depiction of the peptide bond, with a single covalent carbon-nitrogen bond and a double-bonded carbonyl group. b) This representation is more realistic, showing the extra electrons of the carbonyl double bond delocalised and giving a partial double bond character to the peptide bond.

Bond lengths. Listed in Table 3.1 are expected bond lengths for selected covalent bonds, together with those (measured by X-ray crystallography) for bonds in peptides.

Bond	Approx. bond length (nm)
Single bond, C-O	0.143
Double bond, C=O	0.123
C = O, in peptide bond	0.124
Single bond, C - N	0.147
Double bond, C = N	0.127
Peptide bond C ⋯ N	0.132

Table 3.1 Lengths of covalent bonds.

This indicates that the peptide bond is considerably shorter than would be expected for a carbon-nitrogen single bond, and is more satisfactorily represented as in Figure 3.2b. The carbonyl carbon-oxygen bond is slightly longer than would be expected for a carbon-oxygen double bond.

Bond rotation. The sharing of the delocalised electrons of the carbonyl group over the peptide bond prevents rotation around the peptide bond. If this bond was a single bond, rotation around it would be possible (Figure 3.3a), whereas in molecules involving double bonds rotation is not possible (Figure 3.3b).

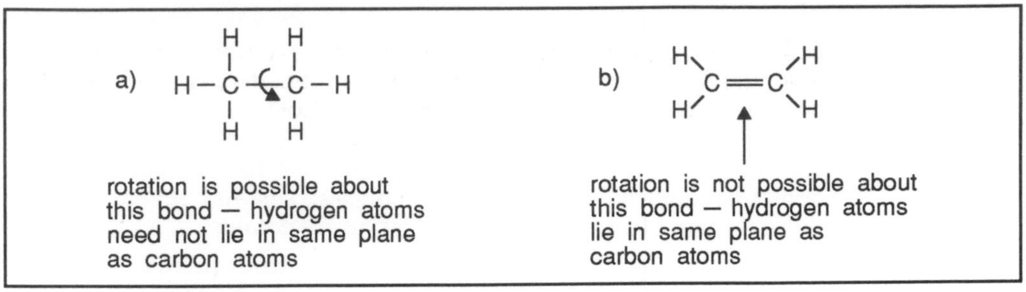

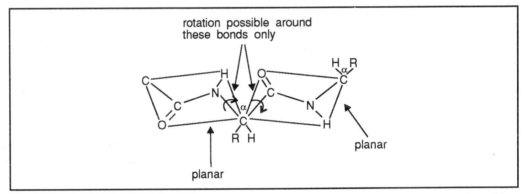

Figure 3.3 a) Rotation is possible about this covalent bond. The hydrogen atoms will not all lie in the same plane, and will not be in a fixed position. b) Rotation is not possible with a double bond. A consequence is that the hydrogen atoms are fixed and lie in the same plane as the carbon atoms.

The partial double bond nature of the peptide bond is sufficient to prevent any rotation. As a consequence, all of the atoms attached to the carbon and nitrogen atoms of the peptide bond are held in the same plane (Figure 3.4).

Figure 3.4 The partial double bond character of the peptide bond (shown in Figure 3.2b) means that all the atoms attached to the carbon and nitrogen atoms of the peptide bond will lie in the same plane.

The peptide bond is said to be planar. This has important consequences for the structure of proteins, since rotation is only possible around certain bonds of the polypeptide 'backbone'. These are the carbonyl carbon to α-carbon bond and the α-carbon to α-amino nitrogen bond, ie those on either side of each α-carbon atom. The α-carbon atom thus acts as a swivel between each planar peptide bond and the next one.

planar peptide bond

trans configuration

Note also that, as shown in Figures 3.2 and 3.4, the carbonyl oxygen atom and the amino hydrogen are positioned on opposite sides of the planar peptide bond, (ie in a trans configuration). The sidechain groups also adopt a trans configuration. When an extended chain is formed, successive sidechains alternate between positions above and below the main backbone of the polypeptide. This is more obvious in the β-pleated sheet forms which we will discuss later (Figure 3.17).

N-terminal amino acid

C-terminal amino acid

In describing the primary structure of a protein, it is customary to write the structure beginning with the amino acid which has a 'free' α-amino group. This is known as the aminoterminal or N-terminal residue of the polypeptide. Correspondingly, the last residue is that with a free α-carboxyl group, which is known as the C-terminal or carboxyterminal residue. Since each polypeptide is an unbranched linear sequence of amino acid residues, there is only one N-terminal and one C-terminal residue per chain. This convention is vital, otherwise ambiguity would occur.

Π Write down, using the 3-letter abbreviations given earlier, the amino acid sequences of the following peptides. You may have to use Figures 2.3 to 2.7 to identify the amino acid residues involved.

1)
$$H_3N^+ - \underset{\underset{H}{|}}{\overset{\overset{H}{|}}{C}} - \overset{\overset{O}{||}}{C} - N - \underset{\underset{H}{|}}{\overset{\overset{CH_3}{|}}{C}} - \underset{\underset{O}{||}}{\overset{}{C}} - N - \underset{\underset{CH_2}{|}}{\overset{\overset{H}{|}}{C}} - \overset{\overset{O}{||}}{C} - N - \underset{\underset{H}{|}}{\overset{\overset{CH_3}{|}}{C}} - COO^-$$
$$COO^-$$

2)
$$^-OOC - \underset{\underset{CH_3}{|}}{\overset{\overset{H}{|}}{C}} - N - \overset{\overset{H}{|}}{\underset{\underset{O}{||}}{C}} - \underset{\underset{H}{|}}{\overset{\overset{H}{|}}{C}} - N - \overset{\overset{O}{||}}{C} - \underset{\underset{(CH_2)_4}{|}}{\overset{\overset{H}{|}}{C}} - {}^+NH_3$$
$$^+NH_3$$

In peptide 1), the amino acid residues are, from the left, gly, ala, asp and ala. By convention, peptides are described from the N-terminal end. Glycine possesses the free α-amino group in this peptide and is therefore the N-terminal amino acid. Hence the sequence is as given above. In peptide 2), the N-terminal amino acid is the third residue as written. Thus the N-terminal amino acid is lysine and the sequence is lys-gly-ala.

numbering of amino acids
With this convention of describing the primary structure of a polypeptide from N- to C-terminal, it is also simple to unambiguously identify a particular amino acid residue. This is done by numbering the residues sequentially from the N-terminal end, the N-terminal residue being number 1. For example the glutamate residue described earlier whilst discussing the effect of pH on the activity of lysozyme (see Figure 2.14) is the 35th amino acid of the peptide chain and can be written as glu-35.

SAQ 3.1

The primary sequence of a polypeptide is as follows:
glu-val-thr-ala-glu-val-ser-asp-pro-arg-val-asp-ala-gly-arg-val-asp-lys.
Which of the amino acid residues:

1) has a free α- carboxylic acid residue;

2) has a free α- amino group;

3) what is the number of the third valine residue?

3.2 Determination of the primary structure of proteins

information needed to determine the primary structure of a protein
Determination of the primary structure of a protein is a complicated procedure. It is first necessary to break the cells up which contain the protein. Typically most cells contain many thousands of different proteins. The protein of interest must be purified from this mixture and single polypeptide chains obtained. Then it is necessary to determine the amino acid composition of the purified polypeptide (this means determining how many of each amino acid are present in the chain). Finally, the sequence of the amino acids in the polypeptide is determined. This involves determination of N- and C-terminal amino

acids and then identifying each residue, sequentially, from the N-terminal end. Cutting the polypeptide at particular positions in the sequence (by use of enzymes which cleave after particular residues, or using similarly specific chemical hydrolyses) is also necessary. In this way the entire primary structure can be determined. The precise strategy and methods are discussed elsewhere in this series. We should also note that it is now possible to deduce the primary structure of a protein from the sequence of bases in the DNA which codes for it. DNA sequencing can thus provide the desired information. We will learn more about this in Chapter 4.

The complete primary structure for insulin was determined, in 1955, by Sanger, for which he received the Nobel prize in chemistry in 1958 (Sanger subsequently received a second Nobel prize, this time for devising and exploiting methods to determine the sequence of bases in DNA; his contribution to our understanding of these macromolecules is very considerable indeed!). Since then the primary structures of many proteins have been described. There have been numerous improvements in methodology and automated 'sequencers' are now widely used. The benefits resulting from knowledge of the primary structure of proteins are various. Here we list a number of them.

disulphide bridge

phosphoproteins

glycoproteins
Rconjugated proteins

In terms of the general structure of proteins, sequence analysis has shown that a protein has a unique amino acid sequence, known to be determined by the base sequence of the gene which codes for it. The only covalent bonds between amino acid residues are peptide bonds and disulphide bridges (we will examine these later) which can form between the sidechains of cysteine residues. Additional covalent bonds occur between the sidechains of some residues and other compounds. These include phosphate (in phosphoproteins), sugars (in glycoproteins) and various cofactors (known generally as conjugated proteins).

As we shall see in Section 3.7, there is ample evidence that the primary structure determines the final 3-dimensional shape of a protein. This, in turn, is responsible for its biological activity. It was hoped that knowledge of the primary structure of many proteins would reveal simple 'folding rules' for proteins. To date these remain to be elucidated; if they exist, they are apparently not simple!

protein engineering

Knowledge of the primary structure is very useful when determining the 3-dimensional structure by X-ray diffraction techniques. Only when the complete 3-dimensional structure is known is it possible to convincingly demonstrate how proteins work, for example, how enzymes achieve catalysis. This information is also necessary before logical and precise changes to the structure of a protein can be made. This is what the technique of protein engineering accomplishes, which is discussed further in the final chapter of this text.

sickle cell anaemia

Determination of the primary structure of proteins demonstrated that some diseases are caused by alterations in the amino acid sequence of particular proteins. One of the earliest to be elucidated, which is of very widespread occurrence, was sickle cell anaemia. This is caused by a single change in the amino acid sequence of the β-chain of haemoglobin. Haemoglobin consists of 4 polypeptides, 2 of one amino acid sequence (called α) and 2 of a second sequence (called β). Normal haemoglobin has a glutamate at position 6 of the β-chain. In individuals suffering from the sickle cell disease, position 6 is a valine. This single substitution is none-the-less sufficient to create a seriously disabling disease. Many disease states are now known to be inherited and 'molecular' in origin.

Determination of the primary structure of proteins reveals how 'related' they are to each other. There are similarities in sequence only if proteins have evolved from a common 'ancestor' gene. The mammalian serine proteases (enzymes which hydrolyse proteins) contain about 40% of the amino acid sequences in identical arrangements. Examples of these enzymes are chymotrypsin, trypsin and elastase. Studies of this type strongly suggest that certain parts of the sequence are absolutely critical and so are conserved: if changes occur, activity is lost. In other parts of a protein, it is less crucial for a particular residue to be at a particular position. Thus for all 3 of the serine proteases mentioned, the sequence around the active site is -gly-asp-ser-gly-gly-pro-. They are called serine proteases because serine is involved in the active site. Conservation of this particular sequence is a powerful indication of its importance.

serine proteases

The 'degree of relatedness' can also be applied to assessing when divergence occurred during evolution. If the amino acid sequence of a particular protein (cytochrome c is a good example) from a large range of organisms is determined, it is found that sequence similarity is high where species are closely related (eg man and monkey). Sequence similarity is less when species are more distantly related (eg mammals and plants). Examples are shown in Table 3.2.

Organisms	Number of differences in amino acid sequence
Man and Rhesus monkey	1
Man and sheep	10
Man and penguin	13
Man and tuna fish	21
Man and wheat	43
Man and yeast	45
Total number of amino acid residues = 104	

Table 3.2 The degree of sequence similarity of cytochrome c in various organisms.

chemical taxonomy

This approach allows for a 'chemical taxonomy' to be constructed, which can be used, together with traditional taxonomies (primarily morphological) to establish how different groups of organisms are related.

Only if the primary structure of a protein is known can it be chemically synthesised, outside a cell. This might have enabled desirable proteins to be made industrially, by organic chemists. In fact, chemical (ie non-biological) synthesis of any but the smallest proteins is technically very difficult, although synthesis of the 124 residue enzyme ribonuclease A was accomplished in 1965. Whilst this demonstrated that the protein product was indistinguishable from the natural molecule (and thus enzymes made by cells do not depend for activity on any mysterious 'life-force'), it does not offer industrial potential. Developments in genetic engineering since the early 1970s provide excellent routes for production of particular proteins in large yields.

You can see that there are varied and important reasons for determining the primary structure of proteins. In summary, a comprehensive understanding of the structure, function and mode of action of a protein requires knowledge of its amino acid sequence.

3.3 Protein conformation and stabilising forces in proteins

Each protein naturally occurs in a single conformation (unless, as a result of binding other molecules, it switches to a second conformation: this occurs, for example, when haemoglobin binds oxygen). The particular primary structure of a protein results in a unique shape which is known as the native conformation. The reason why only one conformation occurs is that it is more stable than other possible shapes. We will consider the reasons for this in this section.

native conformation

3.3.1 Levels of structural organisation

Before examining the overall shape of proteins, we will explain the way in which protein structures are usually described. Four structural levels are used:

Primary structure

This is the sequence of the amino acid residues, together with the position of any disulphide bridges which have formed between two cysteine residues (see later). Primary structure thus includes all the covalent bonds present between amino acid residues, but indicates nothing about the overall shape.

Secondary structure

This involves the occurrence of any regular (repeating) structures or patterns, usually formed by amino acids close to each other in the sequence. These are primarily stabilised by hydrogen bonds and involve either helical or sheet-like structures. We will examine these in Section 3.4.

Tertiary structures

The tertiary structure is the overall shape of the polypeptide; it is usually thought of as the way in which sections of secondary structure (if they occur) are positioned relative to each other. It includes the various bonds formed between parts of the polypeptide which may be far apart in the primary structure.

The distinction between secondary and tertiary structure is not rigid: these terms are simply helpful in describing protein structure. A protein's structure will be the sum of secondary and tertiary structures, since it refers to the overall 3-dimensional shape, covering the path that the polypeptide chain follows as well as the position that the sidechains occupy.

Quaternary structure

This only occurs when more than one polypeptide chain associate together to form the biologically functional molecule. Each polypeptide is described as a subunit, or protomer. Subunits then associate to form the active oligomer. Proteins in which there is only a single polypeptide (such as myoglobin) thus do not have a quaternary structure. Haemoglobin, which consists of 4 polypeptides (2 of one primary sequence, called α-subunits; 2 of a different, although similar, sequence, called β-subunits) is described as having the quaternary structure $\alpha_2\beta_2$.

sub-unit, protomer

3.3.2 Stabilising forces in proteins

We have already said that a protein adopts one particular conformation, which is biologically active, rather than numerous alternative forms (most of which would be inactive). How does this happen? The reason is that the so-called native conformation is more stable than all other possible conformations and is a consequence of a large number of individually weak non-covalent bonds or interactions. Since these are very important, not only for protein structure but also for other macromolecules (eg nucleic acids) and for the interactions between molecules (eg enzymes and substrates; hormones and receptors) we shall review them. Typical strengths of these bonds are shown in Table 3.3, together with that of the peptide bond. Note also that proteins may be stabilised by the formation of disulphide bridges, which will also be discussed.

Bond	Bond strength*, kJ.mol^{-1}
Covalent (peptide, disulphide)	Approx. 400
Hydrogen (C=O....HN-)	up to 20
Ionic (-COO⁻....H$^+_3$N-)	up to 4
Hydrophobic (eg overlapping phenyl rings)	4-8
Van der Waals (transient dipoles)	up to 4

Table 3.3 Strengths of bonds involved in stabilisation of protein structure.
*Bond strength, or bond energy, is the energy required to break the bond.

∏ It would be helpful to you if you drew up a sheet to summarise the properties of each type of bond as you read the next sections.

a) Hydrogen bonds

You may have wondered why secondary structures should occur. Of all the possible directions (ie bond angles) which the polypeptide chain can adopt, why does it follow a particular repeating pattern? The reason is that in so doing, hydrogen bonding is maximised, thereby making such a structure much more stable.

Hydrogen bonds are bonds formed between a hydrogen atom (attached to an electronegative atom, usually nitrogen or oxygen) and a second electronegative atom. By 'electronegative atom' we mean an atom that tends to draw electrons (from surrounding bonds) towards it, giving it an extra, partially negative charge. This leads to partial charge separation in the first group (which is thus a dipole) resulting in the hydrogen atom possessing a partial positive charge (Figure 3.5a). This can be electrostatically attracted to a second atom providing that atom is at least partially negative: this is why the second atom is described as electronegative. The overall effect is that the hydrogen atom is shared between the two other atoms. Hydrogen bonds are highly directional, and are strongest when the three relevant atoms (N-H---O in Figure 3.5b) are in a straight line. The strength of the bond falls rapidly as non-linearity increases.

hydrogen
bonds are
directional

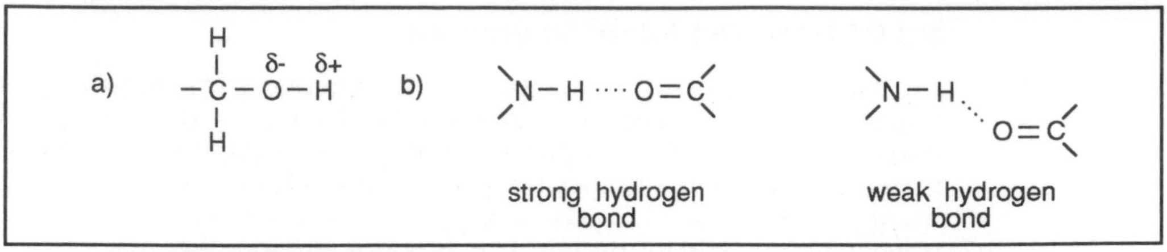

Figure 3.5 a) Partial charge separation enables hydrogen bonds to form. The greater electronegativity of oxygen results in bonding electrons being drawn to the oxygen atom, giving it a partial negative charge (δ-). As a consequence the hydrogen atoms becomes partially positively charged (δ+). b) Hydrogen bonds are strongest when the atoms involved are aligned. Weaker bonds result if these atoms are not all in line.

The atom which formally contributes the hydrogen atom is called the donor and is usually an oxygen or a nitrogen, although sulphur atoms can be donors. The second atom is referred to as the acceptor and can be either partially negative (as described above) or may have a full negative charge. Routinely oxygen or nitrogen is the acceptor, although the sulphur atom in the sidechain of methionine may also be an acceptor. A selection of hydrogen bonds is shown in Figure 3.6.

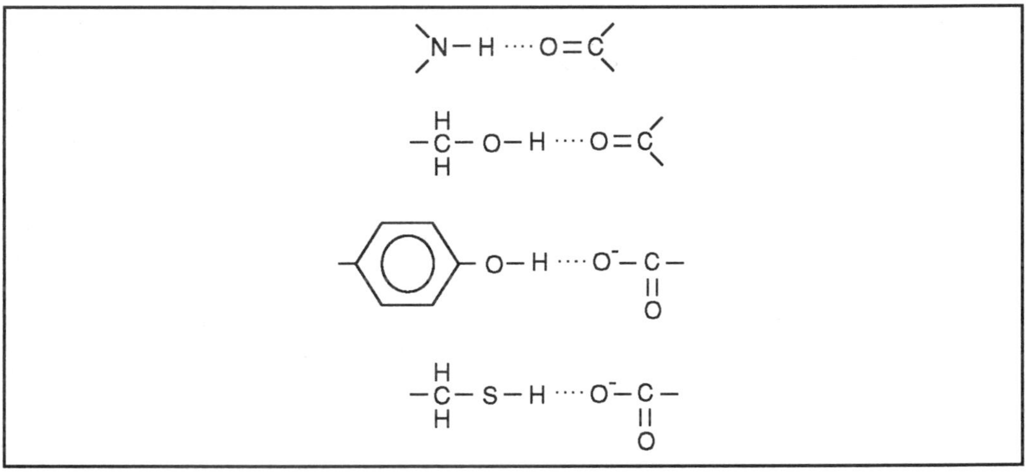

Figure 3.6 Hydrogen bonds between various groups are found in proteins. A selection is shown.

Their lengths and strengths vary according to which atoms are involved. The reason for the widespread occurrence of the α-helix (a type of secondary structure) is that many hydrogen bonds occur and their strength is maximised by the atoms involved being colinear.

b) Electrostatic (ionic) bonds

These are formed between two oppositely charged groups. Their strength depends on the charges (e), the distance between them (r) and the dielectric constant (ε) of the medium. The force (F), ie the strength of the bond, is given by Coulomb's law:

$$F = \frac{e^2}{r^2 \varepsilon} \text{ or } \frac{e_1 e_2}{r^2 \varepsilon}$$

Two factors are clearly vital:

- as the distance between the charges increases, the attraction between them (assuming opposite charges are involved) drops rapidly, since D is a squared term. Note that if the charges are similar (eg both positive) then a repulsion develops. Trying to force two similar charges close together can produce a strong repulsion which will influence conformation;

- a solvent with a high dielectric constant (such as water, $\varepsilon=80$) will make for a weak interaction. Organic solvents (eg chloroform, $\varepsilon=4.8$) will give much stronger electrostatic bonds. Thus these bonds will be weak in aqueous medium; however, as we shall see, the inside of proteins is predominantly non-polar and electrostatic bonds will be much stronger there.

c) Van der Waals bonds

Van der Waals bonds can occur between atoms or groups which are non-polar and electrically neutral. Although in these groups the electron cloud is, overall, symmetrically distributed around the constituent atoms (ie electronegativity of one atom does not lead to electrons being concentrated around that atom), none-the-less local charge differences can transiently occur. Electrons are, of course, mobile and, at a given instant, may not be evenly distributed, hence a charge asymmetry transiently occurs and a dipole is created. Such a dipole can influence a neighbouring atom or group, inducing a corresponding electron asymmetry. This is the basis for a 'bond', since, when the interacting atoms are a particular distance apart, energy would be required to separate them.

The distance between the atoms involved is critical. The attraction is at its greatest when the atoms are separated by the 'van der Waals contact distance'. Each atom has a characteristic contact radius: for carbon it is 0.2nm. This radius can be defined as the van der Waals radius. The van der Waals contact distance is the sum of the relevant contact radii. Atoms which become too close will be strongly repelled; as the distance increases above the contact distance, attraction falls off rapidly (Figure 3.7). Despite their individual weakness (Table 3.3), van der Waals bonds are important in stabilising proteins and in the binding of substrates by enzymes, hormones by their receptors, and antigens by their antibodies. This is because large numbers of van der Waals interactions will occur and the combination of these many interactions have a strong influence on both protein structure and the binding of other molecules by proteins. Anything which prevents their formation (eg mismatches of shape) leads to marked weakening of the overall interaction. This is how enzymes and antibodies obtain their specificity and strength of binding ('affinity') to their substrates or antigens. Note that the other types of weak non-covalent bonds we are discussing are also often involved in the binding of compounds by proteins.

d) Hydrophobic interactions

The hydrophobic interaction is the driving force behind the separation of fluids such as water and cooking oil when they are mixed together. Although a mixture can be made (by vigorous stirring or whisking) they will rapidly separate to form two phases. Thus dispersal of one phase in the other is usually unstable. It spontaneously changes to a more stable system in which contact between the two phases is minimised. This same phenomenon is one of the principal forces responsible for proteins folding to the shapes which they adopt. However, before examining hydrophobic interactions, we must first examine the structure of water.

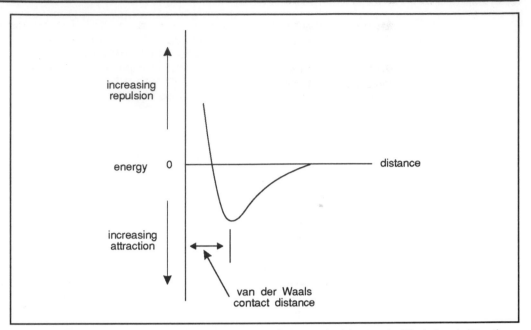

Figure 3.7 The influence of distance between atoms on the strength of the van der Waals bond. Attraction is maximal at a critical distance, the van der Waals contact distance. When closer, strong repulsion occurs (electron orbitals interact); when more distant, attraction falls off rapidly.

Water molecules interact strongly with each other by hydrogen bonds. We have already seen that the oxygen atom is electronegative and draws bonding electrons towards it, leaving hydrogen atoms attached to it relatively deficient in electrons. This partial charge separation leads to extensive hydrogen bonding (Figure 3.8), which explains many of the properties of water (high boiling point compared to other alcohols or other hydrides, eg hydrogen sulphide, H_2S).

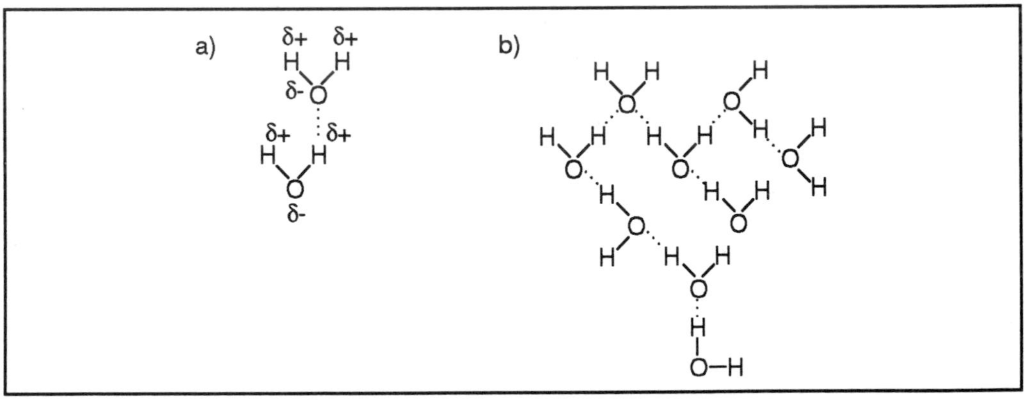

Figure 3.8 a) Due to partial charge separation, hydrogen bonding occurs between water molecules. b) This hydrogen bonding is extensive and is responsible for many of the properties of water, such as high boiling point.

If ionic compounds are introduced into water, they will interact favourably with water molecules. This is why they dissolve. Indeed, this favourable interaction between water and charged molecules means that ionic bonds are very weak in water, because water competes (Figure 3.9).

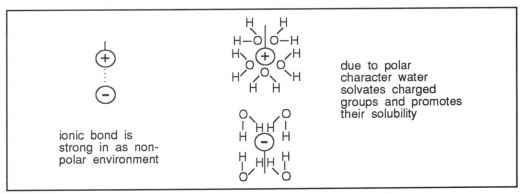

Figure 3.9 Disruption of an ionic bond by water. Water, as a consequence of its dipolar nature, surrounds charged groups, thereby leading to their solution. Thus ionic compounds dissolve in water.

As we have commented, ionic bonds are much stronger when they occur in a non-aqueous hydrophobic environment. This has two consequences:

- atoms held together by ionic attraction (eg NaCl) do not dissolve in non-aqueous solvents, because the ionic bond between the Na^+ and the Cl^- is so strong;

- ionic bonds in proteins become much more important when they occur in hydrophobic surroundings, as is usually the case in the interior of proteins. Such bonds are only strong if water is excluded.

In contrast, non-polar hydrophobic groups do not interact favourably with water: quite the reverse. They disrupt the hydrogen bonding between water molecules (Figure 3.10a). This disruption to the overall hydrogen bonding within the surrounding water is lessened if all the hydrophobic groups coalesce and become close to each other, thereby permitting maximum hydrogen bonding in the surrounding water to occur (Figure 3.10b).

The hydrophobic groups will also form van der Waals bonds. Since the hydrophobic 'effect' is non-specific, it is referred to as an interaction, rather than a bond (van der Waals bonds are also often thought of as 'interactions'). The principal source of its energy (in the sense that, once hydrophobic groups have coalesced, energy will be required to disperse or separate them) is essentially 'negative'. It derives from permitting more hydrogen bonds to form in the surrounding water solvent than if the hydrophobic groups were individually 'sticking out' into the solvent.

As far as proteins are concerned, hydrophobic interactions occur between the non-polar sidechains of the amino acids shown in Figure 2.3, such as valine, leucine and phenylalanine. Note that hydrophobic interactions are also important in stabilising the double helical DNA molecule (Chapter 4) and in the association of lipids (Chapters 6 and 7).

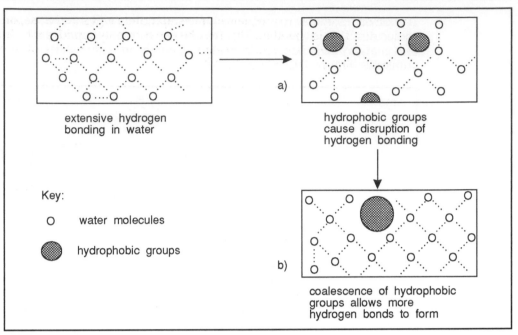

Figure 3.10 a) The presence of non-polar hydrophobic groups disrupts the hydrogen bonding between water molecules. As a consequence, less than the maximum number of hydrogen bonds are formed. b) If several non-polar groups coalesce, with exclusion of water, then the disruption to hydrogen bonding in the surrounding water is lessened. This increased hydrogen bonding (plus van der Waals bonds within the hydrophobic phase) provides the driving force ('energy') for coalesce of non-polar groups.

e) Disulphide bridges

Although these are strong, covalent bonds, they are important in stabilising protein conformation. If upon folding (which is primarily a consequence of the primary structure and the weak bonds we have just reviewed) two sulphydryl residues (from cysteine residues) become positioned close to each other, they may, via oxidation, become covalently joined by a disulphide bridge (Figure 3.11). This is a strong bond requiring about 400 kJ.mol^{-1} to be broken. Thus disulphide bridges, if formed, tend to 'lock' proteins into a particular conformation. This may lead to a more stable protein, whose functional lifetime is consequently increased. Indeed, biotechnologists have used a technique called protein engineering to modify enzymes, such as lysozyme, to increase their stability by increasing the number of disulphide bridges.

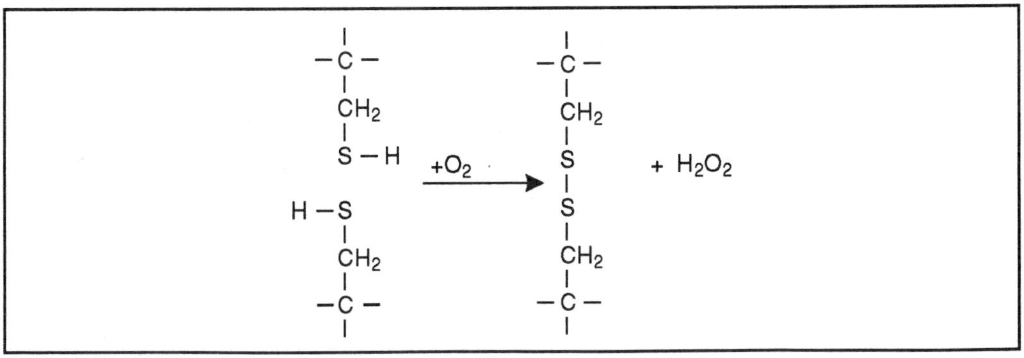

Figure 3.11 Formation of a disulphide bridge between two cysteine residues.

The other particular consequence of disulphide bridges is that separate polypeptide chains can be held together. This may be between polypeptides which were separately synthesised; the disulphide bridges form after association of subunits, as with antibodies of type IgG (Figure 3.12a).

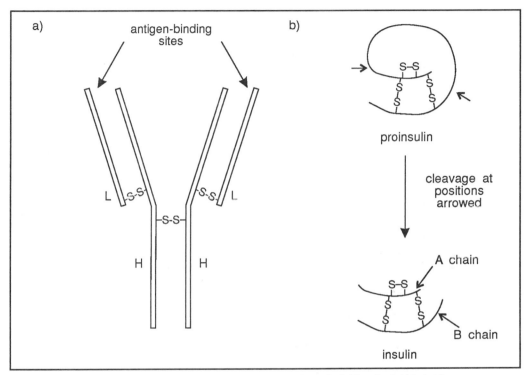

Figure 3.12 a) Simplified structure of an antibody molecule of type IgG. The disulphide bridges ensure that the 4 chains (2H, 2L) remain joined together. Antigen binding sites are arrowed and involve contributions from both an H and a L chain. b) Activation of proinsulin involves proteolytic cleavage and loss of a peptide (the C-peptide). The remaining fragments (the A and B chains) remain connected to each other by disulphide bridges.

Alternatively, a protein may be synthesised as a single polypeptide chain, which folds into a particular conformation and is then stabilised in that conformation by disulphide bridge formation. It may subsequently be modified, by a proteolytic enzyme cleaving between particular residues, thereby cutting the chain; a section of the chain may be lost. Thus insulin (Figure 3.12b) is synthesised as a single chain, which is converted from an inactive form (proinsulin) to its active structure (insulin) by the action of a trypsin-like enzyme. This cuts the original polypeptide twice, leading to removal of a connecting peptide of approximately 30 residues (the C-peptide). The remaining peptides remain connected by disulphide bridges (Figure 3.12b). Note that synthesis of inactive precursor polypeptides (proenzymes) which are subsequently activated by proteolytic cleavage is of widespread occurrence.

SAQ 3.2

Below is a diagrammatic representation of a polypeptide chains showing interactions between some sidechains of its constituent amino acids. Which of the interactions shown are:

1) polar in nature; 2) hydrogen bonds; 3) hydrophobic interactions; 4) disulphide bridges.

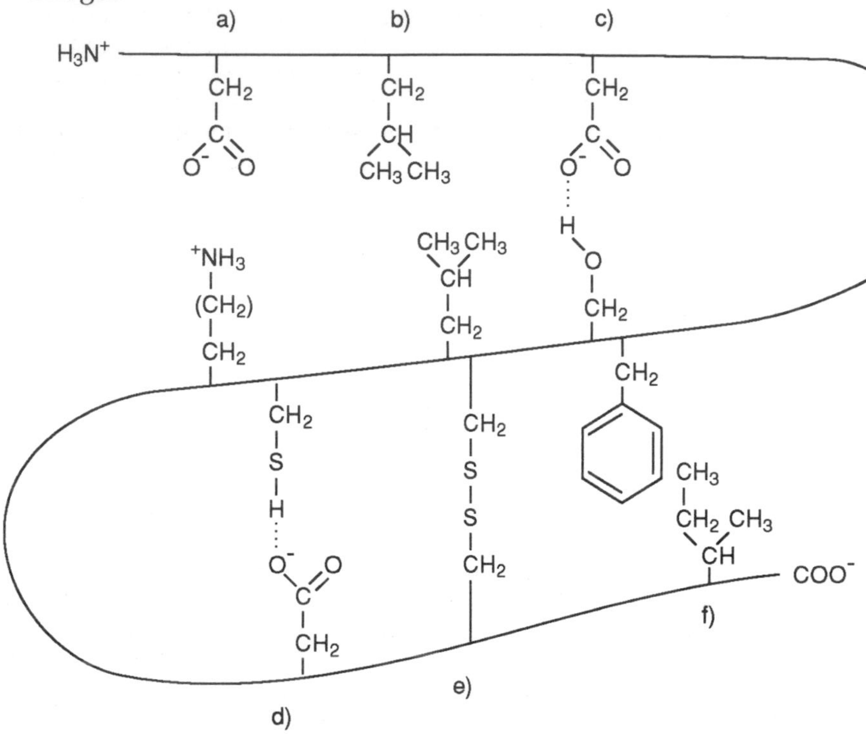

3.4 Secondary structure of proteins

experimental evidence from X-ray diffraction, optical rotatory dispersion, circular dichroism and NMR techniques

Information on folding, the occurrence of secondary structure and on conformation changes can be obtained from a variety of physical methods. The method which has been invaluable in determining protein structure is X-ray diffraction by crystals. This has, in successful cases, provided 3-dimensional structures in which the majority of atoms can be located. Other methods, particularly optical rotation methods (eg optical rotatory dispersion and circular dichroism) and nuclear magnetic resonance (NMR) have also provided useful information.

proteins produce complex diffraction patterns

The principle of X-ray diffraction analysis is that when X-rays are directed at a regularly arranged collection of molecules (as in a crystal), they are scattered by the constituent atoms of the molecules. The structure of the constituent molecules can be worked out from the way in which the rays are scattered. Because of their size, proteins produce an extremely complex diffraction pattern. Proteins composed exclusively of repeating units (periodic structure), such as the fibrous protein of wool, produce X-ray diffraction patterns which reflect these repeating units. Others produce a mass of spots of varying intensity.

To interpret this pattern, the technique of isomorphous heavy metal replacement had to be developed (by Kendrew and Perutz). This involves the introduction of heavy metal atoms at selected sites of the molecule, without distorting the overall structure. The intensity with which X-rays are scattered depends on the electron density of the atoms. The high electron density of the heavy metal (which thus produces strong scattering of X-rays) provides reference points within the molecule. Interpretation then allowed electron density maps to be constructed. From these, highly complicated ball-and-stick models were then constructed. With the enormous improvements in computer graphics, protein structures are now displayed by computer. Analysis of a limited number of X-ray diffraction patterns allows a low resolution image (typically 0.6nm resolution) to be obtained. This shows the position in space of the polypeptide backbone of a protein and the position of some of the sidechains. Much more sophisticated analysis yields a high resolution image (typically 0.15nm resolution), in which all individual amino acid residues can be recognised and the position of the majority of atoms determined. Hydrogen atoms, because of their size, are difficult to locate. Some sidechains, especially those on the surface of proteins, give a less clear-cut image, perhaps because they occupy several positions. Proteins are not rigid and minor changes in position may occur frequently.

The description given above is very much simplified. Note that as well as revealing the 3-dimensional structure of proteins, X-ray diffraction was also vital for the deduction of the structure of DNA.

Early studies, by X-ray diffraction analysis, of fibrous proteins, particularly on one form of the protein of wool and hair (known as α-keratin) suggested that fibrous proteins had a repeating structure. These studies were conducted on fibres: although they were not crystalline, the individual molecules were sufficiently 'ordered' to give an image. The X-ray diffraction pattern indicated repetitive events every 0.15 and 0.5-0.55nm. Analysis of X-ray diffraction patterns of crystals of simple peptides, and careful model building by Pauling and Corey, provided the explanation for this through the occurrence of the

α- helix

α-helix, a secondary structure of widespread occurrence. Pauling and Corey originally proposed criteria which the most stable protein structures should meet. These included

maximum number and maximum strength of hydrogen bonds

formation of the maximum number of hydrogen bonds, which, providing the constituent atoms where in line, would be of maximum strength (you will recall our earlier comments about the directionality of hydrogen bonds). From their X-ray diffraction studies, they also deduced the planar nature of the peptide bond, concluding that a polypeptide must involve planar sections linked by bonds (on either side of the α-carbon atoms) at which rotation was possible. The α-helix fulfils all these criteria: other possible helical structures fail to achieve the maximum possible stabilisation from hydrogen bonding. This is why the α-helix occurs frequently, whilst other configurations do not.

The α-helix is a right-handed helix in which there are 3.6 amino acid residues per turn of the helix (Figure 3.13). This gives a pitch of 0.54nm for one complete turn of the helix.

α helix: carbonyl and amide groups are in line

The α-carbon atom to α-carbon atom distance along the helix axis is 0.15nm. These dimensions are consistent with the repeat distances found for α-keratin. These particular dimensions mean that both the carbonyl and amide groups of the peptide bond point along the helix axis and are almost perfectly positioned to form hydrogen bonds.

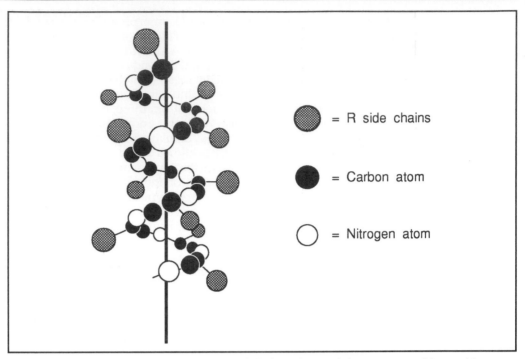

Figure 3.13 The α- helix, found in many proteins. Only the carbon and nitrogen atoms of the peptide core are shown. See text for detailed description.

The repeating structure of the α-helix means that bond angles adopted by those bonds which can rotate are reasonably uniform. Thus the angle of the carbonyl carbon to α-carbon atom bond (known as Ψ) is the same for each residue in an α-helix. Similarly, the angle of the amide nitrogen to α-carbon atom bond (identified as φ) is also constant from residue to residue, although it differs from that of Ψ.

influence of sidechains on secondary structure

Sidechain groups stick out from the helix. Since hydrogen bonding is maximised, polypeptides would be expected to form α-helices, providing other factors do not interfere. These include possible interactions between sidechains or difficulty in adopting the required bond angles on either side of the α-carbon atom. The first point is illustrated by studies on the conformation of synthetic polypeptides composed of a single amino acid such as L-glutamic acid or L-lysine. Conformation changes can be studied by techniques based on optical rotation of plane polarised light. Since polypeptides are composed of L-amino acid residues, each amino acid residue contains an asymmetric centre (except glycine) and, as a consequence, will rotate plane polarised light. Repeating structures, arising as they do from repetition of particular bond angles, result in different optical rotation to that of a protein which shows no regularity of such angles (one in which the polypeptide is in a 'random coil' conformation). Analysis of poly-L-glutamic acid shows a sharp pH-dependent change in optical rotation which corresponds with the pKa value of the sidechain carboxyl group (Figure 3.14).

At low pH (carboxyl group protonated), α-helix is formed; as deprotonation occurs (and negative charges are generated) the α-helix is destabilised and a random structure results. This approach shows that sidechains can influence the stability of α-helices.

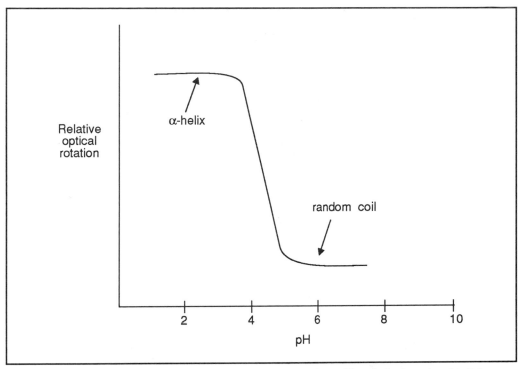

Figure 3.14 Effect of pH on the optical rotation by the synthetic polypeptide poly-L-glutamic acid. At low pH, an α- helix occurs, giving high optical rotation. Deprotonation of the sidechain carboxyl groups causes destabilisation of the α- helix and a transition to a random coil.

proline as a helix breaker

It is impossible for proline to be accommodated within an α-helix. This is because the sidechain of proline is joined to its α-amino group, thereby preventing rotation of the amide nitrogen to α-carbon atom bond. If proline occurs in a polypeptide, it forces a bend to occur: it is thought of as a helix-breaker. This is sometimes used in globular (ie compact, non-extended) proteins to terminate an α-helix and change the direction of the polypeptide.

β-pleated sheets

X-ray diffraction studies of some fibrous proteins, such as silk, revealed a different repeat distance. This was found to exemplify a second periodic or repeating structure, which are known as β-pleated sheets (Figure 3.15). These also arise from extensive hydrogen bonding, but in this case the hydrogen bonds are at right angles to the longitudinal axis of the polypeptide chain and do not usually link residues which are close to each other in the primary sequence (as with the α-helix). Polypeptide chains run parallel to each other. Two forms occur. If the chains run in the same direction (thus the N-terminal ends are together) they are said to be parallel; if they run in opposite directions, they are said to be anti-parallel. The polypeptide chains are more extended than in an α-helix. Some proteins, such as silk, are composed entirely of β-pleated sheets. β-pleated sheets were originally thought to be absent from globular proteins but do occur, for example in the enzymes lysozyme and chymotrypsin. In these cases, hairpin bends allow anti-parallel β-pleated sheets to form.

β-pleated sheets are also present in some globular proteins

∏ Are the β-pleated sheets shown in Figure 3.15 parallel or anti-parallel sheets? (The position of the N-terminals should provide the answer).

You should come to the conclusion that Figure 3.17a is a parallel sheet, while Figure 3.17b shows an anti-parallel sheet.

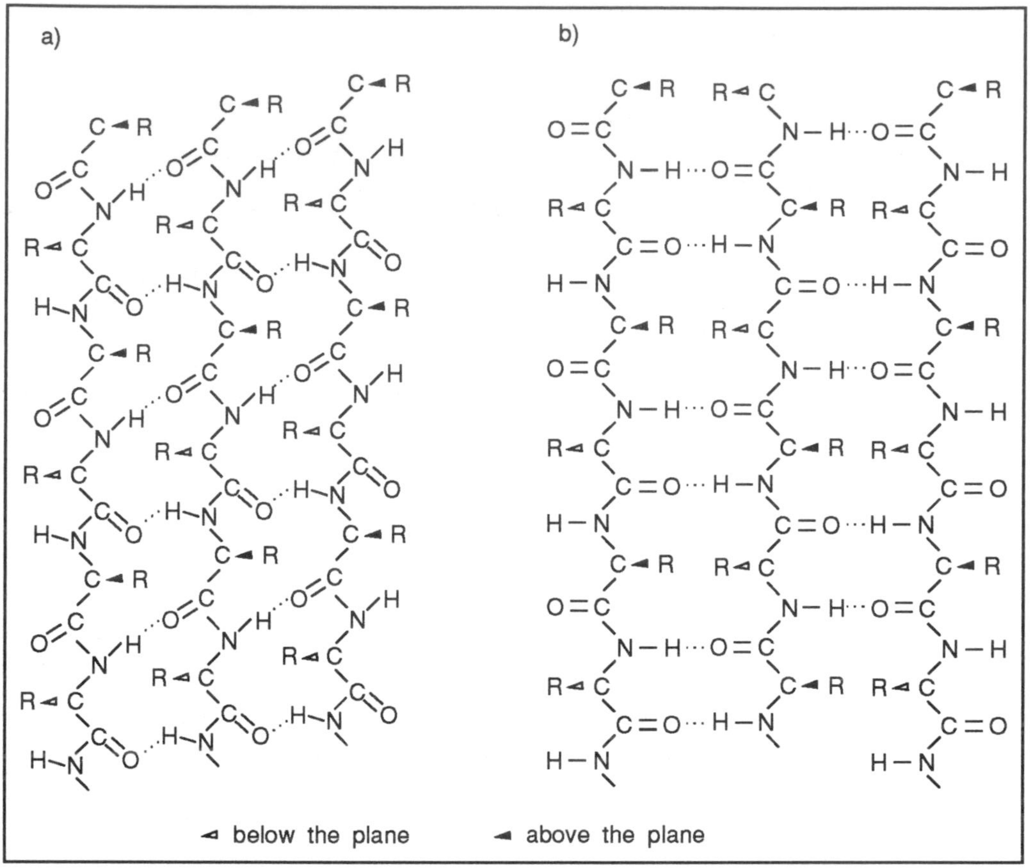

Figure 3.15 β-pleated sheet. Hydrogen bonding between chains leads to a pleated structure in which the sidechains stick out above and below the plane of the 'sheet'. Short sections occur in globular proteins; silk consists entirely of β-pleated sheet.

The α-helix and β-pleated sheets are of widespread occurrence, which is an indication of the stabilisation which results from the maximised hydrogen bonding of the main chain. In globular proteins they are usually short, typically 10-15 residues for α-helices, and less for β-pleated sheets. Both types of secondary structure may occur in one protein, as they do in the enzymes lysozyme and lactate dehydrogenase. The α-helical content varies from essentially 100% in α-keratin to zero; some examples are given in Table 3.4. The secondary structure(s) are then linked by sections of random coil of irregular structure.

extended helices

Other helices can, in principle, occur. Think of either a more extended helix ie greater translation along the helix axis per residue, or a more squashed one. Short segments of a more extended helix (the 3.10 helix) are sometimes seen. The fibrous protein of connective tissue, collagen, forms an extended left-handed helix; three of these strands are combined, rope-like, to form a 'super-helix' which is stabilised by inter-chain hydrogen bonds.

Protein	% α- helix*
α- Keratin	100
Myoglobin	77
Lysozyme	45
Carboxypeptidase	35
Cytochrome c	15

Table 3.4 Percentage α- helical content of some proteins.
*The percentage is the proportion of the total number of amino acids which are found within α- helices.

many fibrous proteins have repeated amino acid sequences

It is notable that some of the fibrous (ie structural) proteins have primary structures which are 'repeats' of a simple sequence. Thus silk largely consists of a 6-residue repeat: gly-ser-gly-ala-gly-ala, whilst collagen has large amounts of glycine and proline (or hydroxyproline, which is a hydroxylated form of proline). Globular proteins, which constitute the enzymes and proteins which interact with other compounds in cells, have much more varied amino acid sequences. As a consequence, their structures are essentially infinitely varied.

3.5 Tertiary structure of proteins

The tertiary structure describes the overall shape of the protein. This overall conformation is the consequence of the presence of any secondary structures and how they and the remainder of the chain are positioned. Each protein adopts a unique tertiary structure. This conformation will be the result of the various weak bonds which can arise. It also depends on interaction with the surrounding solvent (usually water). Thus the unique shape of a protein is that of lowest free energy, which is achieved by minimising unfavourable interactions (eg hydrophobic groups in contact with water, where they disrupt hydrogen bonding between water molecules) whilst maximising favourable interactions.

myoglobin structure elucidated

Myoglobin was the first protein whose structure was determined by X-ray diffraction analysis. Myoglobin is the oxygen-storing protein of muscle and is relatively small (mol weight= 17500 daltons: 153 amino acids), yet it represented a breakthrough in biochemistry. Subsequently, larger proteins and those with subunits (eg haemoglobin) were described. The complete structure of an enzyme (lysozyme) enabled a detailed description of the catalytic mechanism by which an enzyme accelerates reaction rates to be convincingly proposed. Our detailed understanding of the structure and action of proteins relies on application of X-ray diffraction.

So what did the X-ray diffraction analysis of myoglobin reveal? Myoglobin was shown (Figure 3.16) to be very compact, with water almost completely excluded from the interior of the protein.

orientation of polar and hydrophobic sidechains in proteins

It is made up of 8 sections of α-helix (three-quarters of all residues are present in α-helices), which fold to make up a box-like molecule; proline forces a bend at the end of some helices. Strikingly, the interior of the molecule contains almost exclusively hydrophobic, non-polar residues, tightly packed together. Almost all the polar residues are on the surface of the molecule, where the sidechains interact favourably with water (via hydrogen bonding). Thus, where an α-helix runs along the surface of myoglobin, residues whose sidechains point outwards into solvent are polar, whilst those on the inside of the helix are hydrophobic. In this regard (distribution of hydrophilic and

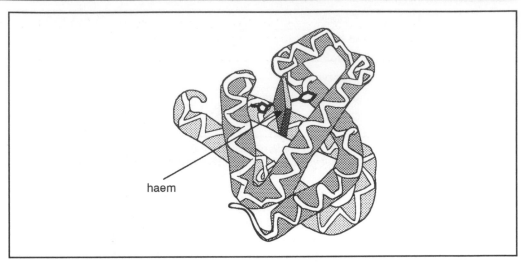

Figure 3.16 The structure of myoglobin, as revealed by X-ray diffraction analysis. The molecule is compact and the hydrophobic residues are internal, whilst hydrophilic residues are on the surface.

hydrophobic groups) myoglobin has been found to be typical of globular proteins normally found in cytoplasm. The aggregation of hydrophobic groups together in the interior of a protein, shielded from water by a layer of hydrophilic groups, is clearly a general feature of proteins. Myoglobin, however, has a very high α-helix content for a globular protein. It also lacks disulphide bridges, which are frequently present in other proteins. Ribonuclease and lysozyme, which are both smaller than myoglobin (124 and 129 amino acid residues respectively), both have 4 disulphide bridges. Myoglobin also contains a haem group (which gives it a characteristic red colour) to which oxygen binding occurs.

In conclusion, the tertiary structure of a protein usually results in a compact structure in which there is hardly any room for water. Nearly all the non-polar sidechains are situated in the interior of the protein and are shielded from the surrounding water, whereas nearly all the polar sidechains are found on the surface. These comments apply to proteins which are normally in solution within the cell or blood stream. Proteins normally located in or on membranes may show altered location of hydrophobic residues. This is developed further in Chapter 7.

3.6 Quaternary structure of proteins

Many proteins consist of a single polypeptide chain. Others are normally found as an association of more than one polypeptide chain and are said to have a quaternary structure. Association of 4 subunits (ie individual polypeptides) to form a tetramer is a particularly common arrangement although dimers, hexamers and more complicated combinations occur. The subunits may be identical to each other (as in the enzyme β-galactosidase) or different. Haemoglobin (the oxygen carrying protein of red blood cells) consists of a tetramer of 2 types of subunits (α and β) of different but related sequences, in equal combination. The quaternary structure is described as $\alpha_2\beta_2$ (Figure 3.17). In other cases, the proportion of 2 types of subunits can differ. Lactate dehydrogenase consists of a tetramer of 2 subunit types (H and M); 5 alternative forms (called isoenzymes) occur, ranging from a tetramer of 4 H subunits (H_4), through

isoenzymes

intervening forms (H3M; H_2M_2; HM_3) to a tetramer of pure M subunits (M_4). These isoenzymes differ in their catalytic behaviour.

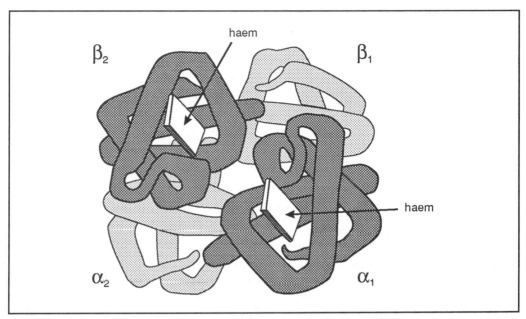

Figure 3.17 The structure of haemoglobin, comprising 2α and 2β subunits.

More complex arrangements involve association of catalytically active and inactive subunits. The enzyme aspartate transcarbamylase consists of 6 catalytic subunits (C-type) associated with 6 'regulatory' subunits (R-type). The R-type subunits do not bind substrates but influence the activity of the C-type subunits. This enzyme is discussed in more detail in Chapter 8.

allosteric properties

Possession of a quaternary structure can be associated with allosteric behaviour. We will discuss this in more detail in Chapter 8. Both haemoglobin and aspartate transcarbamylase are allosteric proteins, whereas myoglobin, with no quaternary structure, is not. One effect of allosteric interactions is that substrate binding to an enzyme is sigmoid. With haemoglobin, oxygen binding displays a sigmoid relationship to oxygen concentration, whereas that of myoglobin is hyperbolic (Figure 3.18). The explanation is that the binding of oxygen by haemoglobin results in a change in the conformation and, as a consequence, a change in affinity of binding of a second (and subsequent) oxygen molecules.

A vital (quite literally!) consequence of this is that haemoglobin binds oxygen in the lungs (because the oxygen concentration is high there) and releases it in the tissues, where the oxygen concentration is lower (Figure 3.18). Myoglobin is able to bind the released oxygen and store it until it is required in muscle. The release of oxygen by haemoglobin would be much less effective without the sigmoid binding relationship; this in turn is a consequence of the interaction between the subunits of the tetramer.

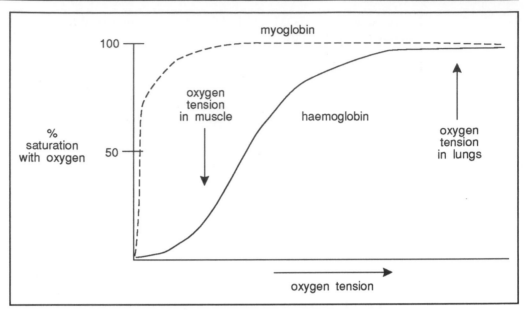

Figure 3.18 Oxygen saturation curves for haemoglobin (solid line) and myoglobin (dashed line).
Haemoglobin displays a sigmoid relationship, whilst that of myoglobin is hyperbolic. Haemoglobin becomes
fully saturated with oxygen in the alveoli of the lungs. At the lowered oxygen tension which exists in resting
skeletal muscle, haemoglobin will release oxygen; myoglobin present in muscle will bind the released
oxygen. As the oxygen tension falls further (eg during exercise), oxygen will be released from myoglobin
and consumed in respiration.

3.7 Factors affecting protein conformation

Influence of primary structure

We have already stated that a polypeptide folds to a particular conformation, known as
the native conformation, rather than adopting an array of shapes. We have seen that this
is a consequence of the formation of stabilising interactions, such that the overall protein
plus surrounding water molecules possess the lowest free energy. We have implied that
all this is a consequence of the primary structure of the protein. This makes sense
because it means that protein folding is an automatic process, following inevitably from
the sequence of the amino acid residues. However, the fact that it seems sensible does
not mean that it occurs. Some evidence is needed.

Anfinsen's experiment urea and 2-mercapto-ethanol

disrupt protein structure

One of the most striking demonstrations of the impact of the primary structure on
overall shape was provided by an experiment conducted by Anfinsen on ribonuclease.
Ribonuclease is a small enzyme consisting of a single chain of 124 amino acid residues
and also containing 4 disulphide bridges. If it is treated with concentrated urea (urea
breaks hydrogen bonds so the protein unfolds) and 2-mercaptoethanol (to destroy
disulphide bridges by reducing them), ribonuclease loses its native structure and
assumes a random coil (Figure 3.19); it is said to be denatured.

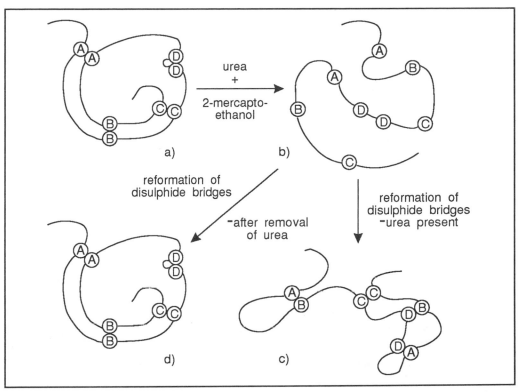

Figure 3.19 Dependence of tertiary structure on the amino acid sequence of a protein. Native ribonuclease (a) is denatured by treatment with urea and 2-mercaptoethanol, resulting in loss of activity and formation of a random coil (b). If reformation of disulphide bridges occurred in the presence of the denaturant urea, then little activity was recovered (c); 'incorrect' conformation(s) and disulphide bridges were occurring. If disulphide bridge formation occurred only after removal of denaturant, then full enzyme activity was recovered; the 'correct' conformation and disulphide bridges were being recovered (d). The incorrect disulphide bridge containing form(s) - (c) can be converted to active form (d) by addition of trace amounts of 2-mercaptoethanol. (Adapted from Stryer, L, (1988). Biochemistry 3rd Edition W.H. Freeman and Co).

Enzyme activity is also lost. If the reducing agent is removed, disulphide bridge formation occurs via air oxidation. This was done in 2 ways:

- when disulphide bridge formation was allowed to take place in the presence of urea (ie the agent which disrupted the weak bonds stabilising the native conformation), essentially no enzyme activity was recovered. The implication is that the native (active) conformation was not being recovered;

- when disulphide bridge formation took place after removal of the urea, enzyme activity was fully recovered. Thus the enzyme was refolding, in a way dictated by its primary structure, to give a single, active conformation.

Another way to look at this is in terms of the reformation of the disulphide bridges. In the native enzyme, there are 4 disulphide bridges, involving 8 cysteine residues. After reduction, how many possible combinations could be made? A given cysteine could combine with any of the other 7 cysteines, leaving 6 - SH groups. Another cysteine can now combine with any one of the remaining 5 cysteines. The next cysteine to reform a disulphide bridge has a choice of 3 others; the final 2 cysteines would then combine. Thus the total number of possible disulphide bridge combinations is $7 \times 5 \times 3 \times 1 = 105$.

Yet of these 105 possibilities, only one actually occurs. This is striking evidence that the primary structure will adopt a particular, precise shape, leading to formation of disulphide bridges only between the cysteine residues at particular positions. This pattern of results has been confirmed with other proteins. These include antibodies of type IgG, which contain 2 types of chain. These are called Heavy (H) and Light (L) chains and they combine into a Y-shaped molecule of 2H and 2L chains (see Figure 3.12). Although these are normally held together by disulphide bridges, they can still be regenerated following denaturation and reduction of disulphide bridges.

SAQ 3.3

Insulin consists of 2 polypeptide chains (A-chain, 21 amino acid residues; B-chain, 30 residues) linked by disulphide bridges. The type of experiment conducted by Anfinsen on ribonuclease was repeated with insulin (ie reduction and denaturation, followed by removal of denaturant and reoxidation). In this case, recovery of activity was only 7%, which is the recovery expected from random pairing of disulphide bridges. Briefly explain how these observations can be accommodated within the concept of primary structure determining overall conformation. (Figure 3.12 might help).

Denaturation of proteins

reversible and irreversible denaturation

The term denaturation is used to describe change(s) in the conformation of proteins leading to loss of biological activity. It excludes minor changes in conformation which are without functional significance and those which occur as part of a proteins' function. Thus the change in tertiary and quaternary structure which occurs in haemoglobin when oxygen is bound is not denaturation. Denaturation normally occurs without cleavage of the polypeptide chain. It involves alterations in the secondary, tertiary and quaternary structure leading to transition from the native conformation to a less ordered shape or shapes (Figure 3.20). Often the most sensitive indicator of denaturation is loss of biological activity, for example enzyme activity. Other indications are a change in optical rotation and alterations in the availability of sidechain groups: particular residues, which are normally 'buried' and not available for chemical reaction, becoming exposed and hence reactive. Denaturation may be reversible or irreversible and may lead to aggregation of protein molecules (through exposure of previously buried hydrophobic groups) and insolubilisation.

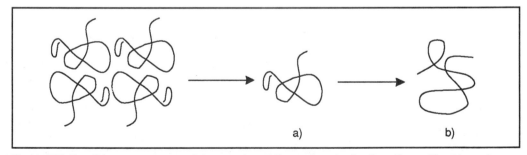

Figure 3.20 Possible consequences of denaturation. a) Separation of subunits, with or without loss of activity. b) Progressive unfolding, with consequent loss of activity. Both processes may be reversible.

Chemical denaturants include high concentrations (6-8 mol.l^{-1}) of urea and guanidinium hydrochloride (these break hydrogen bonds). Renaturation (the reversal of denaturation) is often possible after the removal of the denaturing agent.

thermostability

High temperatures and extremes of pH are effective denaturants, usually irreversibly. The precise conditions required differ from case to case. Enzymes which can withstand high temperatures, and thus exhibit thermostability, may be particularly useful for biotechnologists. Thermophilic organisms growing in hot springs are a useful source of these. Vigorous stirring or shaking will also denature proteins; foam (as produced by whisking an egg white) consists of denatured protein. Certain detergents (eg sodium dodecyl sulphate, SDS) are potent denaturants.

In working with proteins, particularly enzymes, care must be taken to minimise denaturation.

∏ Use the information given above to write a list of some simple precautions you could take to minimise denaturation of an protein.

Simple steps include:

- working at 0-4° whenever possible; this lessens thermal denaturation. It also reduces damage from proteolytic (ie protein-digesting) enzymes which are frequently present in cell or tissue extracts;

- using a buffer to control the pH;

- avoiding shaking solutions unnecessarily.

3.8 Purification of proteins

structural
properties

functional
properties

We conclude this review of the structure and properties of amino acids and proteins by briefly considering how they may be purified. Firstly, why do proteins need to be purified? The simple answer is that most of the structural information which we have discussed can only be obtained once a protein is pure. This includes its amino acid sequence, subunit size and 3-dimensional structure. Similarly, functional properties of proteins such as enzymes can only be unambiguously attributed to a given enzyme if it is known to be pure.

Our purpose in considering how to purify a protein is to provide a general strategy for purification, a framework within which the particular steps necessary for a given protein may be formulated. We will not discuss in detail the particular methods that may be applied; these are the subject of a companion volume of this series.

targeting
differences
between
proteins

measurement
of specific
proteins

A useful strategy is to consider the ways in which the target protein differs from all the other compounds which are also present in the starting material. Providing one can then use separation methods which are based on the ways in which the protein differs, then separation and purification will be possible. An additional prerequisite is that it must be possible to assay the target material to establish how much is present in a sample. Quantitative enzyme assays are described in Chapter 8. For other proteins, appropriate assays, perhaps using antibodies, must be devised.

The ways in which proteins differ can be grouped as follows: the differences may then be exploited for purification procedures.

a) Size

Each protein has a unique structure, resulting from:

- a polypeptide chain of a particular sequence of amino acid residues, and;

- possible association of more than one subunit, which may be of different primary structure.

One consequence of this is the overall size of the protein. Proteins vary enormously in size: a selection to illustrate this is given in Table 3.5.

Protein	Molecular mass (daltons)	Number of subunits
Insulin (bovine)	5780	2
Myoglobin (equine)	16,900	1
Chymotrypsinogen (bovine)	25,700	1
Haemoglobin (bovine)	64,500	4
Lactate dehydrogenase (dogfish)	145,000	4
Glutamate dehydrogenase (bovine)	330,000	6
Pyruvate dehydrogenase (E. coli)	4,600,000	60

Table 3.5 Molecular mass of selected proteins.

gel filtration

Suppose we wish to purify the enzyme lactate dehydrogenase. If we use a method which rejects proteins (and other compounds) of molecular mass of less than 130,000 daltons and more than 160,000 daltons, then the resulting fraction containing proteins of approximately 145,000 daltons will contain partially purified lactate dehydrogenase. Such methods exist. Gel permeation chromatography (also known as gel filtration) is very widely used. Centrifugation methods, such as rate zonal density gradient centrifugation, may also be applicable, although this method is of more limited use.

b) Charge

As a consequences of differences in their amino acid composition, proteins differ in the overall charge they possess. One way in which this can be seen more clearly is by consideration of their isoelectric points. The isoelectric point (pI), as we discussed in Section 2.3.4, is the pH at which a protein (or other compound) has no net charge. It will occur when the various positively charged groups exactly equal the negatively charged ones. Isoelectric points of proteins vary widely. Table 3.6 lists a representative sample.

Protein	pI
Casein	4.7
Serum albumin	4.8
Haemoglobin	6.8
Chymotrypsinogen	9.5
Cytochrome c	10.65

Table 3.6 Isoelectric points of selected proteins.

∏ See if you can write down at least one way in which the differences in isoelectric point of proteins can be useful in their purification? Think about the charge that the above proteins would possess at a given pH eg 7.0.

A generalisation is as follows: if the pH is less than the pI, then the molecules will be positively charged. Conversely, if the pH is greater than the pI, then the molecule will be negatively charged. Generally, as the pH is changed away from the pI, the size of the charge increases (Figure 3.23).

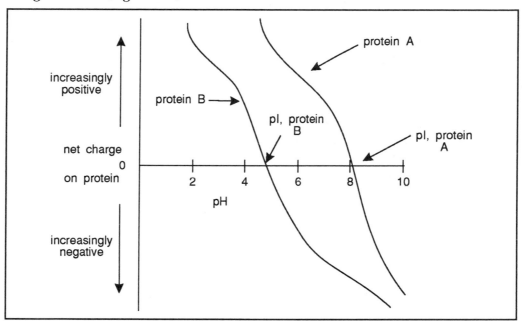

Figure 3.21 Effect of pH on the overall charge of a protein. At pH values below the isoelectric point (pI) of a protein, it will possess a positive charge. When the pH is above the pI, the protein will be negatively charged. Inspection of this figure shows that at pH 6.0 protein A is positively charged whilst protein B is negatively charged. They can therefore be easily separated (look back to Figure 2.13 to see this).

ion exchange chromatography

Thus net charge on a protein varies with pH, and is also likely to vary from one protein to another (since their pI values differ) both in size of charge and nature (+ or -). This is exploited in the technique of ion-exchange chromatography, which you have already encountered in considering amino acid separations (Section 2.3.4). A further purification approach, related to the isoelectric point, is referred to in the next section.

Returning to our question above, at pH 7.0 casein and serum albumin would be negatively charged, whilst haemoglobin would be very close to its isoelectric point and have negligible charge. Chymotrypsinogen and cytochrome c would have a net positive charge.

c) Solubility

For reasons which are less clearly related to structure than the first two categories, the effect of various compounds on protein solubility varies widely from protein to protein. Thus the solubility of proteins is affected to differing extents by various agents. These are:

salting out

Salts. Ammonium sulphate is widely used to selectively precipitate proteins. For example, a 40% saturated solution of ammonium sulphate may precipitate protein A, whilst proteins B, C and D remain in solution. The precipitated protein A can be pelleted by centrifugation, and subsequently resuspended in fresh buffer. It will now have been separated from proteins B, C and D. Further addition of ammonium sulphate, perhaps to 55% saturation, may cause precipitation of protein B, leaving proteins C and D in solution; separation of these proteins can be done as for protein A. This approach is very widely used, especially since ammonium sulphate-precipitated proteins routinely recover full activity when resuspended. We can describe this process as 'salting out' because the addition of salts (eg ammonium sulphate) causes the proteins to come out of solution.

addition of organic solvents

Organic Solvents. Ethanol or acetone may be used in an analogous way to that of ammonium sulphate. Different proteins become insoluble at different concentrations of these organic solvents. These are less widely used than ammonium sulphate, since irreversible denaturation is more frequent.

isoelectric precipitation

Isoelectric precipitation. It is frequently found that proteins (and other zwitterions) display a minium of solubility at the isoelectric point. Thus casein is precipitated from milk at pH 4.5, leaving other proteins (whey) in solution. Since proteins differ in their isoelectric points, such isoelectric precipitation may be a helpful approach. Many amino acids are produced industrially for use as food supplements; they are often concentrated by isoelectric precipitation.

d) Biological specificity

One of the most remarkable properties of proteins is their specific binding ability, ie their ability to discriminate between one compound and many others. This is a consequence of the occurrence, via use of different amino acids, of a virtually infinite range of shapes within proteins. Proteins, more than any other group of macromolecules, are used to interact with small (and large) molecules, for example enzymes with substrates, antibodies with antigens, receptors with hormones. This property of specific ligand-binding underpins many of the roles of proteins.

ligand-binding

∏ Can you devise a method for purifying a protein which utilises this specific ligand-binding property?

affinity chromatography

The high specificity and tight binding of enzyme-substrate interaction is used for purification in the technique of affinity chromatography. For this we could attach the substrate to an insoluble resin. The enzyme would then bind to the substrate and thus become attached to the resin. Other proteins which do not bind the substrate would remain in solution and could be removed. This is one of the most powerful techniques available for purification of proteins.

criteria for purity of proteins

We have discussed electrophoresis of amino acids. Protein electrophoresis is a widely used way of determining the purity of a protein preparation. Gels, usually made of polyacrylamide (a synthetic polymer), are used as supporting medium. A single band following electrophoresis and staining for protein is an indication that the protein is pure. The presence of more than one band indicates lack of purity. Other methods, notably a maximised specific activity (units of enzyme activity per mg protein) of the final preparation, or a constant specific activity across a chromatographic peak, are also useful indicators of purity.

Summary and objectives

In this chapter we have examined the structure and properties of proteins. We have shown how numerous weak interactions between the components of proteins establishes a particular conformation. We have also described the more commonly encountered folding arrangements. We also showed that proteins exhibit different levels of structural organisation and can, through there great diversity, perform a tremendous range of functions within cells.

Now that you have completed this chapter, you should be able to:

* predict the behaviour of proteins in an electric field from stated pI and pH values;

* list the major forces which are involved in maintaining the secondary, tertiary and quaternary structure of proteins;

* identify the types of interactions involved with particular amino acids sidechains involved in maintaining the secondary, tertiary and quaternary structure of proteins;

* outline what features of proteins are important in determining the 3-dimensional structure of the protein;

* list the properties of proteins which may be used in their purification;

* describe what denaturation is and list causes and consequences of denaturation.

Nucleic acids

Nucleic acids

4.1 Introduction

DNA

RNA

ATP

NAD⁺

Nucleic acids include two types of macromolecules with crucial roles within cells. These are deoxyribonucleic acid, DNA, which contains the genetic information of most organisms and ribonucleic acid, RNA, which is involved in the expression of the information contained in DNA. Both DNA and RNA are linear, unbranched polymers of mononucleotides. Mononucleotides, such as adenosine-triphosphate, ATP, and dinucleotides such as nicotinamide adenine dinucleotide, NAD^+, also have important roles in the metabolism of cells.

4.1.1 Information transfer and roles: the central dogma

central dogma

Before discussing the structure of nucleic acids, we will summarise their roles by reference to the so-called 'Central Dogma of Molecular Biology' (Figure 4.1).

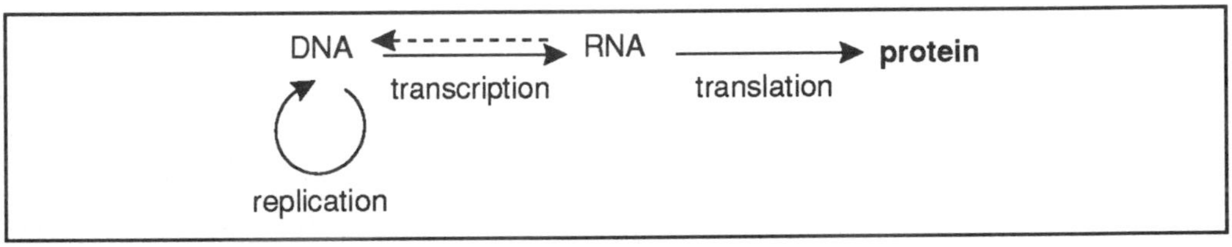

Figure 4.1 The central dogma of molecular biology. Arrows represent the flow of information between the various types of molecules.

This describes the flow of information from DNA through RNA to proteins. The central dogma describes three major processes in which information transfer occurs in cells.

replication

Replication, in which the DNA molecule is copied, with its information content being preserved. This occurs prior to cell division so that each 'daughter' cell receives an exact copy of the DNA present in the original cell. The sequence of the nucleotides (bases) in the DNA represents the information store. This specifies the order of the amino acids in proteins, ie the primary structure of proteins.

transcription

DNA template

Transcription, in which information contained within DNA is converted into the form of RNA, as part of the mechanism by which expression of the genetic message is accomplished. During transcription, one strand of the DNA is used as a template: a molecule of RNA, whose base sequence matches that of the DNA, is constructed using this 'template'.

translation

Translation, in which the information contained within the RNA is used to construct a protein of a particular primary structure. Since both DNA and RNA contain information in a 4-base code (ie 4 different nucleotides are used), the information of the RNA base sequence has to be translated into the 20 letter 'alphabet' of amino acids.

In summary, the sequence of bases in the DNA determines the sequence of bases in RNA transcribed from it, and this in turn determines the amino acid sequence of a protein.

The central dogma has been refined to take account of 'unusual' situations. For example, some viruses contain only RNA as their genetic information. Normally, RNA is not replicated (ie RNA is not usually copied to give other molecules of RNA). So how do RNA - containing viruses replicate? One solution to this problem involves a molecule of DNA first being made using the viral RNA as template. This does occur with some RNA - containing viruses, (hence the dashed arrow from RNA to DNA in Figure 4.1) and requires an enzyme known as reverse transcriptase. (NB this enzyme is a very important tool for genetic engineers). However, you should not lose sight of the fact that the principal information flow, which occurs in the vast majority of cases, is DNA to RNA to protein.

reverse transcriptase

The principal roles of mononucleotides, DNA and RNA are shown in Table 4.1.

Type of molecule	Role(s)
Mononucleotides	Building blocks of DNA/RNA Energy transduction in cell Signalling within cell
Dinucleotides	Serve as cosubstrates (coenzymes) in oxidation/reduction reactions
DNA	Stores genetic information
RNA	
ribosomal	Required for ribosomal structure and function
transfer	Serves as an 'adaptor' during protein synthesis
messenger	Carries genetic information coding for primary structure of a protein
viral	Stores genetic information in some viruses

Table 4.1 Roles of Nucleotides, DNA and RNA.

ribosomal, transfer and messenger, RNA

Note there are 3 types of RNA found in living cells - ribosomal, transfer and messenger; their roles will be discussed in Section 4.6.3. A fourth type of RNA, viral RNA, has already been referred to. All RNA is constructed in the same way, as far as covalent bonds are concerned. These subdivisions thus represent a functional classification.

4.2 Components of nucleic acids

Mononucleotides and the polynucleotides derived from them contain 3 characteristic components. These are:

- a nitrogenous base;

- a sugar;

- a phosphoric acid.

4.2.1 Nitrogenous bases

pyrimidines, thymine, cytosine, uracil

The nitrogenous bases found in nucleic acids fall into 2 categories. They are either pyrimidines or purines. Pyrimidines (Figure 4.2) are structurally related to the heterocyclic compound pyrimidine. Three are commonly found in nucleic acids:

thymine and cytosine are found in DNA, whereas in RNA thymine is replaced by uracil. In order to avoid ambiguity in describing which atoms are involved in bonds, there is a convention involving the numbering of the atoms. This is shown for the compound pyrimidine in Figure 4.2.

a) b) c) d)

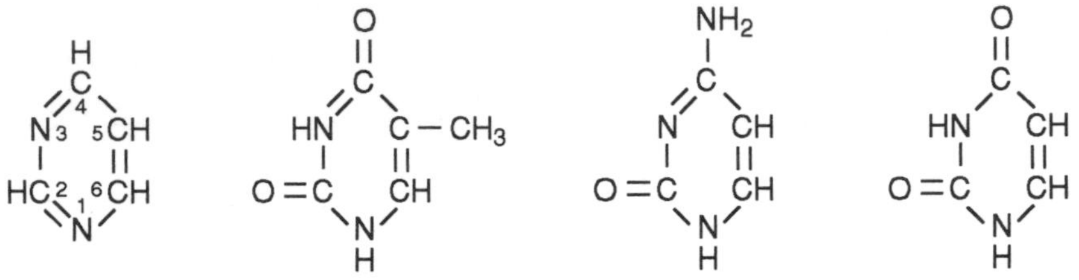

pyrimidine thymine cytosine uracil

Figure 4.2 Pyrimidines. Pyrimidines found in nucleic acids are derivations of pyrimidine. The numbering system used to identify atoms within the heterocyclic ring is shown in structure a). Thymine b), cytosine c) and uracil d) are found in nucleic acids.

purines: adenine, guanine

Purines (Figure 4.3) are derivatives of the fused ring compound purine. Only two are routinely found in nucleic acids: these are adenine and guanine. The numbering system used with purines is illustrated within Figure 4.3.

a) b) c)

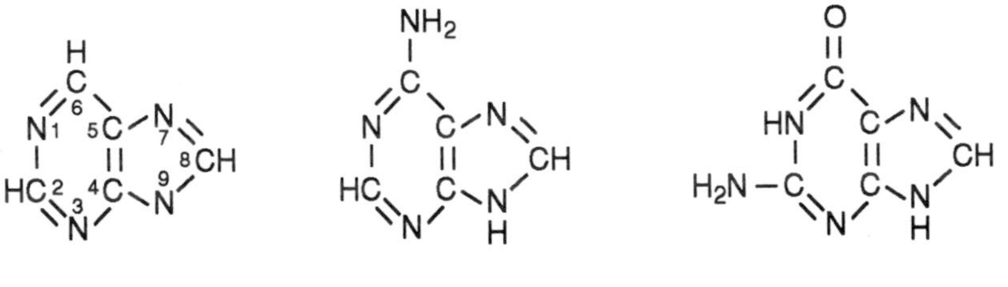

purine adenine guanine

Figure 4.3 Purines. The fused ring nitrogenous bases found in nucleic acids are a) derivatives of purine, b) adenine and c) guanine occur in nucleic acids. The numbering of atoms in the purine ring is shown in structure a).

methylation

Note that unusual bases are sometimes found in nucleic acids, particularly in transfer RNA; these will be briefly discussed in Section 4.6.2. Methylation of bases also occurs as a defence mechanism in bacteria to prevent hydrolysis of DNA by restriction enzymes. Restriction enzymes are the enzymes used in genetic engineering to cut DNA at particular sites (Section 4.7).

Π Nucleic acids can be hydrolysed to their constituent mononucleotides by treatment with acid. Assume that a sample of a nucleic acid is treated in this way, and mononucleotides containing adenine, cytosine, guanine and uracil are recovered. Write down what type of nucleic acid was involved, giving brief reasons.

The fact that a mononucleotide containing uracil was recovered means that the nucleic acid was RNA, and not DNA. Re-read the preceding paragraph if you missed this. We can go a little further. Since no unusual bases were found, this was unlikely to have been transfer RNA (tRNA).

∏ 1) Use the numbering system to name adenine as a derivative of the molecule purine. As a hint, it contains an amino group! 2) Write down the structure of 5-methylcytosine. Look at Figure 4.2 for the structure of cytosine.

1) Adenine is 6-amino purine, since it consists of purine with an amino group attached at position 6. 2) The structure of cytosine is given in Figure 4.2. 5-methylcytosine has a methyl group at position 5:

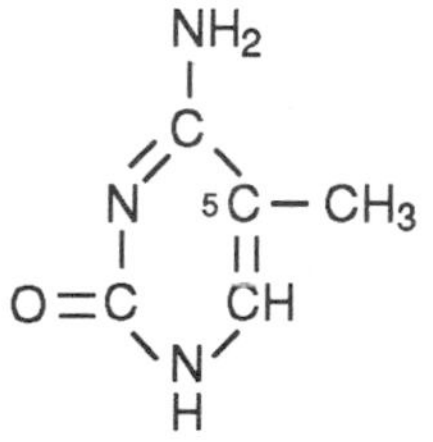

This replaces the hydrogen atom which is present at this position in cytosine.

tautomers   Pyrimidines and purines can exist in alternative forms, called tautomers; these are isomeric forms. These result from redistribution of electrons between the various bonds. For example, thymine can exist in the 2 forms shown in Figure 4.4.

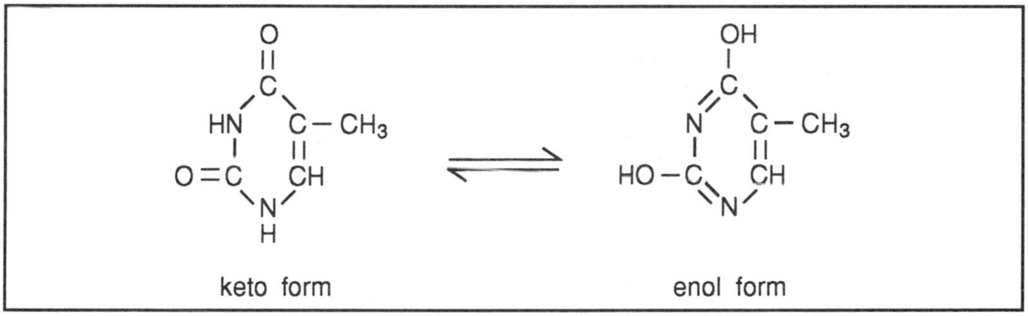

keto form   enol form

Figure 4.4 Tautomeric forms of thymine. Only the form shown on the left (the keto form) is important under physiological conditions.

These alternative forms can occur under particular conditions. Note that the forms shown in Figures 4.2 and 4.3 predominate under physiological conditions and you do not need to worry about the other tautomeric forms. However, it was not always so: in the years leading up to the description by Watson and Crick of the DNA double helix, it was believed that the alternative tautomers were present. It was only by using the 'correct' forms of the bases (as in Figures 4.2 and 4.3) that the necessary bonds between chains in the double helix were made.

SAQ 4.1

Inspect the structures shown below and then write down:

1) Which of them are pyrimidines?

2) Which are purines?

3) Which pyrimidine does not occur in DNA?

4) Which of them is not a nitrogenous base?

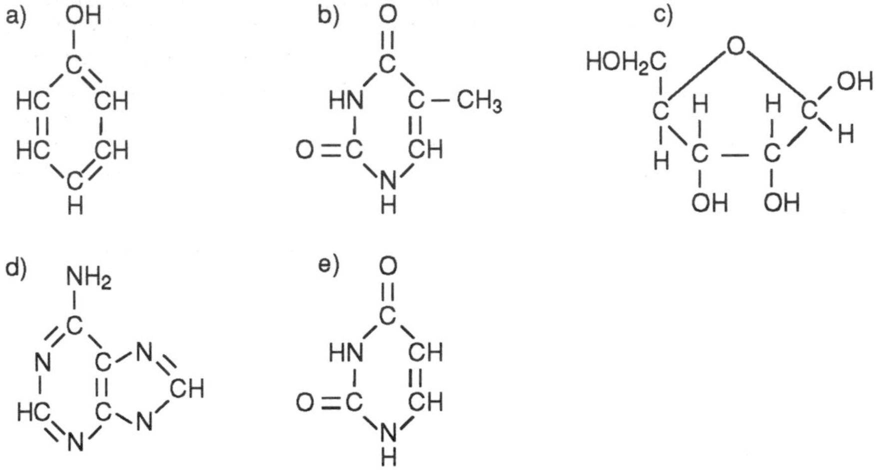

4.2.2 Sugars

D-ribose

2'-deoxy-D-ribose/$lnucleos idesnucleosides

So far we have been discussing the bases found in RNA and DNA, now we turn our attention to their sugar component. One of the them contains the sugar D-ribose, whilst the other contains 2'-deoxy-D-ribose. Ribose is found in RNA; 2'-deoxy-D-ribose is incorporated into DNA. The structures of these sugars as they occur in nucleic acids are shown in Figure 4.5. The sugar units are attached to the purine and pyrimidine bases to form nucleosides and are always present in the ring form. Monosaccharide structure is discussed in detail in Chapter 5.

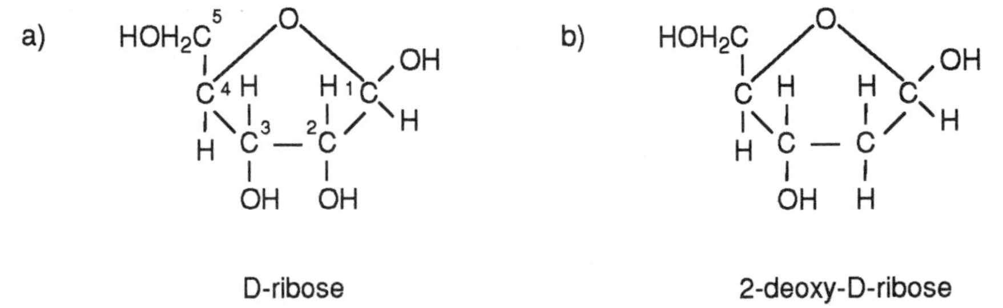

Figure 4.5 Sugars found in nucleic acids. a) D-ribose is found in RNA and ribonucleotides b) 2-deoxy-D-ribose occurs in DNA and deoxyribonucleotides. The numbering of the carbon atoms of these sugars is shown in structure a).

When discussing the nitrogenous bases, we noted that a numbering system is used to avoid ambiguity in describing the position of bonds. A similar problem arises with the sugar moiety. Atoms are again numbered, as shown in Figure 4.5. To avoid confusion between atoms in the base and in the sugar, those in the sugar are also designated 1', 2', 3' etc; these are spoken of as 3-prime or 5-prime. We will return to this in Section 4.4.

4.3 Nucleosides

ribonucleosides

deoxy-ribonucleosides

glycosidic bond

Nucleosides consist of a nitrogenous base (purine or pyrimidine) linked to a 5-carbon sugar (D-ribose or 2'-deoxy-D-ribose). Consequently, nucleosides are described as ribonucleosides or deoxyribonucleosides according to which sugar is present. The linkage is between the C-1' of the sugar and either nitrogen atom N-1 (in pyrimidines) or N-9 (in purines) of the base. Since a sugar is involved, this bond is referred to as a glycosidic bond. Further, the position of the hydroxyl group at the C-1' position of the sugar is always above the plane of the ring ie in the β-position (see Chapter 5). The bond between nitrogenous base and sugar is thus a β-glycosidic bond. Examples of nucleosides are shown in Figure 4.6.

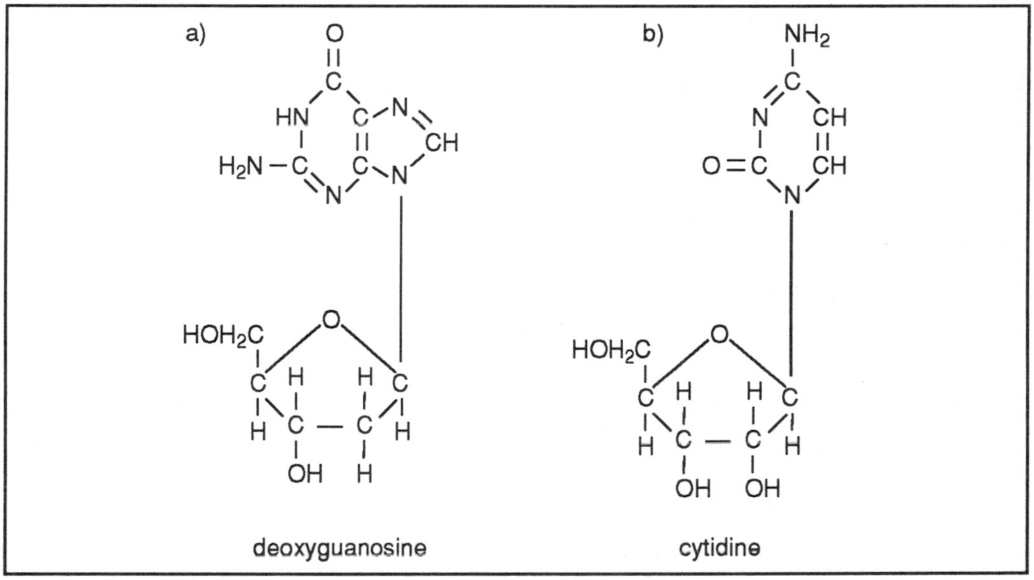

Figure 4.6 Nucleosides. a) deoxyguanosine b) cytidine.

There are also certain conventions involved in naming the various nucleosides. These simply have to be learnt, so that nucleosides (and nucleotides, as the system applies to them as well) can be accurately described. The names used are given in Table 4.2.

Base	Sugar	Nucleoside
Adenine	Ribose	Adenosine
Guanine	Ribose	Guanosine
Cytosine	Ribose	Cytidine .
Uracil	Ribose	Uridine
Adenine	Deoxyribose	Deoxyadenosine
Guanine	Deoxyribose	Deoxyguanosine
Cytosine	Deoxyribose	Deoxycytidine
Thymine	Deoxyribose	Deoxythymidine

Table 4.2 Names of commonly occurring nucleosides.

∏ Write down the structures of a) deoxyadenosine b) thymine c) uridine.
You will need to refer to Figures 4.2 to 4.6 and Table 4.2 unless you have a photographic memory!

Your structures should have been as follows:

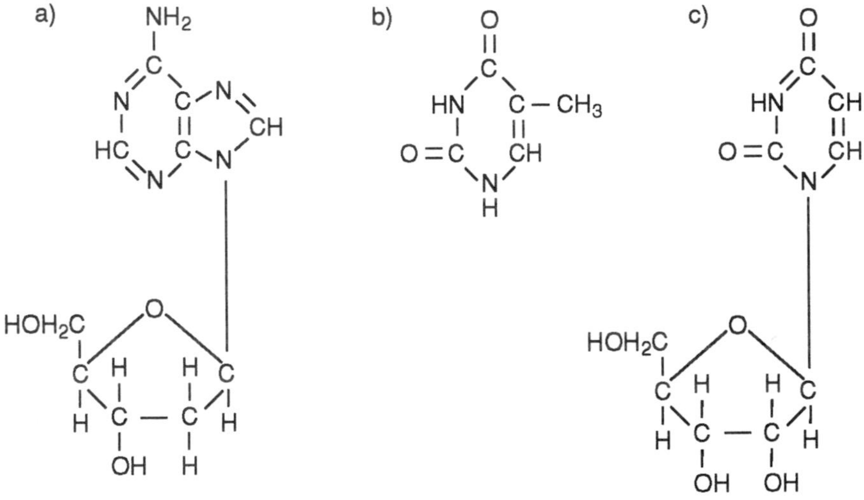

a) This nucleoside has deoxyribose (hence no -OH on C-2′ of the 5-carbon sugar) linked via the C-1′ of the deoxyribose to the N-9 of adenine.

b) Not a nucleoside! Thymine is just the nitrogenous pyrimidine base, thymine.

c) Uridine is the nucleoside compound of uracil and ribose (hence -OH on both C-2′ and C-3′ of the sugar).

4.4 Nucleotides

nucleotide   If phosphoric acid ($H_3 PO_4$) is esterified to the sugar moiety of a nucleoside, a nucleotide is created. The phosphate is always linked to one of the free hydroxyl groups of the sugar ie to C-3' or C-5' of ribose. Nucleotides which contain 2-deoxy-D-ribose are called deoxyribonucleotides; those containing D-ribose are known as ribonucleotides. Deoxyribonucleotides are the components of DNA; ribonucleotides are the building blocks of RNA.

4.4.1 Mononucleotides

AMP

NMP

NDP

NTP

Mononucleotides are named according to the nucleoside which they contain and the position at which the phosphate is attached. Thus a nucleotide containing adenine, ribose and a phosphate linked to the C-5' of ribose becomes adenosine-5'-monophosphate (routinely abbreviated as AMP. The 5' indicates that the phosphate is attached to the C-5' of the sugar). Collectively they are known as nucleoside - monophosphates (NMP). Nucleotides may contain more than one phosphate group, whereupon they become nucleoside-diphosphates (NDP), if two are present; nucleoside-triphosphates (NTP) when 3 phosphates are present. The arrangements for naming nucleotides are summarised in Figure 4.7. Note that at physiological pH (7-7.5), the hydroxyl groups on the phosphates will, to a large extent, be deprotonated. ATP is thus negatively charged and, in the cytoplasm of cells, is routinely complexed with Mg^{++} ions. In enzyme reactions involving ATP, the true substrate is thought to be Mg^{++}-ATP.

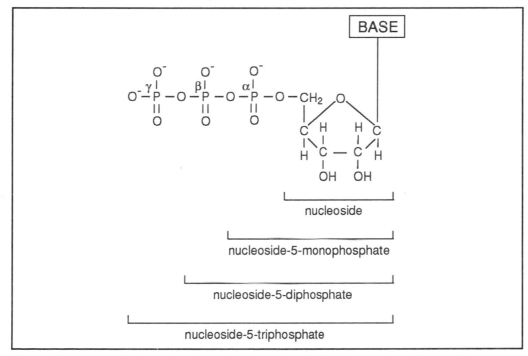

Figure 4.7 The generalised structure of a nucleoside, and nucleoside-5-mono, di- and triphosphates.

The nucleoside di- and triphosphates have several important functions in the cell. These include acting as carriers of chemical energy, carriers of special cellular building blocks, precursors for DNA and RNA and as intracellular messengers. We will briefly examine each of these in turn.

SAQ 4.2

Given below are 6 structures and 5 names. Match up each name (1-5) with the appropriate structure (a-f). Hence which is the structure which is left over, and what is its name?

1) Adenosine-5'-monophosphate.

2) Guanosine.

3) Thymine.

4) Deoxythymidine-5'-monophosphate.

5) Uridine-5'-monophosphate.

a)

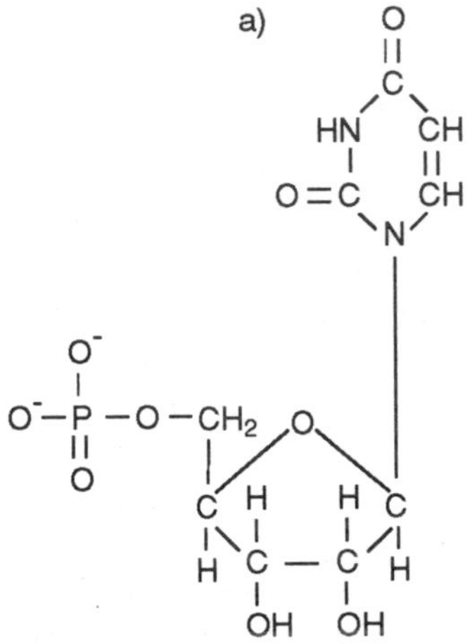

b)

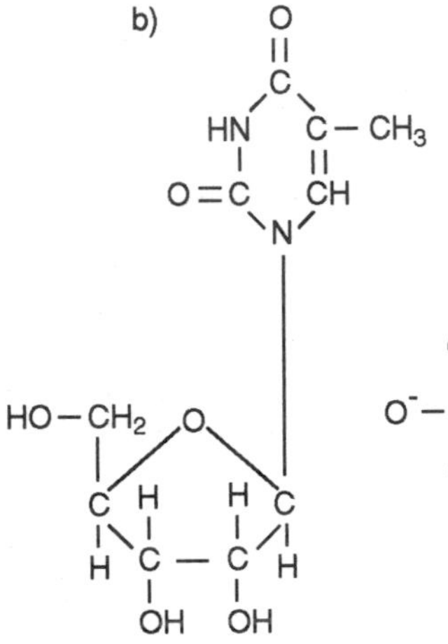

c)

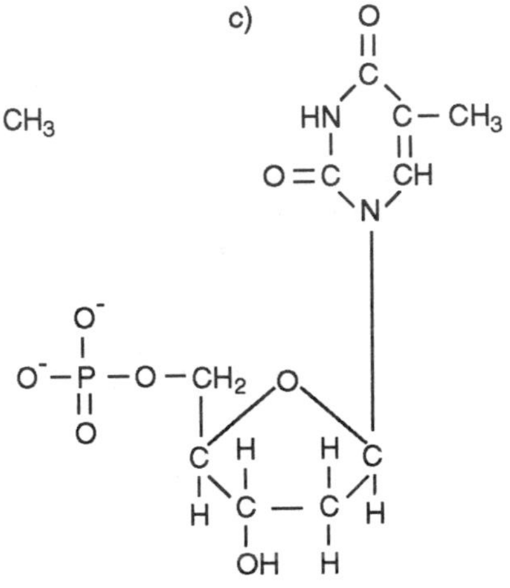

d)

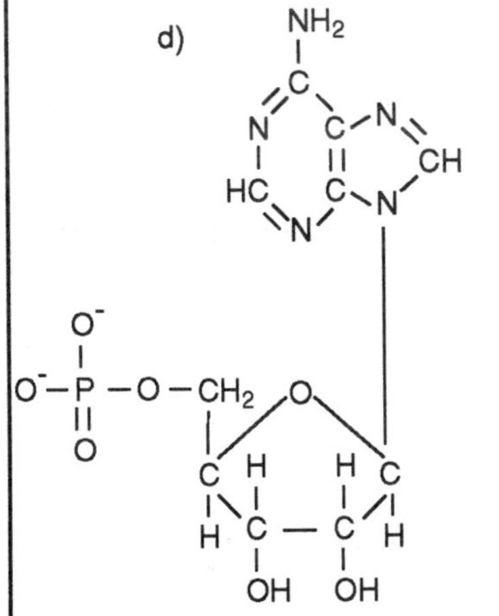

e)

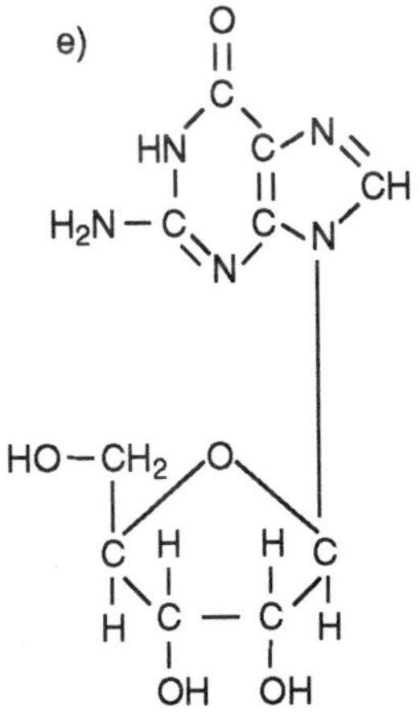

f)

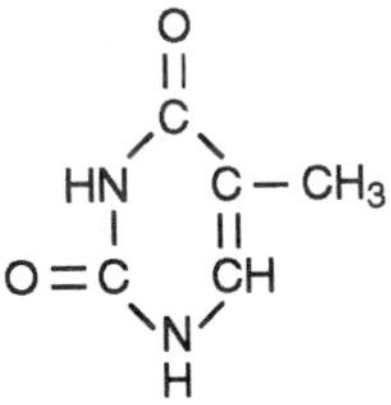

Carriers of chemical energy within the cell: principally adenosine triphosphate (ATP), but also guanosine triphosphate (GTP). ATP is synthesised by coupling phosphorylation of ADP to exothermic (ie energy-releasing) reactions. ATP is then used to supply energy to enable energy-requiring reactions to take place. An example is the phosphorylation of glucose to form glucose-6-phosphate. This is coupled to hydrolysis of ATP, as follows:

1)   Viewed as two separate reactions:

$$ATP \rightarrow ADP + HPO_4^= \quad \Delta G° = -30.7 \text{ kJ. mol}^{-1}$$

$$Glucose + HPO_4^= \rightarrow Glucose\text{-}6\text{-}phosphate \quad \Delta G° = +13.9 \text{ kJ. mol}^{-1}$$

Note that reactions with negative $\Delta G°$ values are exothermic (also known as exergonic, ie energy releasing) and are spontaneous. Reactions with positive $\Delta G°$ values are endothermic (also known as endergonic, ie they need energy).

2)   When these reactions are coupled together:

$$Glucose + ATP \rightarrow Glucose\text{-}6\text{-}phosphate + ADP \quad \Delta G° = -16.8 \text{ kJ. mol}^{-1}$$

**standard free
energy change**
$\Delta G°$ is the standard free energy change for a reaction. Free energy may be thought of as energy available to do work. The greater the value for $\Delta G°$, the greater the change in energy in going from reactants to products. If the value is negative, energy is lost from the compounds (ie energy is released: these are exothermic reactions). If positive, then energy must be supplied for the reaction to occur. Thus synthesis of glucose-6-phosphate on its own is energetically unfavourable. Energy must be supplied; this can be done by coupling the reaction to the hydrolysis of ATP.

**UDP and CDP
carriers**
Carriers of special cellular building blocks eg uridine diphosphate (UDP), which transports sugar groups for polysaccharide synthesis. Cytidine diphosphate (CDP) carries choline for the biosynthesis of choline - containing phospholipids. These types of reactions are exemplified by the synthesis and use of UDP-glucose (Figure 4.8).

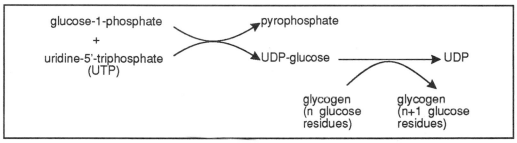

Figure 4.8 Synthesis and use of UDP-glucose. UDP-sugars are also involved in exchange reactions, as described in the text.

UDP-sugars also undergo various interconversions eg:

$$UDP\text{-}glucose + galactose \rightleftarrows UDP\text{-}galactose + glucose$$

High-energy precursors for DNA and RNA. The mononucleotides present in nucleic acids are incorporated from the corresponding nucleoside triphosphate or deoxynucleoside triphosphate eg ATP, CTP, UTP and GTP for RNA and dATP, dTTP, dGTP and dCTP for DNA.

Cyclic nucleotides as intracellular messengers. Cyclic adenosine monophosphate (cAMP, Figure 4.9) is formed within cells by the enzyme adenylate cyclase as follows:

$$ATP \xrightarrow{\text{Adenylate cyclase}} cAMP + Pyrophosphate$$

cyclic AMP

It was discovered during investigation of the mechanism of action of hormones, including adrenaline. Adrenaline (and other hormones) activate adenylate cyclase, resulting in synthesis of cAMP. cAMP in turn activates other enzymes. This particularly occurs with carbohydrate metabolism: through the mechanism described above, adrenaline raises the intracellular cAMP concentration, which leads to release of glucose from glycogen. Interestingly, activation of adenylate cyclase by adrenaline its itself controlled by so-called G-proteins (because they bind guanine nucleotides) which require GTP, GTP is hydrolysed in this activation. cAMP is subsequently inactivated by hydrolysis to 5'-AMP by the enzyme cyclic phosphodiesterase. To fully understand this point you would need to know much more about metabolism and about the action of hormones. At this stage you should remember that nucleotides, especially cAMP, can act as an intermediary messenger through which a hormone can influence the metabolic behaviour of a cell.

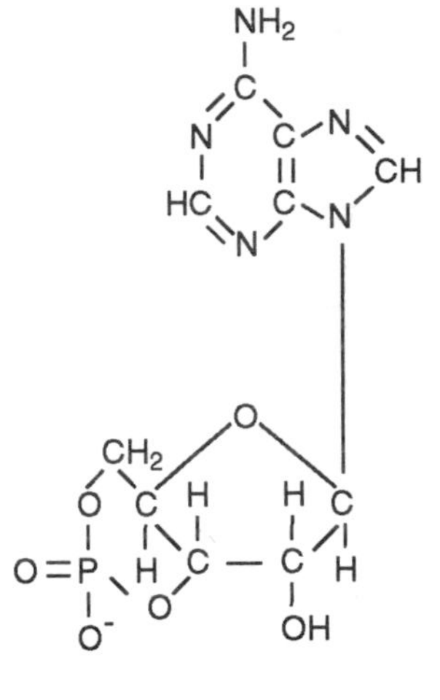

Figure 4.9 cAMP. The complete name is adenosine-3',5'-cyclic monophosphate, indicating that the single phosphate is attached to carbon atoms 3' and 5' of the ribose.

4.4.2 Dinucleotides

dinucleotide
phosphodiester
link

Dinucleotides consist of 2 mononucleotides which are linked by a phosphate group. Many arise through enzymatic hydrolysis of nucleic acids and have a phosphodiester link between the C-5′ atom of one sugar and the C-3′ atom of the second (see Section 4.5.1). An example is shown in Figure 4.10.

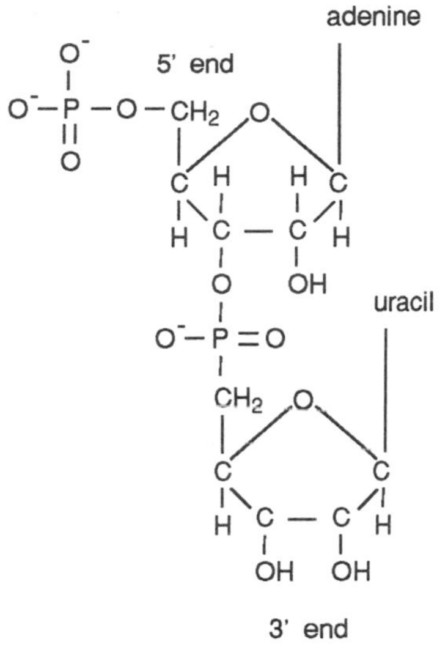

Figure 4.10 A dinucleotide.

Π Carefully inspect the dinucleotide shown in Figure 4.10, which resulted from hydrolysis of a nucleic acid. Was this nucleic acid DNA or RNA? Give brief reasons.

The nucleic acid was RNA. We can deduce this from which bases are present. The uracil is indicative of RNA (the other base, cytosine, occurs in both RNA and DNA, so is of no help). In addition, the sugar involved is ribose - notice that the C-2′ positions have hydroxyl groups attached. This must thus have been RNA.

There are several important dinucleotides which are deliberately synthesised by cells and in which the phosphate bond is not between the C-3′ and C-5′ of identical sugars. Instead they are 2 mononucleotides (ie base-sugar-phosphate) joined together through the phosphate groups. The most important of these dinucleotides are the 3 redox cofactors:

redox cofactors

NAD⁺ • Nicotinamide-adenine-dinucleotide (NAD^+)

NADP⁺ • Nicotinamide-adenine dinucleotide-2′-phosphate ($NADP^+$)

FAD • Flavin-adenine-dinucleotide (FAD).

They are shown in Figure 4.11.

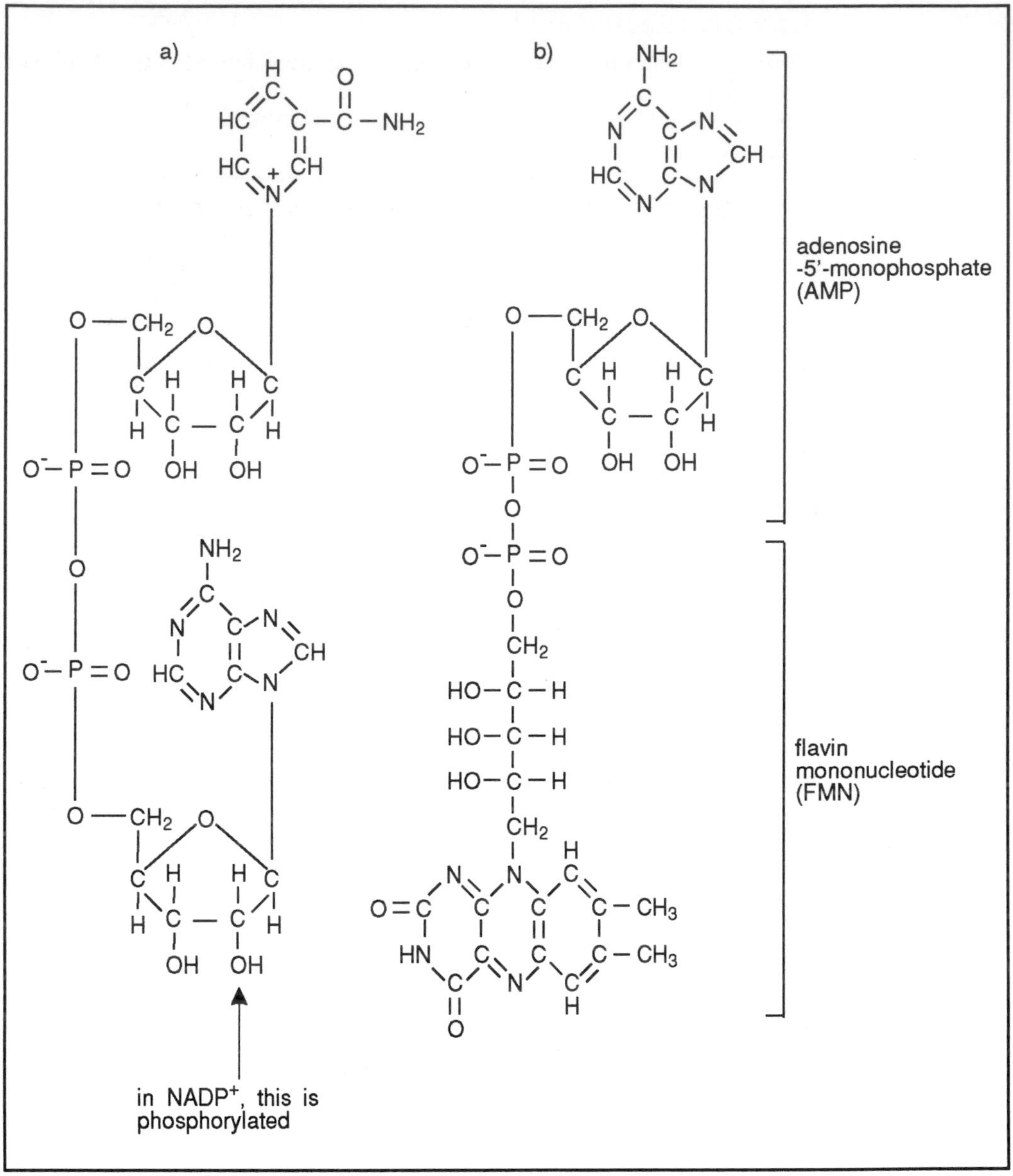

Figure 4.11 Metabolically important dinucleotides a) NAD$^+$. The position at which a phosphate group is attached (thereby creating NADP$^+$) is arrowed b) FAD.

They all contain an adenosine-5'-monophosphate group. NAD$^+$ and NADP$^+$ contain a nicotinamide mononucleotide as the second nucleotide. FAD contains a flavin nucleotide. These compounds are involved in many of the reactions involving reduction and oxidation which occur in cells. An example is the reaction catalysed by lactate dehydrogenase:

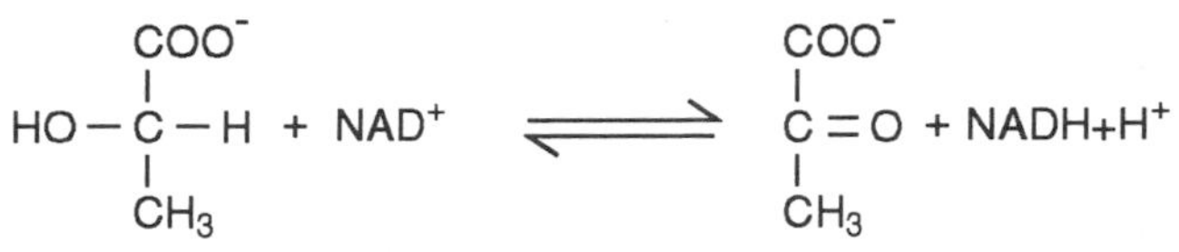

$$\text{HO} - \underset{\underset{\text{CH}_3}{|}}{\overset{\overset{\text{COO}^-}{|}}{\text{C}}} - \text{H} + \text{NAD}^+ \rightleftharpoons \underset{\underset{\text{CH}_3}{|}}{\overset{\overset{\text{COO}^-}{|}}{\text{C}}} = \text{O} + \text{NADH} + \text{H}^+$$

L-lactate   pyruvate

Reduction of NAD$^+$ occurs at the nicotinamide group, as shown in Figure 4.12.

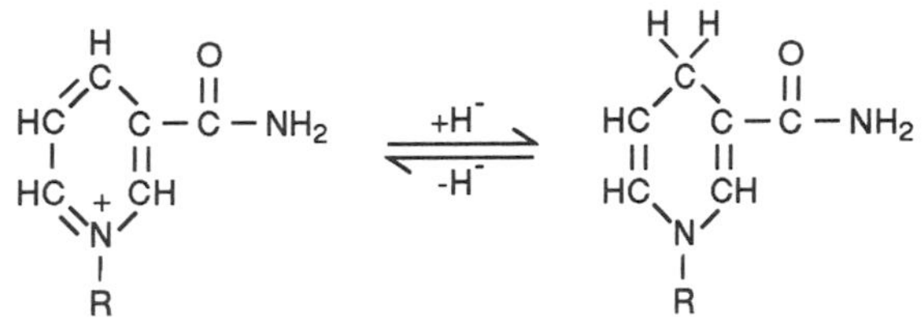

A typical reaction is:

$$\text{SH}_2 + \text{NAD}^+ \rightleftharpoons \text{S} + \text{NADH} + \text{H}^+$$

NAD$^+$ accepts a hydride (H$^-$) ion.

Figure 4.12 Reversible reduction of NAD$^+$, as in dehydrogenase reactions. Only the nicotinamide moiety of NAD$^+$ is shown.

It is not necessary to remember the structures of these dinucleotides. They are mentioned here so that you are aware that mono- and dinucleotides have vital roles in metabolism, and that they are not merely the building blocks of nucleic acids.

SAQ 4.3

Using the constituents given below, construct:

1) a nucleoside containing uracil and ribose, linked as they would be in RNA;

2) a deoxyribonucleotide-5′-triphosphate containing a purine;

3) Guanosine-3′,5′-cyclic monophosphate.

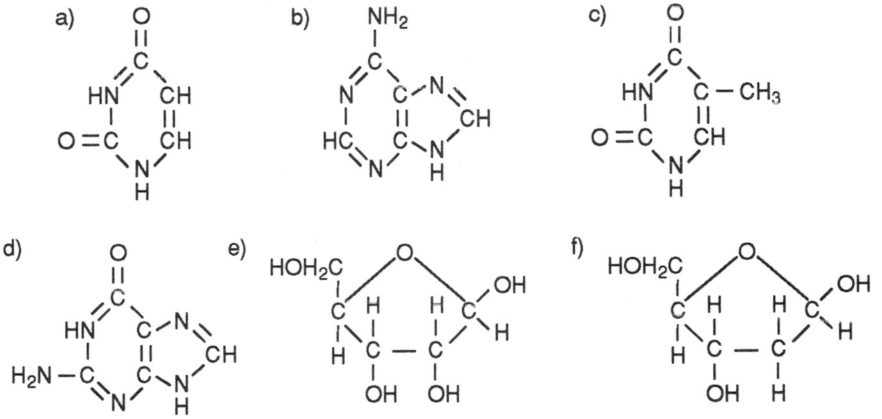

+ as many phosphate groups as you want. [NB The constituents may be used several times if necessary.]

4.5 DNA

Polynucleotides are polymers of mononucleotides. Those consisting of deoxyribonucleotides are called deoxyribonucleic acids, henceforth referred to as DNA. Polymers built out of ribonucleotide units are called ribonucleic acids (RNA).

4.5.1 Structure of DNA

3'5' phosphodiester

The covalent structure of DNA is shown in Figure 4.13a. Each mononucleotide is joined to the next by a 3'5' phosphodiester bond: this is the name given to the linkage, provided by the phosphate group, between the C-3' of one deoxyribose and the C-5' of the next deoxyribose. Notice that the phosphate group carries a negative charge. The pKa of this group is around 6, so at physiological pHs (7-7.5) it will largely be deprotonated. Thus every nucleotide of DNA carries a negative charge. There is no restriction on the order of the bases but genetic information is encoded within their order.

polarity of DNA

Careful inspection of Figure 4.13a shows that the two ends of the strand of DNA are different. The end at the top has a 'free' 5'-hydroxyl group on the deoxyribose; that at the bottom has a 'free' 3'-hydroxyl group. A strand of DNA thus has polarity. By convention, the sequence of bases is always written from the 'free' 5' end towards the 'free' 3' end. The piece of DNA shown in Figure 4.13 is therefore described as dGdAdTdCdA. Usually, the lower case d, denoting deoxy, would also be omitted, giving GATCA as the sequence of this section of DNA.

A convenient shorthand for depicting DNA structure is also shown in Figure 4.13b. In this the deoxyribose is shown as a vertical line with C-1' at the top and C-5' at the bottom. Letters designating bases are written at the top, as they are attached to the deoxyribose C-1'. The 3'-5' phosphodiester bond can be shown as a diagonal line linking the C-3' of one sugar and the C-5' of the next. We will use this shorthand subsequently. As before, the sequence is written in the 5'→3' direction.

X-ray diffraction pattern

Chargaff's rules

The 3-dimensional structure of DNA was described by Watson and Crick in 1953. This was based on X-ray diffraction patterns obtained from DNA by Wilkins and Franklin. The X-ray diffraction pattern obtained was indicative of a helical structure. Also crucial in deducing the structure of DNA were the results of chemical analysis by Chargaff, which became known as Chargaff's rules. On the basis of analysis of DNA from a variety of sources, Chargaff established that:

- the ratio of adenosine (A) to thymidine (T) was 1:1.

- the ratio of guanosine (G) to cytidine (C) was 1:1.

- also the ratio of purine to pyrimidine was 1:1.

However, the ratio of A to G or C, and of G to A or T varied with the source of DNA. The ratio of G + C to A + T was similar within a species, but varied between species. The more closely related the organisms the more similar the G + C: A + T ratio.

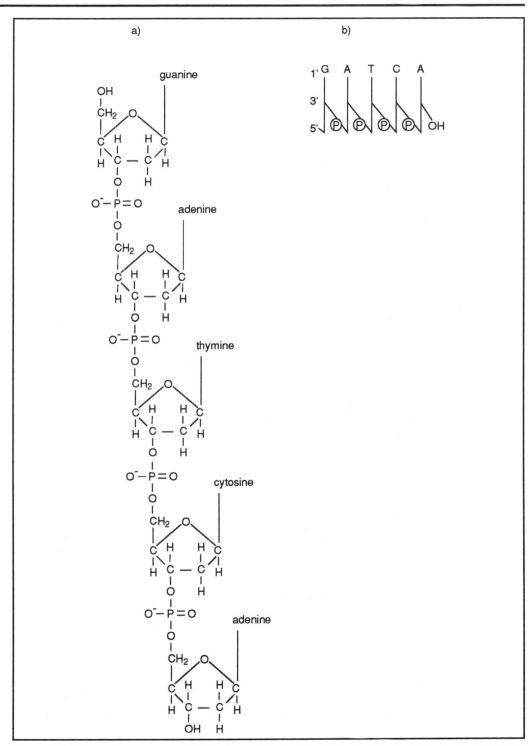

Figure 4.13 The covalent structure of DNA. a) complete structure. b) A 'shorthand' for describing structures of nucleic acids. The procedure is described in the text. The sequence given in b) corresponds to that depicted in a) above.

The double helix structure which Watson and Crick described (Figure 4.14) accommodates the X-ray diffraction analysis and Chargaff's rules, and also revealed an elegant mechanism whereby DNA can be replicated.

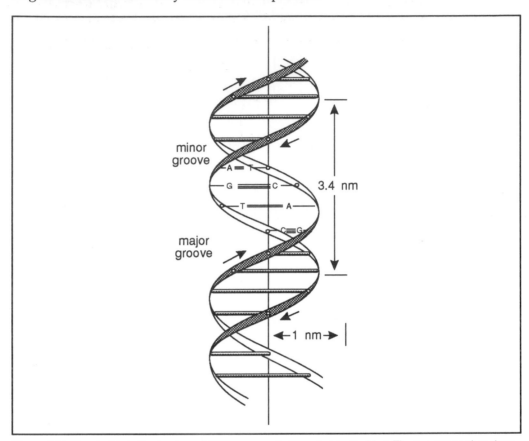

Figure 4.14 The double helix structure of DNA represented by a 'twisted ladder'. The two sugar-phosphate chains are represented by the twisted ladder 'frame'. The bases are represented by the 'steps' of the ladder.

DNA consists of two intertwined strands (of the type shown in Figure 4.13) in which the bases are on the inside. The deoxyriboses, linked by 3'-5' phosphodiester bridges, are on the outside of the double helix. The explanation of the equal amounts of adenosine and thymidine, and also of guanosine and cytidine is that the base adenine can only pair with thymine and guanine can only pair with cytosine. AT basepairs and GC basepairs, when linked by hydrogen bonds, were found to have the same overall size and hence allow a double helix of uniform dimensions to be made. If the double helix is vertical, then the bases are horizontal and flat, stacked one above the next, like a pile of coins. The two strands run in opposite directions (remember what we said earlier about polarity in a single strand of DNA); they are said to be anti-parallel. In moving along the axis of the double helix, there is a 36° rotation from one pair of bases to the next. A complete revolution occurs every 10 bases. This twisting results in the helical structure and creates two grooves, a major and a minor one. These are important for interacting with proteins which recognise and bind to particular sequences of nucleotides.

basepairing

anti-parallel
strands

The role of X-ray diffraction analysis and the significance of Chargaff's rules have already been identified as major contributors to the deduction of the DNA double helix. Also vital was careful model building conducted by Watson and Crick. By using the correct tautomeric forms of the bases, they realised that hydrogen bonding could occur between adenine and thymine and between guanine and cytosine (Figure 4.15).

hydrogen bonding between bases

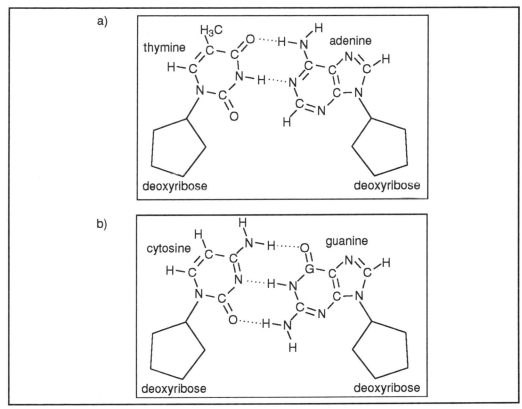

Figure 4.15 Complementary basepairing in DNA. a) Adenine and thymine, forming 2 hydrogen bonds. b) Guanine and cytosine; 3 hydrogen bonds are formed.

Note that 2 hydrogen bonds are formed in an AT basepair, whilst 3 hydrogen bonds occur in a GC basepair. As already noted, these basepairs were of suitable dimensions. With other combinations, relevant groups were incorrectly oriented, or too bulky (2 purines) or too far apart (2 pyrimidines). The anti-parallel nature of the double helix is important here, as it ensures the optimum alignment of bases and thus maximised hydrogen bonding.

Compulsory basepairing is the molecular explanation of Chargaff's rules: whenever adenine occurs in one strand, thymine must occur in the other. The same requirement exists for guanine and cytosine and thus, in a double helix, the number of purines always equals the number of pyrimidines. AT and GC are referred to as complementary base pairs. The 2 strands of a DNA double helix are said to be complementary to each other.

complementary basepairs

Π   Figure 4.13 showed a section of a DNA molecule, whose sequence was given in the text.

1) Given the sequence again (below), write in, below each base, the base you would expect in its complementary strand.

GATCA

Other strand ⎯⎯⎯⎯⟶

2) Now write down the sequence of this second strand (ie the one you have just deduced), bearing in mind any relevant conventions regarding the order of the bases.

Your answer should have been:

1) Other strand ⎯⎯⎯⟶ GATCA   This is the response you
CTAGT ⟵⎯⎯⎯ should have produced.

On the basis of compulsory basepairing (AT; GC), the above pairings would be made. Re-read the previous section if in doubt.

2) This second strand sequence, on its own, should be written as TGATC. This is because single strand DNA sequences are always written from the 5′ end to the 3′ end. Thus we know that the original strand was:

5′ ⎯⎯⎯⟶ 3′
GATCA

In the double helix, the 2 strands are anti-parallel, hence the 2 strands are:

5′ ⎯⎯⎯⟶ 3′
GATCA
CTAGT
3′ ⟵⎯⎯ 5′

This second strand is therefore TGATC when the polarity convention is followed.

Why does a double helix occur? As a consequence of twisting into the double helix, the basepairs are flat and stacked very close to each other (0.34nm apart). The double bonds of the bases give them hydrophobic characteristics. Water is excluded from the centre of the helix and extensive hydrophobic interactions occur between the stacked bases. These, coupled with the hydrogen bonds between the basepairs, make the double helix the conformation of lowest free energy and are responsible for its stability. Quite clearly, DNA has a very pronounced 'secondary structure', since all nucleotides in the molecule are part of a double helix.

secondary structure

A solution of DNA is highly viscous; this is a consequence of the very long strands of double helical DNA, which tend to behave like fairly rigid rods. Molecules of DNA can be extremely long. Table 4.3 gives an indication of the range of sizes which occur.

	DNA length (mm)	No of base pairs (Kbp*)	RMM
Simian virus (SV40)	1.7×10^{-3}	5.1	3.3×10^{6}
Bacteriophage λ	17×10^{-3}	48.6	32×10^{6}
E. coli	1.4	4000	2.6×10^{9}
Drosophila	21	62 000	43×10^{9}

Table 4.3 Sizes of individual DNA molecules.
* Kbp = kilo base pairs = 1000 basepairs. RMM = relative molecular mass.

closed circles

exonuclease

plasmids

In some viruses, the amount of DNA is very limited and small molecules of DNA are found. Many DNA molecules, especially in prokaryotes, are closed circles. The reason is probably that enzymes which degrade DNA, particularly from an end (hence called exonucleases) are widespread in nature. Constructing closed circles means that there are no ends which would be susceptible to attack by exonucleases. The DNA of E. coli displays this property, as do plasmids, the small pieces of DNA which are extensively used in genetic engineering. In eukaryotes, DNA is packaged into chromosomes. One of the chromosomes from Drosophila contains a DNA molecule with nearly 80 million basepairs, an overall length of nearly 3cm. That this is packaged into a very small volume is a remarkable achievement. Most bacterial DNA can be visualised by electron microscopy. Note that all DNA molecules are highly asymmetrical: the diameter of the double helix is only 2nm. As a consequence of their length and asymmetry, DNA is very susceptible to breakage by shear forces. During isolation, DNA must be treated carefully to avoid fragmentation.

chromosomes

histones

chromatin

endonucleases

We have mentioned that in eukaryotes, DNA is present in chromosomes, within the nucleus of the cell (in addition, some DNA is also present in mitochondria and chloroplasts). DNA is complexed with basic proteins called histones. You will recall that DNA is negatively charged (because the phosphate group will be deprotonated, so that the outside of the double helix has negative charges at regular intervals. Histones are highly basic (ie positively charged) proteins. Five types, designated H1, H2A, H2B, H3 and H4 occur and all have high isoelectric points. This is because they contain large amounts of lysine (eg histone H1) or lysine and arginine (the other histones): about a quarter of the amino acid residues are either lysine or arginine. Histones are therefore highly positively charged at neutral pH and bind tightly, through electrostatic attraction, to DNA. Indeed, to remove DNA, strong protein denaturants must be used. The DNA-histone complex is known as chromatin. It can be isolated from nuclei and examined by a variety of methods, including electron microscopy. This revealed that most of the histone is contained in bead-like structures, called nucleosomes (Figure 4.16). Approximately 200 basepairs of DNA are wrapped in and around the histones of the nucleosome. This DNA is partially protected from digestion by endonucleases; the remaining DNA acts as a linker between nucleosomes and is susceptible to degradation by endonucleases (enzymes that hydrolyse DNA within the chains).

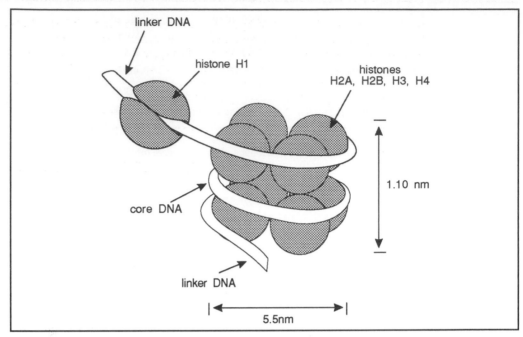

Figure 4.16 A nucleosome. These consist of histones H2A, H2B, H3 and H4 forming an aggregate (a 'bead') around and within which DNA is wound. Some DNA, and another histone (H1) occur as linking sections between nucleosomes.

Π Assume that you are working on an intriguing, newly discovered organism whose DNA is found in 'chromosomes' within a 'nucleus', but in other respects resembles bacteria. You decide to investigate whether the organism has 'histone-like' proteins. What properties would you anticipate such proteins would display, and how could you obtain these proteins?

Histone-like proteins would be expected to bind tightly to DNA. Thus they would be expected to be highly positively charged, resulting from possession of a high proportion (as much as 20-25% of all amino acid residues) of amino acids with positively charged sidechains at neutral pH, ie lysine and arginine. They would have very high isoelectric points (≥ 10).

They would be expected to be bound to the DNA - thus isolation of 'nuclei' and/or chromosomes would be a useful first step. The DNA would then have to be removed. This may well require denaturation of the protein - but its amino acid composition and size could still be determined. Further purification could be based on their high isoelectric point (see Chapter 3). If their isoelectric point was 11, they would still be positively charged at pH 9-10. At this pH the majority of other proteins, whose isoelectric points are typically between 4 and 8, will be negatively charged. Thus the histone-like proteins will bind to a cation-exchange resin (one with fixed negative charges, which binds and exchanges cations), whilst proteins with lower isoelectric points will fail to bind.

4.5.2 Analysis and properties of DNA

absorption spectra of DNA

All nucleotides absorb light in the ultraviolet region, with a maximum at 260nm. This is a consequence of the double bonded ring structures of the bases (thus bases and nucleosides also absorb at 260nm). Typical absorption spectra are shown in Figure 4.17. One consequence of the secondary structure of the DNA double helix is that absorption of light by bases within the double helix is suppressed. What we mean by this is that the absorbance at 260nm of the bases in DNA is less than would be the case if they were present as free nucleotides. As a consequence, when DNA is hydrolysed (whether enzymatically by nucleases, or by treatment with acid - this is how early studies on the base composition of DNA were carried out), the absorbance at 260nm increases. This increased absorbance is useful, as it provides a sensitive way of following changes in the secondary structure of DNA.

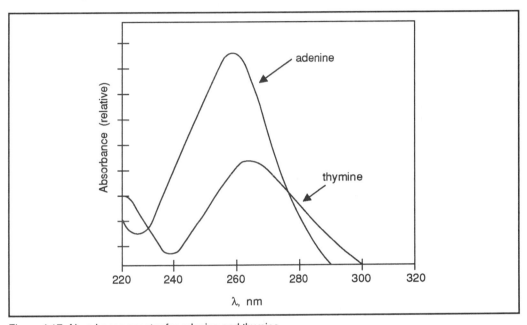

Figure 4.17 Absorbance spectra for adenine and thymine.

denaturation of DNA

If a solution of DNA is heated, the strands of DNA will separate and the double helical structure will be lost. This process is called denaturation and results in single strands of DNA which assume 'random-coil' structures. When denaturation occurs, the absorbance at 260nm of the solution increases by about 40%; this is known as the **hyperchromic effect**. Monitoring the absorbance at 260nm of a solution of DNA is a convenient way of detecting denaturation.

melting temperature, Tm

Denaturation is found to occur over a narrow temperature range and at a characteristic temperature for DNA of a given species (Figure 4.18). This temperature is known as the melting temperature, Tm. The Tm is dependant upon the DNA sample and also upon the physical environment of the DNA (eg the pH of the solution, salt concentration). In the following description assume that the physical environment of the DNA is the same in each case.

The Tm is the temperature of the midpoint of the hyperchromic increase (ie when half the strands have separated). It seems that the 2 strands of DNA remain together, as the temperature is raised, until the double helix, fairly abruptly, 'unzips'. When melting temperatures were analysed for a wide range of organisms, using standardised solvent (buffer) conditions, it was found that the melting temperature was related to GC content. Increase in GC content results in an increase in melting temperature. This

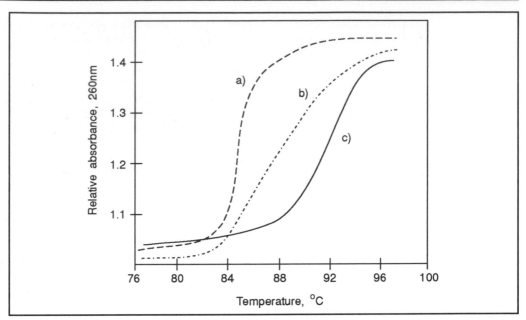

Figure 4.18 Melting curves for DNA from different species. a) *Pneumococcus* b) *E. coli* c) *Serratia*.

enables a calibration graph to be constructed (Figure 4.19) from which the GC content of a test sample of DNA may be deduced. For many years, this was the easiest way of determining the base composition of samples of DNA.

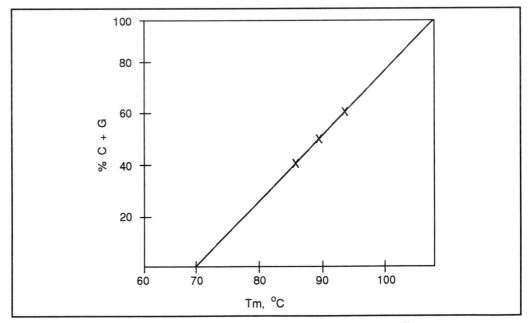

Figure 4.19 The relationship between the G + C content and melting temperature (Tm).

Π Suggest a reason why the melting temperature, Tm, of DNA increases as the GC content is increased. If in doubt, look again at Figure 4.15.

The reason why the melting temperature increases as the GC content increases is that GC basepairs have 3 hydrogen bonds, whereas AT basepairs only have 2. Thus a 100 basepair section of DNA composed solely of GC basepairs possesses 300 hydrogen bonds. The same length of purely AT basepairs would have 200 hydrogen bonds. More energy (= higher temperature) is required to break a larger number of hydrogen bonds.

renaturation of DNA

If samples of DNA which have been denatured, by heating to a temperature above their Tm, are then maintained at a slightly lower temperature (about 20-25°C below Tm), the strands recombine, with correct basepairing. Thus separate strands eventually find their complements and reassociate to form double helices. The rate at which this reassociation occurs depends on the complexity of the DNA in terms of the number of different sequences. The rate of reassociation also depends on the concentration of complementary sequences. DNA from a simple genome will have relatively few different sequences. At a given concentration, it will reassociate more rapidly than DNA from a more complex genome, where the number of different sequences is greater. Analysis of rates of renaturation, as this process is called, provide useful information on the complexity and size of genomes.

Techniques for determining the amino acid sequences of proteins have been available since the early 1950s. Sequence analysis of DNA proved to be much more difficult, although it was always recognised that it would yield enormous benefits. Restriction enzymes (see Section 4.7) enable DNA to be hydrolysed only at particular sites, thereby enabling particular sections of genes to be obtained. In 1977, two groups independently published strategies by which the nucleotide sequence of DNA may be deduced. The method of Maxam and Gilbert involves chemical methods that result in cleavage after particular bases (eg after adenine, or after guanine). That of Sanger and coworkers is very elegant and demonstrates how our understanding of the biochemistry of a process enables it to be exploited. What follows is a very brief description; the technique is discussed in greater detail elsewhere in the BIOTOL series (Biomolecules - Analysis and Properties).

Sanger method

It is well known that DNA chains are extended, during normal replication, by addition of individual bases at the 3' end of a chain. Which base is added at each position is determined by the base in the corresponding position of a template strand (Figure 4.20).

dideoxy analogues

This process is catalysed by the enzyme DNA polymerase (see Section 4.5.3). The incoming base forms a covalent bond with the 3'-OH of the previous base. Sanger and his coworkers realised that if there was no 3'-OH group, then chain growth would be terminated. They used dideoxy analogues (Figure 4.21) of each of the nucleotides in 4 separate incubations, which led to chains being terminated when a particular base was inserted (Figure 4.22).

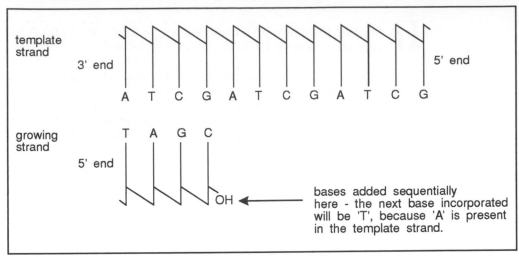

Figure 4.20 During replication, nucleotides are added sequentially to the 3' end of the growing chain. The base at the corresponding position of the template strand determines which nucleotide is incorporated.

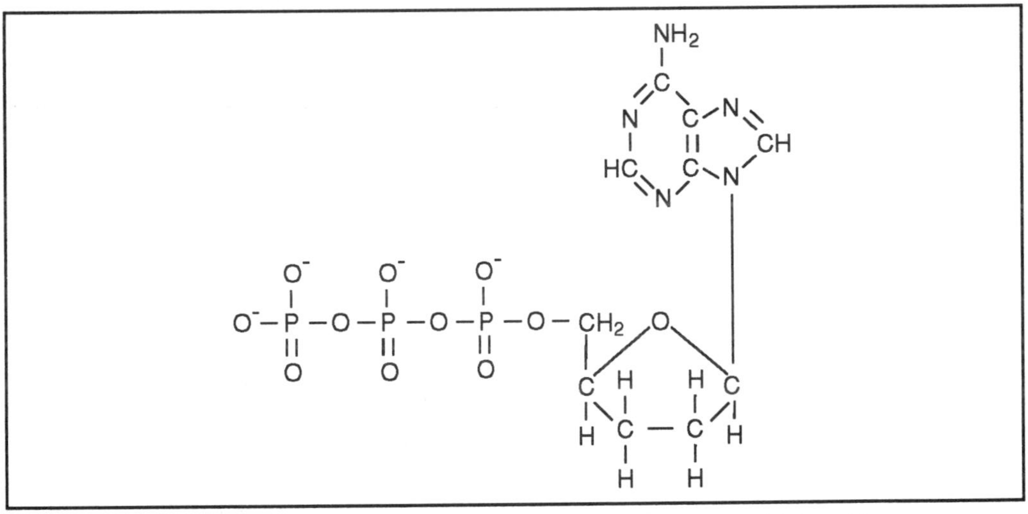

Figure 4.21 Dideoxy ATP. Note the absence of hydroxyl groups at both C-2' and C-3' positions.

For example, if a mixture of normal dATP and dideoxy ATP is supplied, together with normal dNTPs of the other nucleotides, then whenever a nucleotide containing adenine is incorporated, it may terminate the chain.

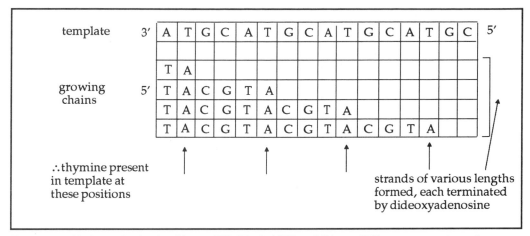

Figure 4.22 Chain termination by dideoxy ATP. If both dATP and dideoxy ATP are present, then when AMP is incorporated it may terminate the chain (if dideoxy ATP is used). This gives rise to strands of different lengths. Their length indicates where thymine must have been present in the template strand.

This leads to the formation of fragments of various lengths; their lengths (which may be determined by gel electrophoresis) indicate precisely where in the template strand a thymine is located. If this is repeated with dideoxy derivatives of the other nucleotides, then the complete sequence can be deduced. A typical sequencing gel, showing bands corresponding to various length fragments, is shown in Figure 4.23.

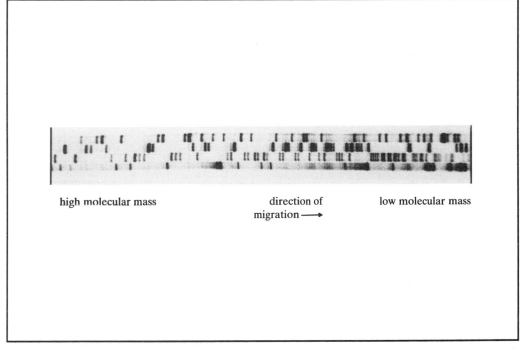

Figure 4.23 An autoradiogram of a section of a DNA sequencing gel. Chains migrate according to size and form discrete bands (small chains migrate further than large chains). ^{32}P- labelled dNTPs are used, thereby enabling visualisation by autoradiography; the actual amount of DNA present per track is very small. In this case, four lanes are shown. (By kind permission of Pharmacia LKB Ltd).

The first successful application was on the DNA of bacteriophage ϕx174, for which Sanger received his second Nobel prize. This DNA is only(!) 5386 bases long. Since then the method has proved invaluable in probing gene structure, the sequences of control regions of genes and in establishing the precise changes in DNA sequence which cause inherited diseases. The project to sequence the entire human genome is now a feasible (although enormous) task.

Π Assume that the dideoxy chain termination method was applied to a sample of DNA (as described in the previous section) using a mixture of normal dGTP and dideoxy GTP, plus normal dATP, dTTP and dCTP. Fragments of the following lengths were obtained, where X is a nucleotide. What can you deduce about nucleotides in the template strand shown at the top.

Template																
Fragments obtained	X	X	X	X	X											
	X	X	X	X	X	X	X	X								
	X	X	X	X	X	X	X	X	X							
	X	X	X	X	X	X	X	X	X	X	X					
	X	X	X	X	X	X	X	X	X	X	X	X	X	X	X	X

5′   3′

Wherever the chain has been terminated, a dideoxyguanosine must have been inserted. For this to occur there must have been deoxycytidine in the template strand at the corresponding position. Thus the last (3′ end) nucleotide in all of the fragments was dideoxyguanosine and deoxycytidine occupies the corresponding position in the template.

3′   5′

Template					C			C	C		C				C
Fragments	X	X	X	X	G											
	X	X	X	X	G	X	X	G								
	X	X	X	X	G	X	X	G	G							
	X	X	X	X	G	X	X	G	G	X	G					
	X	X	X	X	G	X	X	G	G	X	G	X	X	X	X	G

5′   3′

Note that as the strands are anti-parallel, the 5′ end of the template strand is at the right hand end. In the conventional order, the sequence of the template is thus:

CXXXXCXCCXXCXXXX

The identity of bases at remaining positions (ie X) would be deduced from similar analysis using dideoxy derivatives of the other NTPs.

SAQ 4.4

1) Write down the structure of a section of DNA of the following sequence AAGCT. You may just write 'adenine' or 'guanine' etc to indicate where each base would go.

2) Which of the following is the complementary sequence to that given above:

a) 5′ TTCGA 3′

or   b) 5′ AGCTT 3′. Give reasons for your choice.

3) Predict which of the following two double helical deoxyribonucleotides will have the higher melting temperature. Give brief reasons.

a) $^{5'}$ATATATATAT$^{3'}$
      $^{3'}$TATATATATA$^{5'}$

b) $^{5'}$GCGCGCGCGC$^{3'}$
      $^{3'}$CGCGCGCGCG$^{5'}$

4) DNA from a newly discovered bacterium is denatured by heating, using the same conditions as those applying to the data given in Figure 4.19. A melting temperature of 95°C was obtained. What is the G^{C} and AT content of this DNA?

4.5.3 Replication of DNA

The most crucial feature of the structure of DNA for its role as hereditary material is the specificity of basepairing. Watson and Crick recognised the implications of this absolute specificity in their paper describing the structure of DNA. Thus, if the sequence of one chain is given, the sequence of the second may be deduced - as you have done in SAQ 4.4. Further, if the two strands were to separate, then new strands could be assembled, using each existing strand as a template, alongside the existing strands. Bases would be incorporated into the new strand, so as to give only correct AT and GC basepairs. This would give 2 copies of the original DNA molecule, of identical sequence. Watson and Crick predicted this mechanism as the basis of replication. This pattern, in which one of the strands of each 'daughter' molecule is newly synthesised, whilst the other comes from the original DNA molecule, is known as **semi-conservative replication**. Semi-conservative replication was shown to actually occur by Meselson and Stahl.

semi-conservative replication

Meselson and Stahl grew *E. coli* on a medium in which the sole source of nitrogen was NH$_4$Cl in which the nitrogen atom was the heavy isotope ^{15}N. As a result, all of the DNA in the *E. coli* eventually had ^{15}N wherever nitrogen atoms occurred and was therefore more dense than usual. This meant that this 'heavy' ^{15}N-containing DNA could be separated from 'normal' ^{14}N-containing DNA by density equilibrium centrifugation. This is a centrifugation technique, using a concentration gradient of CsCl, by which molecules form bands in the region of the density gradient corresponding to their own buoyant density (Figure 4.24). The purely ^{15}N-DNA is more dense than purely ^{14}N-DNA and distinct bands are formed (Figure 4.24).

density equilibrium centrifugation

Meselson and Stahl then transferred the ^{15}N-grown *E. coli* cells to ^{14}N-NH$_4$Cl containing medium and analysed the density of the DNA present in the cells at various times thereafter. After one doubling time, when DNA had replicated once, a single band,

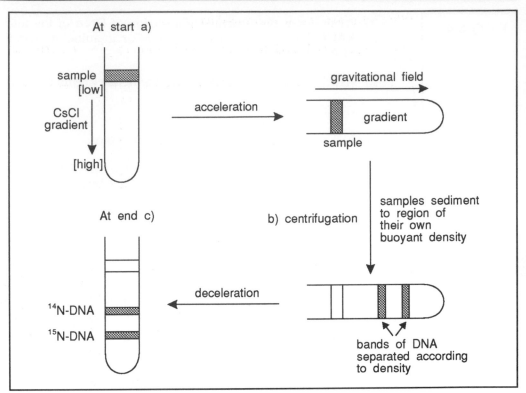

Figure 4.24 Density equilibrium centrifugation of DNA. A CsCl density gradient is constructed a). DNA layered onto this and centrifuged b) will sediment until its buoyant density equals that of the CsCl in that region of the gradient. Samples of DNA of different density (^{15}N-DNA and ^{14}N-DNA) form discrete bands c).

precisely intermediate in density between pure ^{15}N-DNA and pure ^{14}N-DNA was found (Figure 4.25).

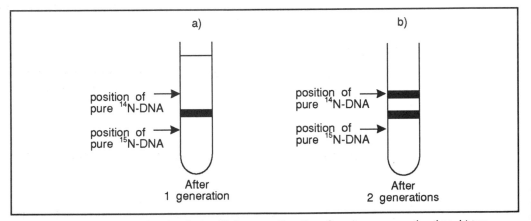

Figure 4.25 Band patterns after density equilibrium centrifugation after a) one generation time; b) two generation times.

After 2 generation times, 2 bands were seen, one corresponding to the intermediate density form seen after one generation, the second to pure ^{14}N-DNA. With increasing time, the ^{14}N-DNA came to predominate.

What is the explanation for this data?

Three generalised models for DNA replication can be proposed, as shown in Figure 4.26.

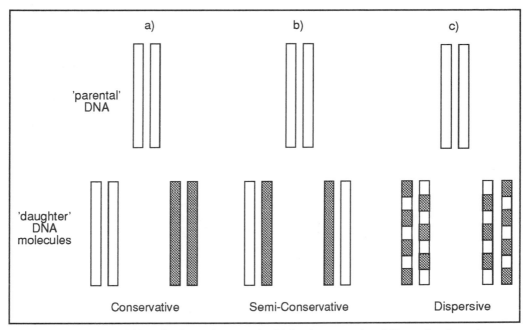

Figure 4.26 Models for DNA replication a) Conservative b) Semi-conservative c) Dispersive. In each case parental strands (or parts of them) are shown as unshaded; newly synthesised DNA is shown by the hatched pattern.

Conservative

In this model the 2 'parental' strands remain together; two newly synthesised strands together form one of the 'daughter' helices.

Semi-conservative

Here the 'parental' strands separate, with each strand remaining intact, and a new strand is synthesised alongside each 'parental' strand. Each daughter helix had one old and one new strand.

Dispersive

In this model the 'parental' strands are fragmented and sections of the 'parental' strands are found in all 4 strands of the 2 daughter helices.

Π   1) Analyse the results obtained by Meselson and Stahl in terms of the 3 models described above. Which model best fits the observed data? 2) Now predict the results that would have been obtained if each of the other models actually was correct and represented replication.

1)   The semi-conservative model is the only one which is consistent with the data, After one generation, all of the DNA should have one ^{15}N-containing and one

^{14}N-containing strand, and therefore be of intermediate density. After 2 generations half of the DNA will be (1 x ^{15}N- + 1 x ^{14}N-strands), the other half will be entirely ^{14}N-containing strands (see Figure 4.26). This is precisely the pattern that was observed. The other models cannot give these results.

2)   With the conservative model, after one generation there will be 2 types of DNA: one composed entirely of ^{15}N-DNA, the other composed entirely of ^{14}N-DNA (as in Figure 4.24). This did not occur. After 2 generations, of the 4 double helices, 3 will be pure ^{14}N-DNA; one will be pure ^{15}N-DNA. With the dispersive model, all strands would have some ^{15}N-sections and some ^{14}N-sections. Thus after one generation the actual result is consistent with this model. After two generations all strands will still have some ^{15}N-DNA, but there will be less ^{15}N atoms per strand (and hence per double helix) and more ^{14}N atoms. By now we would expect about 25% of the N atoms to be ^{15}N and 75% to be ^{14}N. Thus there should be a single band intermediate between ^{15}N/^{14}N and pure ^{14}N band positions. With further generations the band should slowly shift to pure ^{14}N.

If you had any difficulty with this analysis, re-read the main text section again.

template   The semi-conservative model for replication was that envisaged by Watson and Crick. Each strand acts as a template, with a new strand being polymerised alongside it, one nucleotide at a time (Figure 4.27). The base in each position in the template strand determines which nucleotide is incorporated at each position of the growing strand. Replication is carried out by the enzyme DNA polymerase, as shown in Figure 4.27.

Template $^{3'}$A-T-G-C-A-T-G-C-A-T-G-C

Primer 5'   T-A-C-G ———————→ Primer extended in this direction

DNA polymerase
+dATP +dGTP +dCTP +dTTP

Figure 4.27 DNA replication. DNA polymerase extends the primer strand in a 5' → 3' direction. The nucleotides in the template strand determine which nucleotide is inserted at each position.

primer   DNA polymerase can only extend chains, so it requires a primer. This is a short chain which is then extended at its 3' end by successive addition of bases. New DNA is always synthesised so that the strands are anti-parallel and the primer strand always grows in a 5' → 3' direction.

∏   Predict the sequence of the primer strand in Figure 4.27 after 4 more nucleotides have been incorporated.

The primer strand will now be $^{5'}$TACGTACG$^{3'}$. This is the sequence required by AT/GC complementary basepairing with the template strand.

replication fork   Analysis of the replication of the single DNA molecule (the chromosome) of *E. coli* revealed that a replication fork occurs, which moves in one direction along the DNA (Figure 4.28a).

This implies that primer extension in the 5' → 3' direction occurs along one strand, but 3' → 5' extension occurs along the second strand (bearing in mind that the parental strands of the double helix are antiparallel). How can this be, especially as no enzyme capable of 3' → 5' chain extension has ever been found? The answer is that DNA is synthesised continuously along one strand (the leading strand), but in short fragments along the second (the lagging) strand. At all times, chain extension is 5' → 3' (Figure 4.28b). This gives a series of fragments (Okazaki fragments) which are subsequently joined together by another enzyme, DNA ligase. DNA polymerase needs a primer to synthesise DNA. In producing the Okazaki fragment RNA polymerase first makes short RNA molecules which are then used as primers by DNA polymerase. The short sections of RNA are subsequently removed and the gap filled with DNA.

leading and lagging strand

Okazaki fragments

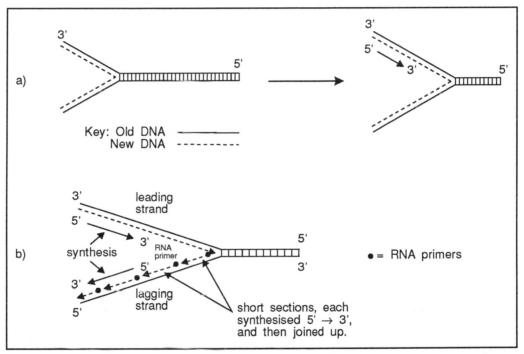

Figure 4.28 a) A replication fork moves along the DNA molecule. b) Primer extension is continuous along the leading strand. For the lagging strand, DNA is synthesised in short fragments, which are subsequently joined up. In all cases chain growth is 5' → 3'.

stringency of nucleotide incorporation

The description given above is a very simple account of a complex event. Many more enzymes and other proteins are required to accomplish replication. The key to avoiding any error occurring (inaccurate copying) is the very high stringency which occurs in selecting each nucleotide to be incorporated. Further, DNA polymerase 'checks' for incorrect nucleotides - a form of proof-reading! - and replaces any incorrect nucleotide incorporations. The error rate (ie frequency with which incorrect, in the sense of AT and GC compulsory pairings, nucleotides are inserted) is extremely low. This means that exact copies of DNA molecules are produced, with precisely the same base sequence. Note that errors (ie the wrong base being incorporated) represent, at the molecular level, a mutation, since it will be maintained during subsequent replication and may lead to an incorrect amino acid being incorporated into a protein.

mutation

Π   Chargaff's rules are consistent with the compulsory basepairing of A and T, and G and C in the DNA double helix. The DNA of a small bacteriophage, called φx174, was found to have bases in the following ratio:

$$A = 24.5$$
$$G = 24.7$$
$$C = 18.4$$
$$T = 32.4$$

In addition, φx174 DNA was less viscous than other DNA of similar length, and behaved more like a randomly coiled polymer. Further, the bases of φx174 DNA were more reactive with chemicals, whereas in DNA from other sources the bases failed to react. Suggest reasons for these anomalies.

single stranded DNA

The bacteriophage φx174 is very unusual in having single-stranded DNA. Chargaff's rules are obeyed by all double stranded DNA molecules, but in a single strand of DNA (ie in one of the 2 strands of 'normal' DNA) there is no requirement for A to equal T or for G to equal C. The hydrodynamic behaviour is consistent with single-stranded DNA, as occurs, for example after denaturation (Section 4.5.2). In the double helix the bases are in the middle of the helix and inaccessible to chemicals eg those reacting with amino groups. In a single-stranded DNA, there is no comparable protection.

Incidentally, φx174 DNA does form a double stranded DNA during its replication - a so-called replicative form. It is this double stranded form which serves as template for production of more φx174 DNA. Thus the principle of DNA strands acting as templates during replication to enable exact copies to be made is universal.

4.6 RNA

4.6.1 Structure and properties

RNA is similar to DNA in its covalent linkages. It is a polymer of nucleotides linked by 3′, 5′ phosphodiester bonds, and 4 different nucleotides are usually involved. It differs from DNA in that the sugar involved in RNA is ribose, and uracil replaces thymine. Uracil basepairs with adenine in the same way that thymine does. A section of an RNA molecule is shown in Figure 4.29.

The structures of the constituents of RNA were described in sections 4.2 to 4.4.

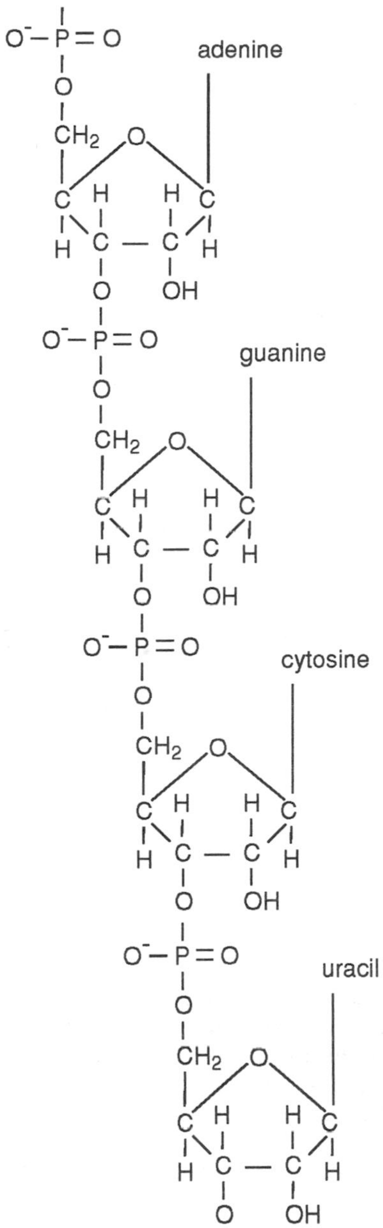

Figure 4.29 The structure of RNA.

One consequence of the presence of ribose is that RNA (unlike DNA) is susceptible to hydrolysis by sodium hydroxide. Both DNA and RNA can be hydrolysed by enzymes; various endo- and exonucleases, which act internally or at the ends of nucleic acids chains, respectively. DNA is stable to treatment with NaOH, whereas RNA is readily hydrolysed. Two nucleotide products are formed, in which the phosphate is attached to either the 3'- or the 2'-position of ribose. The reaction is shown in Figure 4.30.

The reason for the rapid base-catalysed hydrolysis of RNA and the alternative products is now apparent. Reaction involves an attack by the free 2'-OH group, resulting in a cyclic-2',3'-diester. This is then hydrolysed to give either a 2'- or a 3'-monophosphate.

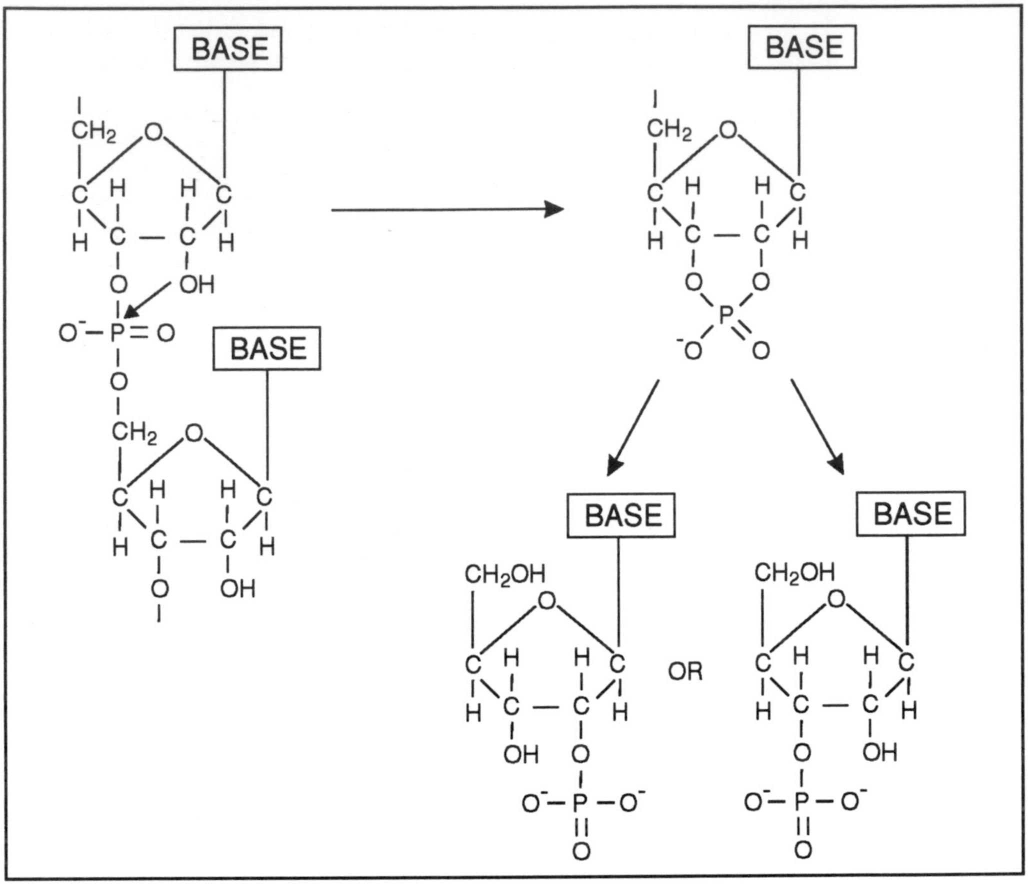

Figure 4.30 Hydrolysis of RNA. An intramolecular attack by the C-2' hydroxyl leads to a cyclic intermediate. This is hydrolysed to give a mixture of nucleoside-2'-phosphates and nucleoside-3'-phosphates.

This reaction provides the basis for another means of distinguishing RNA from DNA. Note that there are also separate chromogenic reactions for DNA (the diphenylamine reaction) and for RNA (reaction with orcinol).

limited secondary structure of RNA

RNA is predominantly present in cells as a single-stranded polynucleotide. However, it can fold so as to form double helical regions in which adenine - uracil and guanine - cytosine basepairs are made. It thus has limited secondary structure. As a consequence, there is much less of a hyperchromic effect when RNA is heated: the transition is neither so sharp or so extensive as is the case with DNA. Different types of RNA (Section 4.6.2) are likely to have different secondary and tertiary structures. It seems likely that, when double helical regions occur, they are interspersed by regions of more random coil structure.

4.6.2 Types of RNA

Three types of RNA are found in a typical cell. These are ribosomal (rRNA), transfer (tRNA) and messenger (mRNA) RNA. All of them are synthesised in the nucleus (in eukaryote cells), as a transcript from a DNA template, but they are used in the cytoplasm. Their structures and roles will be briefly summarised. In addition, we will also comment briefly on viral RNA. Note that transcription and translation are described in greater detail in the BIOTOL text 'Cell Infrastructure and Activity'.

rRNA

Ribosomal RNA is found in ribosomes and typically represents 70-80% of the RNA of the cell. Several types are present in ribosomes. They can be distinguished by the rate at which they sediment under centrifugal force. These are routinely identified by their

sedimentation coefficients

sedimentation coefficients. Sedimentation coefficients are measured in Svedberg units which are given the symbol S. In bacterial ribosomes (70S), three forms of rRNA occur: 5S, 16S and 23S. The relative molecular mass of the larger forms of rRNA is about 10^6, but the 5s RNA species only has about 120 nucleotides.

Ribosomes are the sites where proteins are synthesised. Ribosomal RNA probably has a structural role, ensuring that the various components are correctly positional for protein synthesis. Ribosomes consist of 2 subunits. For prokaryotes these are 30S and 50S subunits, together forming the 70S ribosomes (no misprint; it really is 70S). The smaller subunit contains the 16S RNA and some 21 proteins. The larger subunit contains the 23S and 5S RNA and a further 34 proteins. They are obviously complex structures! These details of 70S ribosomes and comparable information for the ribosomes of eukaryotes (which are larger, 80S) are shown in Table 4.4.

70S -Prokaryote		80S-Eukaryote	
Subunits	Constituents	Subunits	Constituents
30S	16S rRNA	40S	18S rRNA
	21 proteins		33 proteins
50S	23S rRNA	60S	28S rRNA
	5S rRNA		5.8S rRNA
	34 proteins		5S rRNA
			49 proteins

Table 4.4 Components of 70S and 80S ribosomes.

These are given purely for information and to show how complex ribosomes must be - do not try to memorise the various components in detail.

leader sequence

One particular role of the 16S rRNA is that it has a sequence, near its 3' end, which is complementary to a sequence in mRNA molecules. In the mRNA sequence, this is 'upstream' of the coding region of the mRNA molecule, and is part of the mRNA known as the leader region. It ensures that mRNA binds to, and is correctly positioned in, the 30S subunit, in order for it to be 'read' (ie translated) during protein synthesis.

mRNA

Messenger RNA carries the coded information from DNA to the ribosomes and determines the sequence of amino acids of a protein. We explain this more fully in Section 4.6.3. Messenger RNA only represents a small proportion (3-4% in *E. coli*) of the cellular RNA. As it carries the genetic information for proteins (which differ widely in size) and as several proteins may be coded within one mRNA molecule, mRNA varies widely in size. It tends to be short lived, at least in prokaryotes. A typical mRNA has a

Shine-Dalgarno sequence

leader sequence (which contains the Shine-Dalgarno sequence, which is complementary to the 16S rRNA and was referred to earlier) and a coding region.

Eukaryote mRNA also has a 'cap' and a poly-A 'tail'. The 'cap', at the 5' end, involves methylated bases and is added after transcription of the intact mRNA. The poly-A 'tail' of adenosine nucleotides (150-200 long) is of unknown function but is useful for isolating mRNA molecules. This is important for genetic engineering of eukaryote genes (Section 4.7), because eukaryote genes contain coding regions (called exons) separated by non-coding regions, called introns. By 'coding regions', we mean that this part of the DNA specifies the amino acid sequence of the protein. In contrast, the 'non-coding' introns do not specify the amino acid sequence of a protein - their function is unknown. In eukaryotes there is a complex 'processing' of the primary transcript during which the introns are removed (Figure 4.31) and the cap and tail are added.

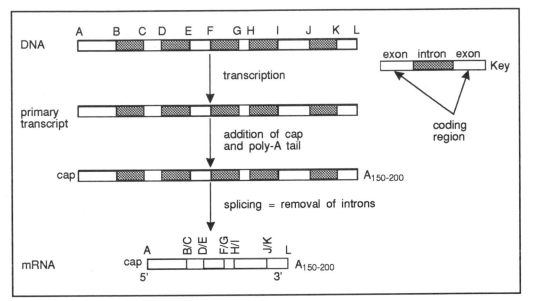

Figure 4.31 Processing of RNA in eukaryotes. The primary transcript produced by RNA polymerise is modified by addition of a cap at the 5' end and a poly A 'tail' or the 3' end. Introns are removed and the remaining coding regions (exons) spliced together.

Transfer RNA is the smallest of the various RNA molecules. It comprises about 20% of the RNA of cells and serves a vital role in protein synthesis as an adaptor. It enables the correct amino acid to be incorporated according to the base sequence of the mRNA. Before describing the precise role of tRNA, we will briefly digress and examine the genetic code. This will enable the importance of tRNA molecules as adaptors to be appreciated.

The genetic code describes the relationship between the sequence of bases in mRNA and that of the amino acids in the protein coded for by the mRNA. On theoretical grounds, 3 or more bases are needed to code for each amino acid. Given that there are only four different bases in mRNA, if a single base dictates which amino acid is incorporated, only 4 amino acids can be specified. If it were a 2-base code, then 4^2 amino acids could be specified (choice of 4 bases at position 1 and at position 2; total 16 combinations). This is insufficient to code for the 20 amino acids found in proteins. Three bases permits up to 64 different coding possibilities (ie 4 x 4 x 4). Careful studies led to the elucidation of the genetic code, shown in Table 4.5.

First Position*	Second Position				Third Position
	U	C	A	G	
U	Phe	Ser	Tyr	Cys	U
	Phe	Ser	Tyr	Cys	C
	Leu	Ser	Stop	Stop	A
	Leu	Ser	Stop	Trp	G
C	Leu	Pro	His	Arg	U
	Leu	Pro	His	Arg	C
	Leu	Pro	Gln	Arg	A
	Leu	Pro	Gln	Arg	G
A	Ile	Thr	Asn	Ser	U
	Ile	Thr	Asn	Ser	C
	Ile	Thr	Lys	Arg	A
	Met/start	Thr	Lys	Arg	G
G	Val	Ala	Asp	Gly	U
	Val	Ala	Asp	Gly	C
	Val	Ala	Glu	Gly	A
	Val	Ala	Glu	Gly	G

Table 4.5 The genetic code.
* All codons are read from the 5' end of the mRNA.

codon

The genetic code is a triplet code, with no punctuation signals. Each group of 3 bases specifies a particular amino acid, and is referred to as a codon. Certain codons also serve as initiation ('start') signals (eg AUG) or as polypeptide chain termination ('stop') codons (UAG, UAA, UGA). The mRNA is read from the 5' end, and the polypeptide chain grows from the N-terminal towards the C-terminal. For many amino acids, several (up to 6) codons can be used to specify that amino acid. Since more than one codon specifies a given amino acid (but not in all cases - methionine is specified by only

degenerate
code

one codon), the code is said to be degenerate. The genetic code is universal: although it was initially established for *E. coli*, all organisms are believed to use the same code.

∏ What is the amino acid sequence coded by a section of mRNA whose sequence is: 5' UUGCACUCCGGGAGA 3' (Examine Table 4.5).

The amino acid sequence will be: leu-his-ser-gly-arg.
The mRNA is read from the 5' end and amino acids are inserted from the amino terminus. Thus if this was a penta-peptide, the leucine residue would have a free α-amino group and the arginine would be the C-terminal amino acid residue.

∏ Write down the advantages that arise from the degeneracy of the genetic code. Hint: one way to approach this is to consider possible outcomes if only 20 codons were used to specify amino acids and the remaining 44 codons were 'stop' signals (ie do not code for an amino acid).

The principal advantage of degeneracy lies in what happens if an error (eg a mutation) occurs in the DNA, such that the RNA has a 'wrong' base in the sequence. If, for example, only 20 codons specified amino acids, whilst the remainder were stop signals, then, on a random chance basis most base substitutions would lead to chain termination. This would give prematurely shortened polypeptides, which are unlikely to be functional (or would display an impaired function). With only 20 codons specifying amino acids, some substitutions would result in a different amino acid being inserted. Whilst the polypeptide would be of the correct length, the altered amino acid might affect the protein's performance.

In contrast, with degeneracy there is an excellent chance that a full length polypeptide will be formed, despite a base substitution. Indeed, for many amino acids there are several codons: some substitutions will not even result in a differemt amino acid being incorporated. For example, a change in the mRNA from CUU to CUC, CUG, CUA, UUA or UUG has no effect on the sequence of the polypeptide: leucine would still be incorporated.

What role does tRNA play in enabling proteins to be synthesised according to the codons of the mRNA? Transfer RNA solves a problem of recognition. It is most unlikely that an amino acid could recognise and specifically bind to a particular codon. However, a complementary base sequence - an anti-codon - would only bind to a given codon, given the specificity of AU and GC basepairing.

anti-codon

Molecules of tRNA have a cloverleaf structure (Figure 4.32).

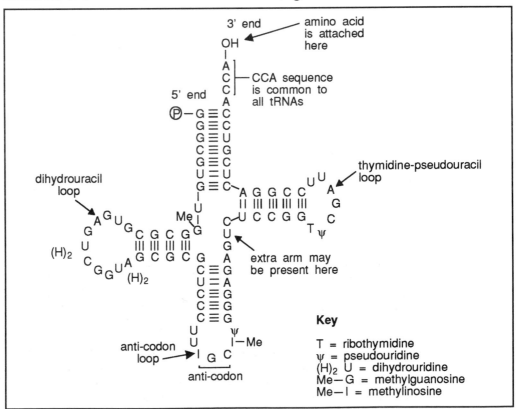

Figure 4.32 Cloverleaf structure of a tRNA molecule. Various features which are important for tRNA function are identified. These are described more fully in the text.

There is at least one tRNA molecule for each amino acid. All tRNA molecules are of similar size (about 80 nucleotides) and all have the features shown in Figure 4.32. These are:

- Various loops, linked by short sections of double helix in which basepairing occurs. The cloverleaf structure was originally proposed on the basis of the base sequence of tRNA. It represented the structure with maximum basepairing (and therefore hydrogen bonding) and, as a result, was expected to be the most stable. The actual 3- dimensional shape was elucidated much later and showed that the loops of the cloverleaf structure were essentially correct. The loops are given various names, as shown in Figure 4.32. In space, tRNA molecules adopt an L-shaped structure in which the anti-codon site is at the end of one arm of the L and the amino acid attachment site is at the end of the other arm. All tRNA molecules adopt this same general shape.

- Occurrence of unusual bases, such as inosine, pseudouracil, and methylated or acetylated bases. Inosine is particularly important in the anti-codon.

amino acyl-tRNA synthetase

- The 3′ end of the tRNA always ends in the sequence CCA and it is to this terminal adenosine that the amino acid is attached. This is accomplished by an amino acyl-tRNA synthetase and involves hydrolysis of ATP, which represents activation of the amino acid prior to protein synthesis. There is a different amino acyl-tRNA synthetase for each amino acid: this ensures that the correct amino acid is attached to the correct tRNA (eg valine only becomes activated and attached to tRNAval).

amino acid activation

Enzyme specificity in only binding the correct tRNA and amino acid is vital. It is the tRNA which subsequently determines which amino acid is incorporated. If the amino acid, after activation and attachment, is chemically converted to a different amino acid, as in the following example, then the incorrect amino acid is inserted.

$$\text{tRNAcys + Cys} \longrightarrow \text{Cys-tRNAcys} \xrightarrow[\text{Nickel}]{\text{Raney}} \text{Ala-tRNAcys}$$

Clearly it is vital that only the correct amino acid is attached to each tRNA. Enzyme specificity is often important: never more so than here!

wobble

- The anti-codon loop contains a trinucleotide which is the anti-codon. By basepairing with the codon, only the appropriate tRNA should be utilised. Interaction between codon and anti-codon is anti-parallel: the anti-codon base which pairs with the third base is often inosine and the stringency of basepairing is less at this position. This lessened discrimination is known as wobble and means that one tRNA (eg tRNAala) can bind to 3 codons (all of which specify alanine) ie GCU, GCC and GCA. You may have deduced from looking at the genetic code that the third position is often less important than the first two. In some cases either a purine, or a pyrimidine is sufficient to specify a particular amino acid (eg asparagine is coded by both AAU and AAC); in other cases the identity of the third base is irrelevant (GGU, GGC, GGA, GGG all specify glycine). Wobble provides the molecular explanation for degeneracy.

We can now clearly see the importance of tRNA molecules. Providing the correct amino acid is attached to each tRNA molecule, the tRNA will, through anti-codon to codon basepairing, ensure that the correct amino acid is incorporated. You should now understand why tRNA is described as an adaptor.

∏ Predict a base sequence for the anti-codon region of a tRNA for alanine.

The codons for alanine in mRNA are GCU, GCC, GCG and GCA, written in conventional 5′ → 3′ terms. Basepairing with the anti-codon is anti-parallel, therefore suitable possibilities will be:

Codon ^{5}'GCU^3' and ^{5}'GCC^3' and ^{5}'GCA^3'and ^{5}'GCG^3'

Anti-Codon $_3$'CGA$_5$' and $_3$'CGG$_5$' and $_3$'CGU$_5$' and $_3$'CGC$_5$'

Thus tRNA molecules with the anti-codon sequences AGC, GGC, UGC and CGC (in 5′ → 3′ notation) would be satisfactory. If inosine was present in the first position of the anti-codon (giving IGC), it could interact with 3 of these codons, since inosine can basepair with A, U or C. The first position in the anti-codon is the 'wobble' position. The advantage of inosine in this position is that only one tRNA is needed for several codons for a given amino acid.

viral RNA | **Viral RNA.** Some viruses do not contain DNA. Instead, their genetic information is carried by the base sequence of an RNA molecule. For example, tobacco mosaic virus (TMV) is composed of 2,130 identical protein subunits and a single-stranded RNA molecule of 6,400 nucleotides. In some cases, viral RNA can be double-stranded, although double-stranded RNA is less stable than double-stranded DNA. Viruses code for very few proteins and rely on using the host cells biosynthetic machinery following infection.

∏ The last sentence of the previous paragraph began with the words 'viruses code for very few proteins....'. What information, within this paragraph, is consistent with this statement?

Look at the total number of bases in the TMV RNA. Even if all of the bases coded for amino acids (which is highly unlikely - there must be 'control' regions), then the maximum number of amino acids which can be coded is 6400/3 or approximately 2130 amino acid residues. A typical sized protein of 50 000 daltons will contain about 450 amino acid residues. Thus the TMV RNA could only code for about 5 proteins of this size!

If RNA represents the genome of RNA-containing viruses, then some special strategy for producing further copies of the viral RNA is required. We earlier noted one solution to this: the use of reverse transcriptase to make DNA, using the viral RNA as a template. The DNA can then be transcribed, using the cells normal transcription apparatus, to produce more copies of the viral RNA. Hence the viral RNA can be reproduced.

∏ The enzyme reverse transcriptase does not occur in host cells prior to infection by an RNA-containing virus. What must happen for it to be present?

The viral RNA (or part of it) must be treated as a mRNA and used to direct protein synthesis. The viral RNA must therefore code for reverse transcriptase; unless this enzyme is produced, no further copies of the viral RNA will be made.

4.6.3 Transcription and translation: an overview

In order for the information contained in the sequence of bases of DNA to be used, ie for genes to be expressed, an RNA transcript has to be made. In eukaryotes, this occurs in the nucleus (where the vast majority of the genome is located) but the RNA product, which contains the information from the DNA (ie mRNA), migrates to the cytoplasm.

In the cytoplasm, protein synthesis occurs on ribosomes, whereby the information contained in the base sequence of the mRNA is 'decoded' and determines the sequence of amino acids in the protein product. These processes are summarised in Figure 4.33.

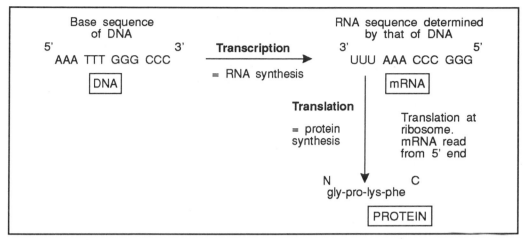

Figure 4.33 Transcription and translation.

RNA polymerase

Transcription is catalysed by the enzyme RNA polymerase (also known as DNA-dependant RNA polymerase or transcriptase). This synthesises RNA, in a 5' → 3' direction, inserting bases according to a template DNA strand. The DNA is read in an anti-parallel direction. As far as the double-stranded DNA is concerned, only one strand is transcribed and is known as the 'sense' strand. The process, shown in outline in Figure 4.33, is shown more fully in Figure 4.34.

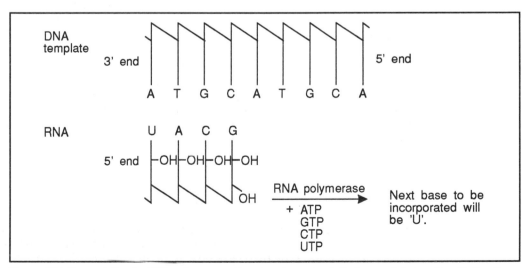

Figure 4.34 Transcription by RNA polymerase. No primer is required; chain growth is in a 5' → 3' direction, with bases incorporated according to the base sequence of the DNA template. DNA is read in an anti-parallel direction.

RNA polymerase does not require a primer (as DNA polymerase does). It can start an RNA molecule *de novo*. We remind you that during DNA replication, RNA polymerase is first used to synthesis a short section of RNA, which is then used as a primer by DNA polymerase; the short section of RNA is subsequently cut out and the gap filled with DNA).

RNA
polymerase
holoenzyme

sigma factor

In bacteria a single type of RNA polymerase catalyses the synthesis of all types of RNA. It has a complex structure, consisting of 4 subunits (α, β, β' and σ) which are present in the ratio 2:1:1:1, respectively. This assembly is known as the holoenzyme. The role of the σ (sigma) subunit is to enable the RNA polymerase to find the correct starting point at which to begin transcription - it would clearly be a waste of energy and material resources to start mRNA molecules midway through the coding region. The start signal is a sequence of bases upstream from the coding region of the sense strand of the DNA (the strand which will be used as template during transcription); this sequence is called

promoter

core RNA
polymerase

rho

the promoter. Once the RNA polymerase has bound to the promoter and started transcription, the σ subunit is no longer required. It can and does dissociate, leaving the 'core' RNA polymerase to complete transcription. Although the core RNA polymerase will stop transcription when it reaches a termination sequence in the sense strand, correct termination is sometimes aided by a termination protein called rho.

Eukaryotic transcription, which occurs in the nucleus, involves 3 different RNA polymerases, which transcribe different types of genes:

- RNA polymerase I transcribes rRNA genes;

- RNA polymerase II transcribes mRNA genes;

- RNA polymerase III transcribes tRNA and the 5S rRNA genes.

All of these enzymes are complex, with a number of subunits.

Translation of the mRNA into a polypeptide takes place on ribosomes. Ribosomal RNA ensures that the mRNA binds correctly (via the leader region) and tRNA molecules bring activated amino acids. The mRNA is read from the 5' end. Correct basepairing of codon and anti-codon (as described in Section 4.6.2) ensures that the correct amino acid, as specified by the genetic code, is inserted in each position. The overall effect is as shown in Figure 4.33.

Both transcription and translation are discussed in greater detail in the BIOTOL text 'The Infrastructure and Function of Cells'. These brief descriptions are intended to show the roles of the various nucleic acids we have discussed and to clarify their involvement in gene maintenance and expression.

SAQ 4.5

Earlier in this chapter we described the so-called central dogma of molecular biology. Without looking back, choose appropriate words from those given below and write them into appropriate spaces in the diagram:

Protein; ATP; Guanine; Ribose; Replication; DNA; RNA; Endonuclease; Transcription; Complementary basepair.

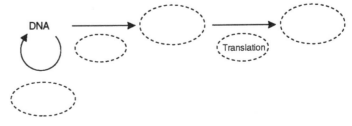

SAQ 4.6

In Section 4.5.3, we said that the error rate of DNA polymerase during replication was extremely low. Predict the consequence of errors during replication by DNA polymerase, and what would happen if the error rate was higher.

SAQ 4.7

Scattered throughout this chapter we have indicated differences between DNA and RNA. Write down, as a table, as many differences as you can which could, at least in principle, be useful in identifying a sample of nucleic acid. Thus, the fact that DNA is normally present as a double helix, whereas RNA is not, is perfectly true. It is not, in itself, sufficient to distinguish DNA from RNA. However, certain consequences of double helical structures are helpful!

SAQ 4.8

1) Deduce the base composition of a sample of DNA if adenine is known to represent 30% of the bases. The DNA shows a sharp Tm, with an increase in absorbance at 260nm of about 40%.

2) What is the base composition of a sample of RNA whose uracil composition is also 30%? Think carefully!

SAQ 4.9

Which of the following is the correct sequence of a peptide which is coded for by a section of the coding region of a gene containing the nucleotide sequence: 5' AGACCTGACGGAATT 3'?

1) ser-gly-leu-pro

2) asn-ser-val-arg-ser

3) Some different sequence.

4.7 Genetic engineering - an outline

The discovery in the early 1970s of restriction enzymes paved the way for genetic engineering, which is generally viewed as being of central importance in biotechnology. The crucial effect of genetic engineering is that novel combinations of DNA, which could never arise naturally, can be produced, with the outcome that organisms can produce proteins which they would not normally make.

genetic
engineering

The process of genetic engineering is conveniently described in four stages. These are:

• the preparation of the gene which is to be transferred;

• the incorporation of the gene into a suitable vector;

- the introduction of the vector into the new host cell (a process called transformation);

- demonstration that successful incorporation of the gene has taken place and that the product coded by the gene is being made (gene expression).

Before considering each of these stages, the properties of restriction enzymes will be considered.

Restriction enzymes are enzymes which hydrolyse double stranded DNA at particular points in the DNA. Each restriction enzyme (several hundred have been described) recognises a particular sequence of bases in DNA and only cuts the DNA at this site. For example, the restriction enzyme Eco R1 (obtained from *E. coli*) recognises the sequence GAATTC and only cleaves at this sequence. Use of restriction enzymes enables DNA to be reproducibly cut at a relatively few sites (since each recognition sequence only occurs infrequently). Restriction enzymes are essential for the production of particular genes, and are also indispensable in other areas of applied molecular biology (DNA sequence analysis, DNA fingerprinting, DNA probes).

4.7.1 Preparation of the gene

sticky ends

Bacterial genes can usually be obtained by digesting chromosomal DNA preparations with restriction enzymes. In most cases the restriction enzyme used generates fragments with overlapping (or sticky) complementary single strand ends (Figure 4.35).

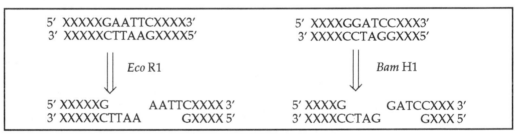

Figure 4.35 Restriction sites for Eco R1 and Bam H1. Note the overlapping (sticky) ends. These facilitate joining of pieces of DNA, providing their ends are complementary.

cDNA

Genes coding for small gene products (eg peptide hormones) may be chemically synthesised, providing the amino acid or DNA sequence is known. Eukaryotic genes (eg those from humans) cannot be produced in a form which is suitable for genetic engineering by restriction enzyme digestion of chromosomal DNA, because of the non-coding regions (introns) which they contain. These have to be removed before protein synthesis takes place. As bacterial genes do not contain introns, bacteria have no system for recognising and removing introns after transcription. Unless the introns present in eukaryotic DNA are removed prior to introduction into a bacterial host, a protein with incorrect amino acid sequence will be produced. A gene which will be correctly expressed in bacteria is produced by making a DNA copy (cDNA) of the mRNA present in eukaryote cells (Figure 4.36); this cDNA is used for genetic engineering. If you examine this figure carefully, you will see that by making a DNA copy of the mRNA isolated from eukaryotes, we can produce a new version of the gene without intron sequences.

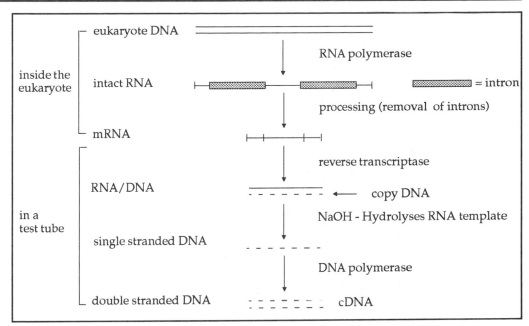

Figure 4.36 Preparation of cDNA (in outline). The processed mRNA (which now codes for the desired protein) is used as a template by reverse transcriptase to produce a single-stranded copy DNA. This can be used to generate double-stranded cDNA.

4.7.2 Incorporation of the gene into a vector

vector

plasmids

In order to ensure that the foreign gene is taken up and expressed by the host cell, a vector is required. Foreign DNA is not normally incorporated into the bacterial chromosome but must none-the-less divide (replicate) when the host cell divides. The vector has to act as a replicon in order that the novel gene is maintained in the host cell. Very often so called plasmids are used as vectors. Plasmids are circular pieces of double-stranded DNA which are replicated within the host cell. Useful features of plasmids used for genetic engineering include the following:

* single sites for several restriction enzymes;

* presence in multiples copies within the host cell (thus one obtains many copies of the novel gene);

* plasmid codes for selectable markers, such as antibiotic resistance: this permits selection of transformants.

Widely used plasmids, such as pBR322 (Figure 4.37), contain these features.

For use, the vector is cut with the same restriction enzyme as that used to produce the gene for insertion: thus gene and vector will have complementary sticky ends (Figure 4.38).

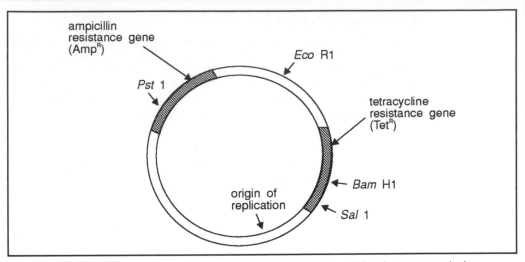

Figure 4.37 pBR322. The positions of certain restriction enzyme sites, which only occur once in the plasmid, are indicated. Also shown are the two genes whose products enable selection of transformed *E. coli* cells to be made. This type of diagram in which the position of restriction sites is marked, is known as a restriction map.

The prepared gene and vector are incubated with DNA ligase, which forms covalent bonds where complementary ends overlap.

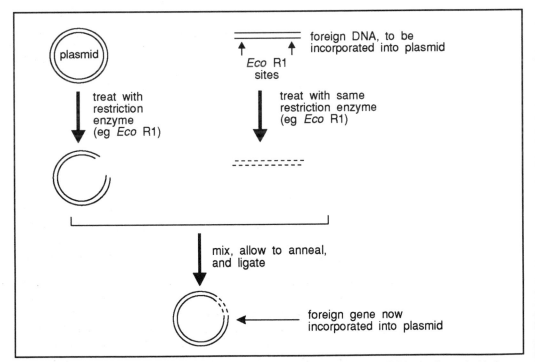

Figure 4.38 Introduction of foreign DNA into a vector. The plasmid and foreign DNA are cleaved with the same restriction enzyme: this ensures that they both have complementary sticky ends (note that there are other ways of achieving complementary ends). Upon mixing, some of the fragments will re-anneal to give a plasmid containing an insert; DNA ligase forms the necessary covalent bonds.

4.7.3 Transformation of host cells

A suitable vector containing the inserted gene is then introduced into a bacterial host cell. When *E. coli* is the host, treatment with calcium chloride results in easily transformed cells. Transformed cells may be selected by their ability to grow in the presence of an antibiotic, resistance against which is coded by the vector (eg for *E. coli* transformed with pBR322, the plasmid confers resistance to the antibiotics ampicillin and tetracycline). Selection by antibiotic resistance is highly effective (Figure 4.39).

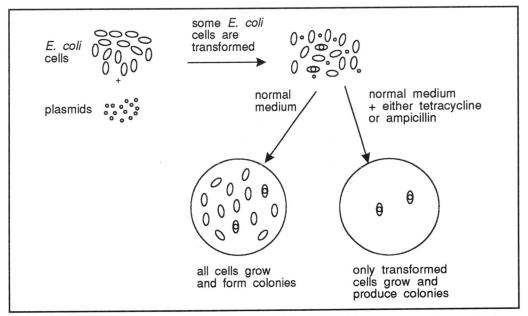

Figure 4.39 Selection of transformed cells. *E. coli* cells lacking pBR322 fail to grow on medium containing either ampicillin or tetracyclin. The presence of the plasmid confers resistance on the *E. coli* cells.

This means that only a small proportion of bacteria present need to be transformed since they are easily selected.

Π Assume that some foreign DNA is incorporated into the single *Bam* H1 restriction site of pBR322, which is then used to transform cells of *E. coli*. Predict the phenotype of the transformed cells with regard to the antibiotics ampicillin and tetrancycline. (By phenotype we mean the characteristics displayed by the organism; in this case, are they sensitive to or resistant to these antibiotics?).

insertional inactivation

The phenotype will be Amp^R, Tet^S, meaning that the *E. coli* cells are resistant to ampicillin but sensitive to tetracycline. Look at the restriction map of pBR322 (Figure 4.37). Successful incorporation of extra DNA into the *Bam* H1 site of pBR322 means that, although the tetracycline gene will still be transcribed, it will produce a protein with an extra section in the middle of it. It is most unlikely to be functional; this is known as insertional inactivation. Thus the gene product conferring resistance to tetracycline is not produced in a functional form. Hence Tet^S. The penicillinase which confers resistance against ampicillin is not affected, hence the transformed cells would be Amp^R.

Note that if foreign DNA was not incorporated into pBR322 and it simply re-annealed without an insert (eg in Figure 4.38, intact pBR322 is reformed), or if the pBR322 was not cut by the *Bam* H1 at all, then the phenotype would be Amp^R, Tet^R. *E. coli* cells which are not transformed will be Amp^S, Tet^S. These phenotypes can all be distinguished by plating onto appropriate media. Incorporation of foreign DNA into the Pst 1 site of pBR322 would give an Amp^S Tet^R phenotype. These insertional inactivations are very useful as a means of monitoring the success of incorporation of foreign DNA and transformation.

4.7.4 Detection of the cloned gene and gene product

in situ hybridisation

The presence of the gene within transformed cells may be demonstrated by the use of a radioactively labelled DNA probe which, by complementary basepairing, only binds to the complementary sequence (Figure 4.40).

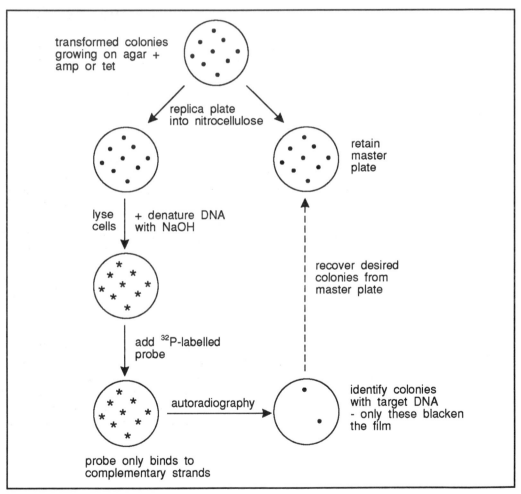

Figure 4.40 Colony hybridisation (also called *in situ* hybridisation). DNA from transformed cells is denatured and bound to a nitrocellulose support. ^{32}P-labelled DNA (the probe) only binds to complementary sequences. Use of an appropriate probe reveals which colonies contain the desired DNA sequence. Adapted from Old, R.W.and Primrose, S.B. (1985) 'Principles of Gene Manipulation', 3rd Edition, Blackwell.

This enables colonies of cells which contain the gene to be identified. Correct expression of the cloned gene (ie synthesis of a protein of the correct amino acid sequence) may be demonstrated by detection of enzymic activity or by an immunological method using specific antibodies to identify the presence of particular proteins.

Our intention with the description of genetic engineering given above was to identify the various stages of the process. Numerous problems are encountered and a variety of ingenious strategies have been developed to resolve them and to attain the desired objectives. The techniques of genetic engineering and the outcomes of the application of these techniques are developed in the BIOTOL texts 'Techniques for Engineering Genes' and 'Strategies for Engineering Organisms'.

4.8 Hybridisation in molecular biology and biotechnology

The importance of specific binding by one piece of nucleic acid to another containing a complementary sequence has been discussed. Interaction between mRNA and the Shine-Dalgarno sequence of 16S rRNA ensures correct binding of the mRNA to ribosomes. Specific interaction between codons of mRNA and the anti-codons of the tRNA molecules underpins the accuracy of protein synthesis. Exploitation of specific interaction between complementary base sequences has had a major impact on molecular biology and biotechnology.

hybridisation

Hybridisation is the term used when separate strands of nucleic acid combine through complementary basepairing. It was originally applied to the formation of DNA RNA hybrids, which provided the first direct evidence that the sequence of mRNA was complementary to that of DNA. The term hybridisation is now routinely used when any complementary base paired nucleic acid is formed, irrespective of the types of strands involved. The techniques and strategies which are now briefly described all involve hybridisation. Whilst these methods were first carried out with labelled RNA or DNA as the probe, they are now frequently conducted with labelled oligonucleotides. An

oligonucleotides

oligonucleotide of 18-20 bases in usually adequate to give highly specific binding. Oligonucleotides of defined sequence can now be synthesised very easily.

4.8.1 Colony and *in situ* hybridisation

During genetic engineering, it is necessary to establish whether a particular DNA sequence is present in bacterial cells, and which cells contain it. We described colony hybridisation, which accomplishes this objective, in Section 4.7.4 and Figure 4.40. Similar probing strategies can be used to detect histochemically the presence of particular DNA or mRNA sequences in human cells or tissue sections. This can be useful in establishing the presence of viruses, such as HIV or Epstein-Barr virus.

4.8.2 Southern blotting

gel
electrophoresis

polyacrylamide
gel

agarose gel

ethidium
bromide

Gel electrophoresis is a powerful technique for separating charged molecules, including DNA. Speed of migration of DNA fragments during electrophoresis is inversely proportional to size; different sized fragments can easily be separated, as was seen with the DNA sequencing gels described earlier. For small pieces of DNA (up to 1000 basepairs) polyacrylamide gels give the best resolution, and these are routinely used for DNA sequence analysis. For the larger fragments which typically result from digestion of DNA with a restriction enzyme, a more porous gel is needed and agarose is used. Following electrophoresis, DNA can be visualised by treatment with ethidium bromide. This binds to double-stranded DNA and fluoresces. Photographs of the flourescent bands can then be taken.

The strategy described above is sufficient to reveal the number and position of DNA fragments following restriction enzyme digestion. Southern devised the technique which bears his name, Southern blotting, which enables a fragment containing a particular sequence to be located (Figure 4.41).

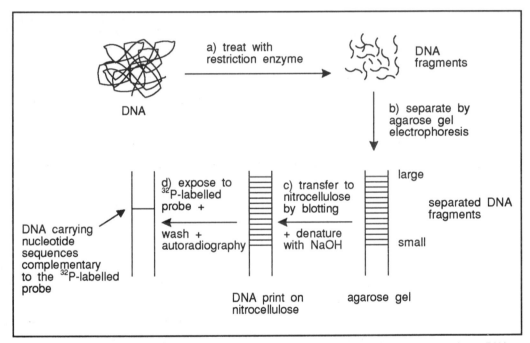

Figure 4.41 Southern blotting. a) Treatment of a sample of DNA with a restriction enzyme produces DNA fragments. b) These are separated, on the basis of size, by agarose gel electrophoresis. c) The separated fragments are blotted into nitrocellulose and denatured to give single stranded DNA. d) When ^{32}P-labelled probe (DNA or an oligonucleotide) is added, it only binds to fragment(s) containing the complementary sequence; this fragment is revealed by autoradiography.

Following agarose gel electrophoresis, the separated DNA fragments are transferred to nitrocellulose (this can be done by placing the nitrocellulose over the agarose gel, and then covering the nitrocellulose with several layers of absorbent paper and a weight. The absorbent paper draws fluid, containing DNA, from the gel and through the nitrocellulose; DNA is bound by the nitrocellulose; this is the origin of the 'blotting' of the method's name). The DNA, now stuck to the nitrocellulose, is denatured to create single strands and then incubated in a solution containing a ^{32}P-labelled probe. The ^{32}P-labelled probe only binds to complementary sequences. After washing, autoradiography reveals which bands contain the particular sequence. Non-radiolabelled probes are increasingly widely used: the basic principle remains the same. Analysis of samples of RNA, using similar separation and hybridisation principles, is rather coyly known as Northern blotting.

Northern blotting

Areas in which the use of Southern blotting has made major contributions include the following.

Preparation of DNA fragments for genetic engineering

Southern blotting allows the single restriction digest fragment which contains the required gene/sequence to be identified. This fragment can then be isolated for incorporation into other systems.

Preparation of DNA fragments for sequence analysis

A similar strategy can be used to prepare a particular DNA fragment for DNA sequence analysis.

Diagnosis of inherited conditions and infections

sickle cell anaemia

Sickle cell anaemia is caused by a mutation which results in valine replacing glutamic acid at position 6 of β-globin. The alteration in the DNA (a thymine occurs in the sense strand instead of an adenine) fortuitously removes a recognition site for the restriction enzyme Mst II. As a consequence, treatment of human DNA with Mst II gives a different band pattern (following electrophoresis and probing with a ^{32}P-labelled β-globin DNA probe) for normal- and sickle cell anaemia -coding DNA (Figure 4.42).

With normal DNA, the β-globin gene is recovered on two fragments after Mst II digestion. For sickle cell anaemia DNA, the β-globin containing fragment is on a single, larger fragment. This analysis can be carried out on DNA from any cells.

cystic fibrosis

Cystic fibrosis is a debilitating and usually fatal disease. The mutant gene is inherited as a recessive gene, with carriers being unaffected. The identification and cloning of the cystic fibrosis gene in 1989 means that diagnosis *in utero* of those who are homozygous for the mutant gene will now be possible.

Gene probes are also available for detection of various pathogenic organisms, such as *Salmonella* and *Legionella* and viruses.

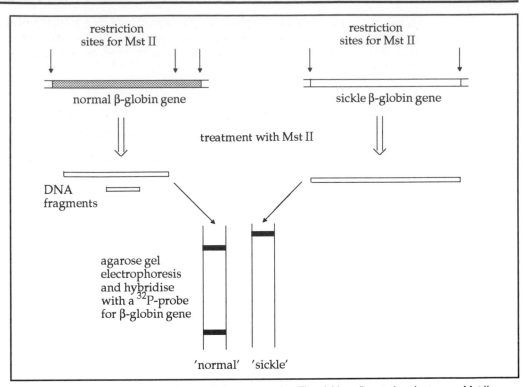

Figure 4.42 Diagnosis of sickle cell anaemia by gene probe. The sickle cell mutation destroys a Mst II recognition site, thus a large β-globin gene-containing fragment is produced. This is detected by agarose gel electrophoresis and Southern blotting with labelled β-globin gene as probe.

Restriction fragment length polymorphism (RFLP)

We have just seen that mutations within restriction sites can destroy a recognition site for a particular restriction enzyme, and thus cause fragments of a different size to be produced. Similarly, a mutation may create an additional restriction site, causing smaller fragments to be produced. Fragments of different length, generated in this way and revealed by probing with a particular ^{32}P-labelled probe, are known as RFLP and are very useful in mapping genetic disease. The diagnosis of sickle cell anaemia described above is an example of the technique.

In many cases, the genes causing serious hereditary disease are not yet identified. However, probes for particular sequences may reveal a particular RFLP which shows a close linkage with the disease: this is established by the careful study of families in which the inherited condition occurs. Until the discovery of the cystic fibrosis gene, RFLP analysis was the only available method for screening: it remains the only method for other diseases, such as Huntington's chorea. Since the gene which the DNA detects is only linked to the disease - causing gene (and is some distance away from it) diagnosis with RFLP analysis in this way cannot be absolute, since recombination could occur.

DNA fingerprint analysis

Many organisms, including humans, contain DNA known as mini-satellite DNA. This DNA consists of multiple repeats of a so-called 'core' sequence. Jeffries devised a hybridisation technique, using suitable ^{32}P-labelled sequences, which revealed that the distribution of these core sequence patterns of each individual is different. An

RFLP

individuals 'fingerprint' is related to that of his/her parents and to other close relatives. Given the uniqueness of the pattern in individuals, and its inheritance, this technique has enormous implications in the investigation of serious crime and in paternity disputes.

4.8.3 Isolation of mRNA

poly-A tails

The occurrence of a poly-A tail at the 3' end of eukaryote mRNA enables it to easily be purified. The technique of affinity chromatography exploits specific interaction between 2 molecules, as occurs between complementary basepairs. If an oligonucleotide consisting of deoxythymidine is attached to a resin and packed into a column, then mRNA will bind to it via complementary basepairing with its poly-A tail. Other types of RNA would not bind. This technique can be made more specific. If the base sequence of the mRNA is known, an oligonucleotide complementary to part of its coding region can be constructed: purification of a single mRNA should then be possible.

Summary and objectives

In this chapter we have learnt about the structure of DNA and how, through basepairing, it can be replicated. We have also learnt of the major groups of RNA and how they are involved in interpreting the nucleotide sequences of DNA into the amino acid sequences found in proteins. The final part of this chapter showed how knowledge of DNA structure enables it to be manipulated and used in genetic engineering. The importance of correct interactions between and recognition of nucleic acids, via basepairing, was also stressed.

Now that you have completed this chapter you should be able to:

- identify the nucleotide bases found in RNA and DNA and be able to distinguish between nucleosides and nucleotides;

- draw the structure of the major nucleic acids;

- list a variety of methods that can be used to distinguish between RNA and DNA;

- describe why the structure of DNA and the bonds within it are responsible for properties such as UV absorbance and the hyperchromic shift and use data relating to the melting temperature of DNA to determine its GC content;

- explain the roles of mRNA, tRNA and rRNA in the translation of a nucleotide sequence into a sequence of amino acids;

- use knowledge of nucleotide sequences and the genetic code to predict the order in which amino acids will be joined to form a peptide;

- describe how nucleotide sequences can be used as probes for identifying the presence of genes and for identifying changes in DNA which occurs in some inherited diseases;

- describe, in outline, how genetic engineering may be accomplished.

Carbohydrates

Carbohydrates

We have already referred to the fact that cells are made of a restricted number of major classes of molecules. As we have progressed through the examination of these classes of molecules we have commented that, in all aspects of cell function, there seems to be a close relationship between the structure of a molecule and its biological function. In this and the next chapter, we discuss carbohydrates and lipids. Unlike proteins and nucleic acids, neither of these classes of molecules possess information within their structures (as the amino acid sequence of a protein or the base sequence of a nucleic acid does). Carbohydrates and lipids tend to have less precise structures than proteins and nucleic acids. This is presumably because their roles within cells can be adequately fulfilled without precisely controlled (or uniform) structures. They none-the-less have vital roles, as follows:

(margin: roles of carbohydrates and lipids)

Carbohydrates

- Biological fuel
- Storage form of food reserves
- Structural components of cell walls

Lipids

- Fuels and storage forms of fuel
- Major components of cell membranes

We shall return to these functions later in this and the next chapter. For the major part of this chapter, we will examine the structure and properties of carbohydrates.

5.1 Classes of carbohydrates

Carbohydrates are conveniently subdivided into three groups:

Monosaccharides

These consist of single 'units' and are important in metabolism and as the building blocks from which the remaining three groups are constructed.

Disaccharides

(margin: glycosidic bond)

These consist of two monosaccharides linked through a glycosidic bond. **Oligosaccharides** contain a few (3-6) monosaccharide units linked by glycosidic bonds.

Polysaccharides

Polysaccharides are long chains (branched and unbranched) of monosaccharide units. They may consist of only one type of monomer (hence homopolysaccharides) or of more than one monomer type (heteropolysaccharides). They do not have a precise, uniform structure or size.

Whilst a general formula $(CH_2O)x$ is often given for carbohydrates, note that:

- this precise ratio of numbers of atoms frequently does not hold;
- some carbohydrates contain other atoms (especially nitrogen).

5.2 Monosaccharides

aldoses
ketoses

Monosaccharides are all polyhydroxy compounds with either an aldehyde or a keto group. If they contain an aldehyde group they are called aldoses, whilst those containing a keto group are ketoses (Figure 5.1).

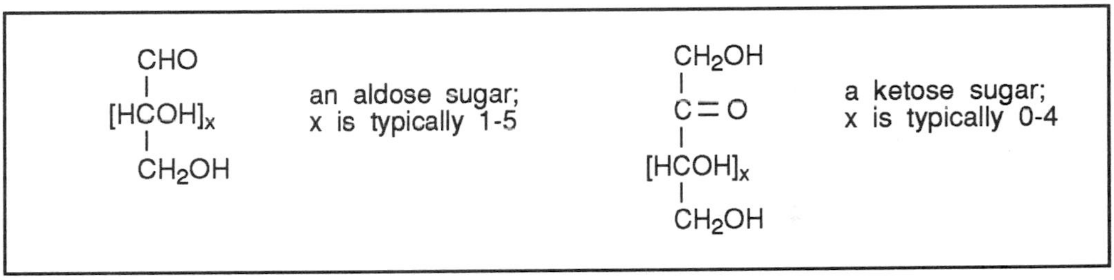

```
CHO                               CH2OH
 |         an aldose sugar;        |        a ketose sugar;
[HCOH]x    x is typically 1-5     C=O       x is typically 0-4
 |                                 |
CH2OH                           [HCOH]x
                                   |
                                 CH2OH
```

Figure 5.1 General formula of aldose and ketose sugars. X is typically between 0 and 5.

Our interest lies with the more common monosaccharides which have between 3 and 7 carbon atoms, especially those with 3, 5 and 6 carbon atoms. These sugars are known as trioses, pentoses and hexoses, respectively.

Match the following monosaccharides (1-4), for which certain information is provided, to the structures a-e. Hence which is the odd sugar out?

1) a 5 carbon ketose sugar.

2) a hexose with an aldehyde group (also known as an aldohexose).

3) a 6 carbon ketose.

4) an aldotriose.

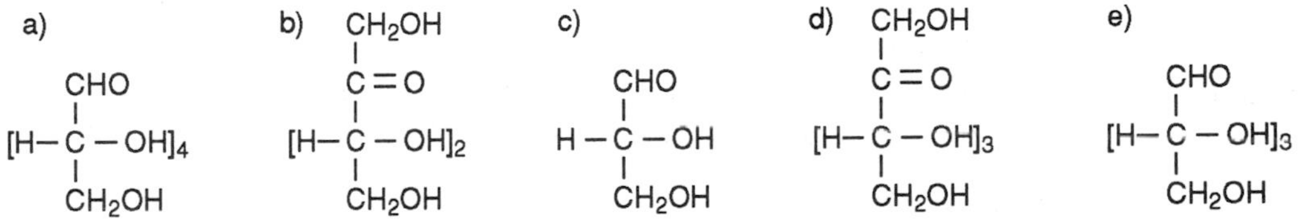

```
a)              b)  CH2OH       c)            d)  CH2OH       e)
    CHO             |               CHO           |              CHO
     |              C=O              |            C=O             |
[H-C-OH]4       [H-C-OH]2        H-C-OH       [H-C-OH]3       [H-C-OH]3
     |              |              |             |               |
   CH2OH          CH2OH          CH2OH         CH2OH           CH2OH
```

We would expect you to have spotted the following:

1) The only 5 carbon ketose sugar is structure b). Ketose sugars have a keto group, in which the carbonyl (-C=O) group is attached to two other carbon atoms. Whilst structure d) is also a ketose, it has 6 carbon atoms.

2) Structure a) is the answer. In aldose sugars the carbonyl group is terminal and is an aldehyde group (ie -CHO). Structure e) is a 5 carbon aldose.

3) The correct answer is structure d), being the only compound with 6 carbon atoms and a keto group.

4) Structure c) is an aldotriose: thus it has 3 carbon atoms (the 'triose' part of the title) and an aldehyde group. This leaves structure e) as the remaining compound: it is a 5 carbon aldose sugar.

5.3 Isomerism in monosaccharides

5.3.1 Structural isomers

structural isomers

If we examine 3 carbon monosaccharides, we find that both an aldose (glyceraldehyde) and a ketose (dihydroxyacetone) exist (Figure 5.2). Close inspection reveals that these molecules have identical atomic composition and hence molecular mass, but differ in functional groups (one has an aldehyde, the other a keto group). They are therefore said to be structural isomers (isomers being compounds of identical atomic composition) Thus within monosaccharides there are numerous structural isomers, as a consequence of the occurrence of either aldehyde or keto (also known as ketone) groups.

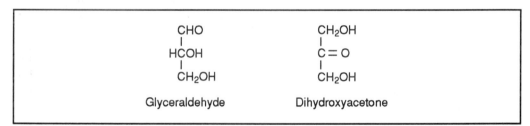

Figure 5.2 Structural isomers: carbohydrates are either aldehydes (carbonyl group is terminal) or ketones (carbonyl group is attached to two other carbon atoms).

5.3.2 Stereoisomers and stereochemistry or monosaccharides

configuration

stereoisomers

optically active

chiral or asymmetric atom

The simplest aldose, glyceraldehyde, contains an asymmetric carbon atom. This arises because the central carbon atom has 4 different substituent groups (H; OH; CHO; CH_2OH) attached to it (Figure 5.3). A consequence of this is that glyceraldehyde exists in two forms. Whilst these have the same molecular mass and the same functional groups, they differ in the arrangement of these groups in space, (ie in their configuration). Such molecules are termed stereoisomers. Possession of an asymmetric carbon atom results in a compound being optically active. This means that a solution of the compound will rotate the plane of plane-polarised light. (Another term which is used to describe an asymmetric carbon atom is chiral. The particular atom is then a chiral carbon atom or chiral centre; compounds containing an asymmetric centre are said to display chirality - from the Greek, meaning hand).

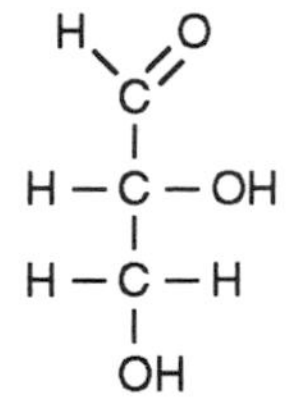

Figure 5.3 The simplest aldose, glyceraldehyde.

Our diagram of glyceraldehyde shown above (Figure 5.3) is 2-dimensional. But what is the real molecule like? It is much more realistically shown as a 3-dimensional structure (Figure 5.4) in which the central carbon atom is at the centre of a tetrahedron, with each of the four substituent groups at a different corner of the tetrahedron.

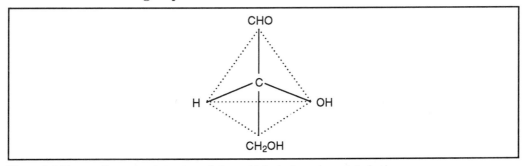

Figure 5.4 Three - dimensional structure of glyceraldehyde, based on a tetrahedron.

Viewed in this way, we can construct two different compounds by repositioning substituent groups, as shown in Figure 5.5. These compounds are different, in that they cannot be superimposed on each other. However, they are closely related to each other, in the same way as a right glove or hand is related to the left glove or hand; indeed, they are said to possess 'handedness'. These compounds are mirror images of each other: if one structure is placed in front of a mirror, the image in the mirror resembles the other molecule. Two structures which are related to each other in this way are termed enantiomers   enantiomers.

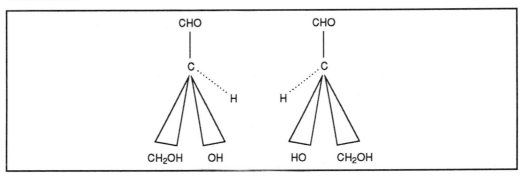

Figure 5.5 Alternative three - dimensional structures for glyceraldehyde.

∏ If you can get hold of two four cornered pyramids of equal faces or can make them out of card you could do the following little demonstration which will help you understand this principle. Label the corners of one pyramid A, B, C and D. Now label the other one as though it is a mirror image. Thus

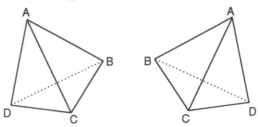

Now try as hard as you may; you will find that you can never arrange the pyramids so that they are labelled in an identical way.

Fischer projections

How do these 3-dimensional molecules relate to the first (2-dimensional) glyceraldehyde structure we presented (Figure 5.3)? The simplest way to convey the 3-dimensional image on paper is to use the projection formula introduced by Fischer. These are also known as Fischer projection. The 3-dimensional picture, with the central atom in the plane of the paper is displayed so that the substituent groups in a vertical line (through the central carbon atom) are behind the plane of the paper; those in a horizontal line are in front of it. This gives the projection formula for glyceraldehyde shown in Figure 5.6b; note how this relates to the 'ball and stick' model shown beside it (Figure 5.6a). Also shown is a so-called perspective formula in which dashed lines indicate bonds behind the plane of the paper, whilst wedges depict bonds projecting in front of the paper (Figure 5.6c).

a)
$$CHO$$
$$C$$
$$H \quad OH$$
$$CH_2OH$$

b)
$$CHO$$
$$H - C - OH$$
$$CH_2OH$$

c)
$$CHO$$
$$H - C - OH$$
$$CH_2OH$$

Figure 5.6 Alternative ways of depicting molecular structure a) Ball and stick b) Fischer projection c) Perspective formula.

∏ Inspect the molecules shown below and decide whether the statements that follow are true or false.

a)
$$CHO$$
$$H - C - OH$$
$$HO - C - H$$
$$CH_2OH$$

b)
$$CH_2OH$$
$$C = O$$
$$HO - C - H$$
$$CH_2OH$$

c)
$$CHO$$
$$HO - C - H$$
$$HO - C - H$$
$$CH_2OH$$

d)
$$COOH$$
$$CH_2$$
$$HO - C - H$$
$$CH_2OH$$

e)
$$COOH$$
$$CH_2$$
$$CH_2$$
$$CH_2OH$$

f)
$$CHO$$
$$H - C - OH$$
$$H - C - OH$$
$$CH_2OH$$

1) a) and c) are stereoisomers.

2) b) and d) are stereoisomers.

3) b), c) and d) are structural isomers.

4) a) and e) are structural isomers.

5) c) and f) are enantiomers.

1)   True: the only difference is in the configuration of the hydroxyl at carbon -2 (ie the second carbon when numbered from the aldehyde group). Both compounds contain the same numbers of each atom and the same functional groups.

2)   False: stereoisomers differ only in distribution (ie configuration) of groups. Compounds b) and d) differ in what groups are present; b) contains a keto group, d) contains a carboxylic end group. However, they are still isomers, since they contain the same numbers of each atom. They are structural isomers.

3)   True: the 3 compounds contain quite different groups (aldehyde; keto; carboxylic acid) and are thus not stereoisomers. They do, however, all have the same numbers of each atom, and hence the same molecular mass. They are thus structural isomers.

4)   The answer is false. In this case, compound a) has 4 oxygen atoms, whereas compound e) only has 3 oxygen atoms. They are thus not isomers at all.

5)   Enantiomers are a special case of stereoisomers in which the compounds are mirror images of each other. c) and f) are enantiomers, so this statement is true. They are also stereoisomers. Note, however, that neither pair a) and c), nor a) and f) are enantiomers, a) is not the mirror image of c), although they are stereoisomers.

Let us re-examine the two forms of glyceraldehyde we identified earlier (Figure 5.5). Glyceraldehyde is the simplest monosaccharide which possesses an asymmetric carbon atom and hence displays stereoisomerism. By convention, the aldehyde group is written at the top (and is thus behind the plane of the paper). If the hydroxyl group is to the left of the central carbon atom, it is designated L - glyceraldehyde (Figure 5.7); if the hydroxyl group is to the right of the central carbon atom, it is D-glyceraldehyde. D- and L-glyceraldehyde are used as reference compounds for other carbohydrates and other groups of molecules, especially amino acids. By convention we number the carbon atoms from the aldehyde group.

D and L forms

Figure 5.7 D- and L- glyceraldehyde.

absolute configuration

The D- and L-designation is described as the absolute configuration. Note that it tells us nothing about in which direction plane polarised light is rotated. It so happens that D-glyceraldehyde rotates plane - polarised light in a clockwise direction. This is designated as (+) and also described as dextro (or d); you may see this information incorporated into the description of glyceraldehyde as D (+) - glyceraldehyde. Similarly, L-glyceraldehyde happens to rotate plane - polarised light anticlockwise, (ie to the left) and is designated (-) or 'l' (for laevo): this would be summarised as L (-) - glyceraldehyde. To conclude, do not confuse the capital D/L, indicating absolute configuration, with lower case d/l, indicating dextro/laevo rotation.

differences between D, L and d l

Note also that the simplest ketose sugar, dihydroxyacetone, does not exist as stereoisomers, since the central carbon atom is not an asymmetric centre (there are only three substituent groups).

What happens with larger monosaccharides? If we examine 4 carbon aldose sugars, we find that we can draw 4 structures, since there are now 2 asymmetric carbon atoms, at each of which 2 alternative configuration are possible (Figure 5.8). Five carbon aldoses will have 8 possible forms since there are 3 asymmetric centres (ie 2^3 structures). For 6 carbon aldoses 2^4 or 16 forms occur. The general rule is that there are 2^n possible forms where n is the number of asymmetric carbon atoms.

numbers of forms $=2^n$

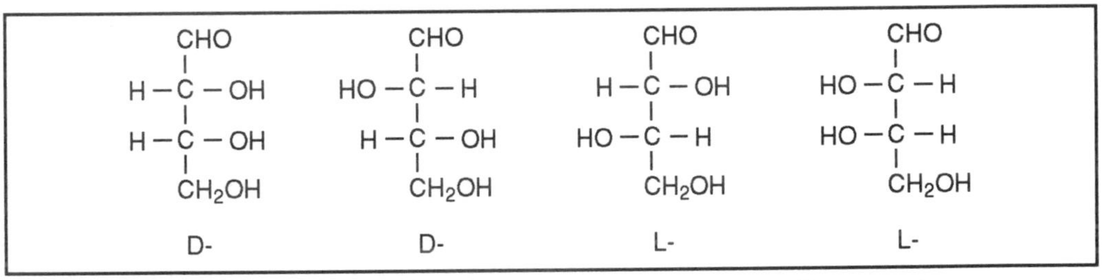

Figure 5.8 Possible 4 carbon aldose sugars. Two have absolute configuration D-; two are L- sugars.

For ketoses, it is with 4 carbon sugars that an asymmetric centre first appears (Figure 5.9) and thus two 4 carbon ketoses exist, as shown: there are thus only half the number of stereoisomers in the ketose 'family'. Again the rule that the number of forms $=2^n$ where n is the number of asymmetric carbon atoms applies.

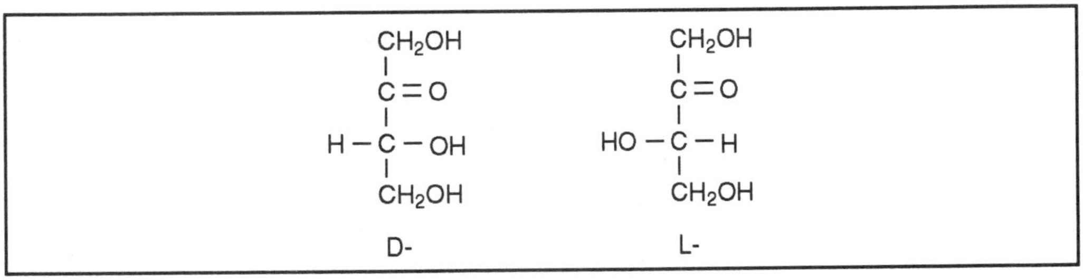

Figure 5.9 D- and L- 4 carbon ketose sugars.

reference to glyceraldehyde

How do we assign an absolute configuration to those sugars with more than one asymmetric centre? The designation of absolute configuration of sugars having at least two asymmetric centres is based on the configuration at the asymmetric carbon furthest from the carbonyl group. Their conformation is then designated by reference to that of glyceraldehyde.

SAQ 5.1

Construct a 'family' of 5 carbon aldoses, by analogy with the way in which the 4 carbon aldoses in Figure 5.8 have been derived from D- and L-glyceraldehyde. Identify which are D-sugars and which are L-sugars. Do this before you read on, as it will help to convince you that you have understood this rather complex section.

You may well be thinking that this topic is getting rather complex! Do not worry, since for most occasions the situation is much simpler for two reasons. Firstly, most naturally occurring sugars belong to the D-series. Secondly, we routinely encounter only a few of the hexose and pentose sugars. Nonetheless, it is important that you thoroughly understand why stereoisomerism occurs and the basic rules of configuration.

SAQ 5.2

Inspect the following 6 carbon compounds and then answer the questions which follow.

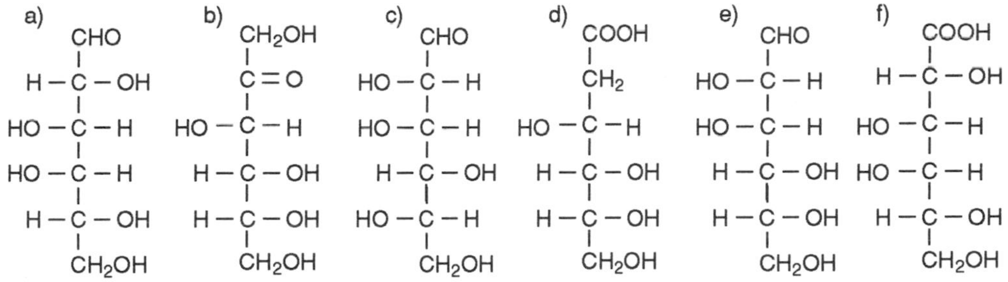

a)	b)	c)	d)	e)	f)
CHO	CH$_2$OH	CHO	COOH	CHO	COOH
H–C–OH	C=O	HO–C–H	CH$_2$	HO–C–H	H–C–OH
HO–C–H	HO–C–H	HO–C–H	HO–C–H	HO–C–H	HO–C–H
HO–C–H	H–C–OH	H–C–OH	H–C–OH	H–C–OH	HO–C–H
H–C–OH	H–C–OH	HO–C–H	H–C–OH	H–C–OH	H–C–OH
CH$_2$OH	CH$_2$OH	CH$_2$OH	CH$_2$OH	CH$_2$OH	CH$_2$OH

1) Which sugar(s) are stereoisomers of D- glucose (Look ahead to Figure 5.10a to find out the structure of D- glucose. You might like to draw it out so that you can compare it with the above structures).

2) What is the absolute configuration of compounds a), b) and c)?

3) Which compounds are not aldoses?

4) Which compounds are structural isomers of compound c)?

5.3.3 Cyclisation and the formation of anomers

glucose in solution is not in a linear form

We have seen that, through stereoisomerism, 16 six-carbon aldose sugars can exist, 8 of which will be L- sugars and 8 of which will be D-sugars. Of these, only the D- sugars are naturally widespread and important to us. D-glucose is the most common of these: its structure is shown in Figure 5.10a. As you can see, it possesses a terminal aldehyde group, which is responsible for many of its properties. Analysis of the behaviour of D-glucose in solution has shown, however, that most of the molecules do not occur as linear structures but as ring structures. How does this occur?

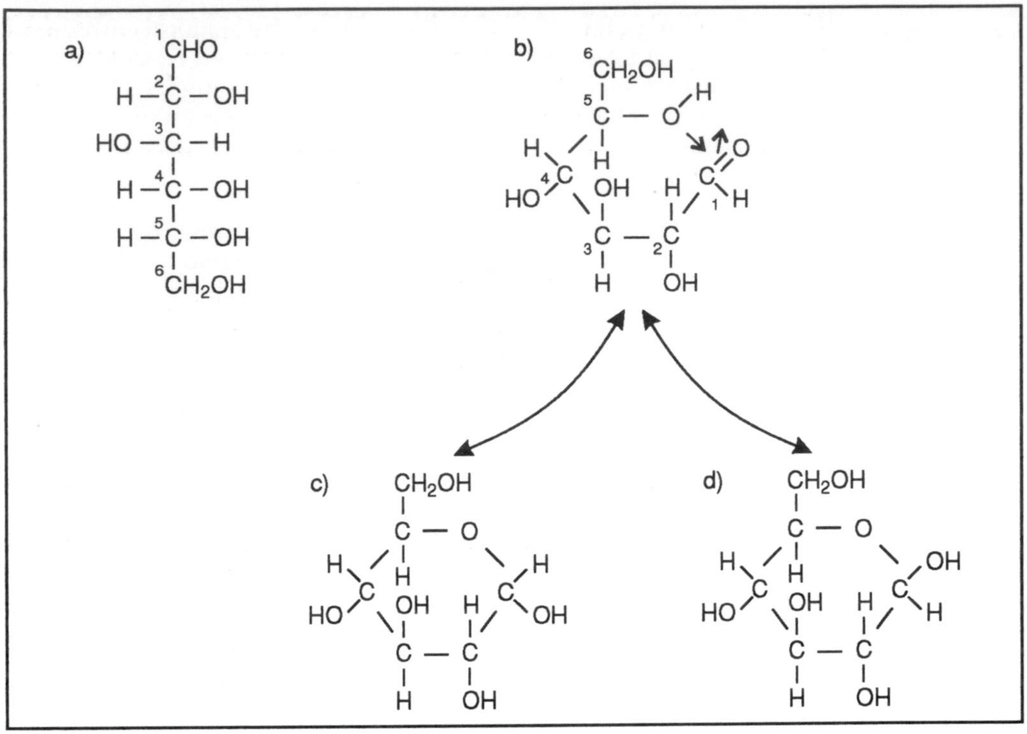

Figure 5.10 The structure of D-glucose. a),b) linear chain form of glucose. b) shows intramolecular reaction of C-5 hydroxyl group with the C-1 carbonyl group, forming a ring structure. c),d) two alternative ring forms of D-glucose may be made. c) is α-D-glucose. d) is β-D-glucose.

numbering of sugars

The mechanism is that the straight chain form naturally bends into a C- shape (Figure 5.10b). For convenience in discussing sugar structures, we identify each carbon atom by a number, starting with the aldehyde carbon atom as C-1. This numbering is also included in Figure 5.10. By rotation of the bond between carbon atoms 4 and 5, the C-6 -CH₂OH group moves away from the C-1 carbon, whilst the hydroxyl group on C-5 comes close to the C-1 aldehyde group.

This allows an intramolecular reaction of the hydroxyl group at C-5 with the carbonyl (C=O) group of the C-1 (Figure 5.10b), resulting in the ring forms of glucose as shown in Figures 5.10c, d. This process is reversible: an equilibrium exists with small quantities of the open chain form present. Note the position of the hydroxyl group on position C-1 in α-D-glucose and β-D-glucose.

∏ In Figure 5.10, we have seen that 2 cyclic structures can be formed. Do you think this will have any effect on the optical properties of D-glucose?

The answer is 'Yes', it will, because from a single pure compound, D-glucose, two alternative new structures are being made. The difference arises through the generation of another asymmetric centre. The C-1 carbon atom now has 4 different substituent groups, allowing the alternative forms of D-glucose to occur. These are called α and β-D glucose. These have markedly different optical rotation properties, summarised in Table 5.1.

α and β- D glucose

Compound	Specific Optical Rotation
α - D - Glucose	+ 112°
β - D - Glucose	+ 19°

Table 5.1 Optical rotation properties of D- glucose.

Π   If you dissolve pure α-D-glucose in water, its specific optical rotation changes with time from + 112° to + 52°. How can this be explained? Hint: Look carefully at the C-1 carbon atom in the ring forms of glucose (Figure 5.10 c and d); how many different groups are now attached to it?

We have said that α-, β- and straight chain forms of D-glucose exist in solution in an equilibrium mixture. If you were to start with any of these 3 forms (although only α- and β- forms can be obtained pure), it would slowly form the other structures. Thus the α-D glucose we started with opens out into the straight chain form and then recyclises, eventually reaching an equilibrium mixture of approximately 33% α-D-glucose, 67% β-D-glucose, and less than 0.1% straight chain D-glucose. This ratio gives the equilibrium specific rotation value of 52°.

anomers

We have seen that from a single pure compound, D-glucose, two alternative ring forms α- and β-D-glucose can be formed and an additional asymmetric centre is created. These two forms are called anomers. Pairs of anomers will be formed whenever cyclisation of linear forms occurs. Whether the hydroxyl group is above the plane of the ring of the sugar (β- form) or below it (α- form) has an important impact on the shapes of any polysaccharides which may be formed from them. We will examine this later. Cyclisation of monosaccharides is important in 5 carbon aldoses (eg ribose) and 6 carbon sugars (aldoses such as glucose, galactose and mannose; ketoses such as fructose). In the cases of aldoses, 6-membered rings are formed. In the case of ketoses, 5-membered rings are formed (Figure 5.11).

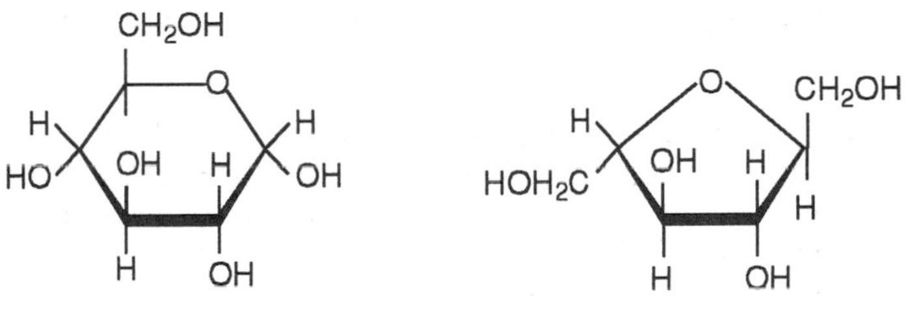

α-D-glucose or glucopyranose

β-D-fructose or fructofuranose

Figure 5.11 Ring forms of aldoses and ketoses. The ring formed by aldoses is similar to the molecule pyran: hence cyclic forms of 6 carbon aldoses are often called 'aldopyranose' (eg glucopyranose). Similarly, the 5-membered rings adopted by ketoses are similar to the chemical furan and hence are often called 'ketofuranose' (eg fructofuranose). For clarity, the carbon atoms within the ring are frequently omitted, as here.

epimers

A final term which is used in describing some stereoisomers in sugars is epimers. This is used when sugars only differ in the configuration around a single carbon atom. Looking at Figure 5.12, D-glucose and D-mannose only differ in their configuration at the C-2 position: thus D-mannose is an epimer of D-glucose.

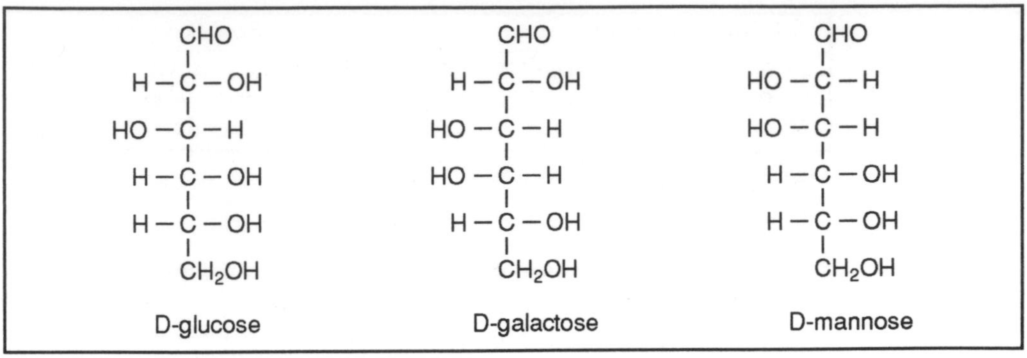

Figure 5.12 Open chain formulae of D -glucose, D-galactose and D-mannose.

Π Examine Figure 5.12. Are D-glucose and D-galactose epimers? Are D-galactose and D-mannose epimers?

D-glucose and D-galactose are epimers, since the only difference lies in the configuration at the C-4 position. D-galactose and D-mannose are not epimers: they differ in configuration at both C-2 and C-4. Note that although cyclisation occurs with all 3 of these 6-carbon aldoses to produce α and β forms, this does not stop D-mannose and D-glucose being epimers.

5.4 Properties of monosaccharides

5.4.1 Reducing properties

aldoses are reducing sugars

Carbohydrates are routinely classified as either reducing or non-reducing compounds. The basis for acting as a reducing agent comes from a free, or potentially free, aldehyde group. What is meant by this is that, with glucose for example, the aldehyde group is potentially free even in the ring forms since it is in equilibrium with the open chain form (free aldehyde group present). Thus aldoses are all reducing sugars and this property is very widely used in testing for monosaccharides. The product of the reaction will be an acid such as D-gluconic acid (Figure 5.13). As we shall see shortly in some cases further bonding between monosaccharides may prevent the straight-chain aldoses forming; when this happens, the reducing property is lost.

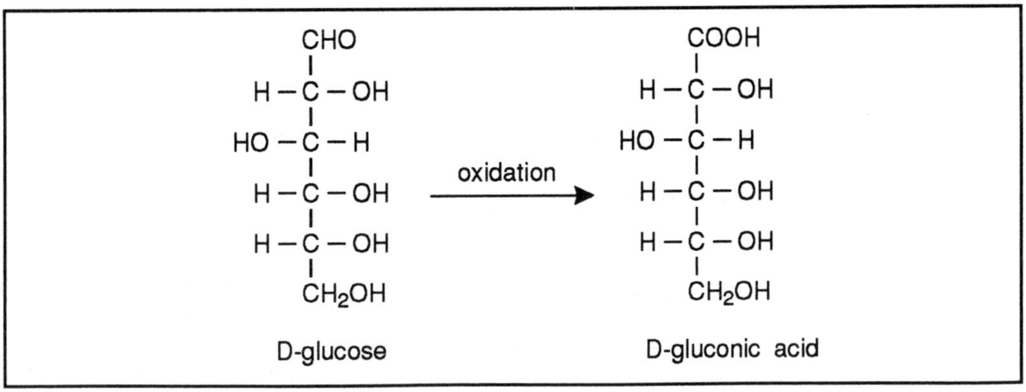

Figure 5.13 The reducing property of monosaccharides is related to the ability of the aldehyde group to be oxidised.

5.4.2 Mutarotation

the
interchange of
α and β forms:
mutarotation

The occurrence of two alternative ring forms of D-glucose (α-and β-D-glucose) has already been described. α-and β-D-glucose display marked differences in their optical rotation. The existence of an equilibrium between these forms, via an open chain intermediate, means that α-D-glucose will slowly partially convert to β-D-glucose and vice versa. This is accompanied by a change in optical rotation: this is known as mutarotation.

5.4.3 Important monosaccharides and their roles

importance of
glucose

A number of 3 to 7 - carbon monosaccharides are important intermediates in the central pathways of metabolism within cells. They usually occur in a phosphorylated form, which prevents their leakage from the cell. D-glucose is an important source of energy for both prokaryote and eukaryotic organisms. Although most of the carbohydrate present in living organisms is in the form of polymers (polysaccharides), D-glucose is the primary product of polysaccharide breakdown. Following hydrolysis of starch (eg in humans) or cellulose (in herbivores), glucose is absorbed from the gut, transported in the bloodstream and either used directly by cells or stored as glycogen in the liver. Thus glucose is an important transportable fuel in the body.

importance of
fructose

Fructose can also be important as a source of energy. It may occur as a constituent of foodstuffs (for example, in honey) or be produced by breakdown of the disaccharide sucrose or the polysaccharide inulin (this occurs, for example, in the roots of the Jerusalem artichoke).

important
pentoses

As we have already learnt, five - carbon sugars have important roles as components of nucleic acids and nucleotides. Ribose (Figure 5.14a) is found in the nucleotides used in metabolism (eg ATP, NAD$^+$) and is present in ribonucleic acid (RNA). 2-deoxyribose (Figure 5.14b) replaces ribose in deoxyribonucleic acid (DNA).

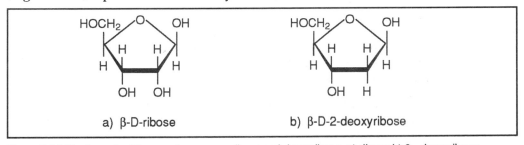

a) β-D-ribose   b) β-D-2-deoxyribose

Figure 5.14 The important five - carbon sugars ribose and deoxyribose, a) ribose b) 2 - deoxyribose.

Various derivatives of the commonest monosaccharides described here also occur, either free or as components of polysaccharides. The C-6 carbon atom in aldose sugars can be oxidised to form a carboxylic acid group, -COOH. Such acids are called uronic acids. The one derived from glucose is called glucuronic acid (Figure 5.15a). Alternatively, the aldehyde group of a monosaccharide may be reduced, forming a sugar alcohol. These are also referred to as polyols. When this occurs with glucose, sorbitol (found in many berries) is formed (Figure 5.15b). Finally, a common and important monosaccharide is ascorbic acid or vitamin C (Figure 5.15c). This is an essential carbohydrate that virtually all organisms (except primates and guinea pigs) can make from glucose. The inability of humans to synthesize vitamin C means that it must be obtained in the diet or scurvy, caused by its deficiency, will occur.

glucuronic acid

sorbitol

ascorbic acid

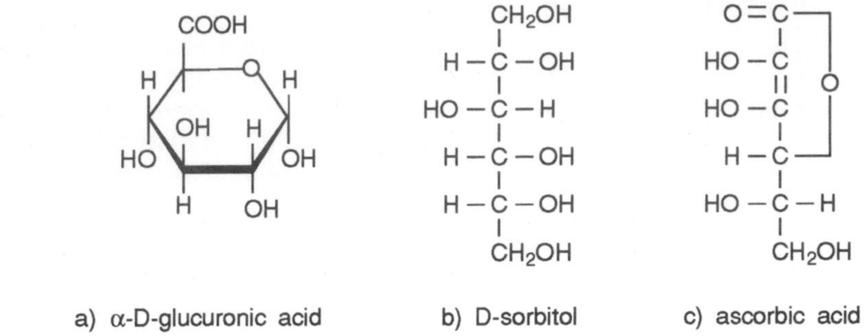

a) α-D-glucuronic acid   b) D-sorbitol   c) ascorbic acid

Figure 5.15 Oxidation and reduction products of D - glucose, a) D - glucuronic acid, b) D - sorbitol, c) ascorbic acid (vitamin C).

5.5 Disaccharides

glycosidic bonds

Disaccharides are formed by the linking of two monosaccharides with the elimination of a molecule of water (Figure 5.16). This reaction is therefore a condensation and the bond thus formed is called a glycosidic bond. Note that Figure 5.16 is intended to show the overall result of the condensation. The mechanism of condensation of monosaccharides to form disaccharides in cells is more complicated.

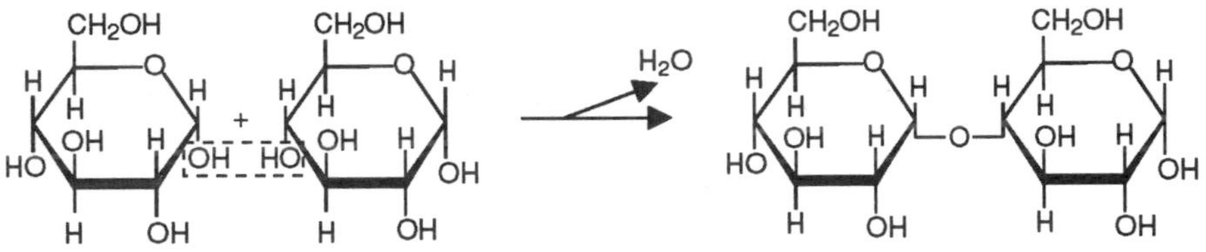

Figure 5.16 Scheme to illustrate the overall condensation reaction by which disaccharides are formed.

5.5.1 The glycosidic bond

The nature of the glycosidic bond between sugar units has a profound effect on the structure of the resulting molecule. It is therefore important to be able to describe it adequately. Glycosidic bonds are therefore routinely designated according to which carbon atoms of the constituent monosaccharides are involved. Thus the bonds shown

1 - 4 linkages

in Figure 5.17 are 1 - 4 or -1,4- glycosidic bonds. If the anomeric carbon atom (number 1) involved in the bond has the α configuration, then the bond is depicted as shown in Figure 5.17a and is designated α-1,4. Conversely, if the C-1 carbon atom involved was in the β-configuration, a β-1,4- bond results, as shown in Figure 5.17b. Such differences have considerable consequences for polysaccharide structure and properties. Bonds can and do occur between other carbon atoms (eg 1,6), resulting in branching (if it occurs in addition to -1,4- bonds).

5.5.2 Common disaccharides

Disaccharides may consist of either two identical or two different monosaccharides.

maltose

Maltose (Figure 5.18a) consists of 2 molecules of D-glucose linked by an α-1,4 glycosidic bond. It is found naturally and is also formed during the enzymatic hydrolysis of starch. Although the anomeric carbon atom of the left hand glucose unit is involved in the

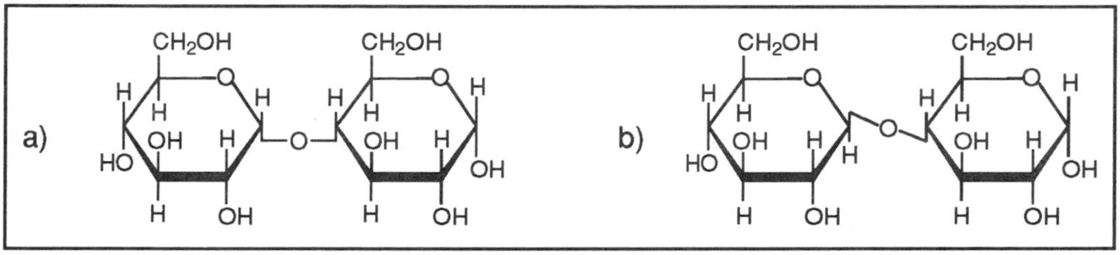

Figure 5.17 Alternative formulation of a glycosidic bond between the 1-carbon of one glucose and the 4-carbon of a second. a) α-1,4- glycosidic bond, as found in maltose; b) β-1,4- glycosidic bond, as found in cellobiose.

glycosidic bond the right hand glucose unit can open into the linear form. This means that the -CHO group on C-1 is potentially 'free'. Since the reducing property, and mutarotation, result from a free C-1 group, maltose is a reducing sugar and displays mutarotation.

cellobiose

lactose

Cellobiose (Figure 5.18b) also contains two molecules of glucose but linked by a β-1,4 glycosidic bond. It is formed during acid hydrolysis of cellulose. Lactose (Figure 5.18c) consists of D-glucose and D-galactose joined by a β-1,4 bond. Since it also contains a free C-1 carbon atom, it is a reducing sugar and displays mutarotation.

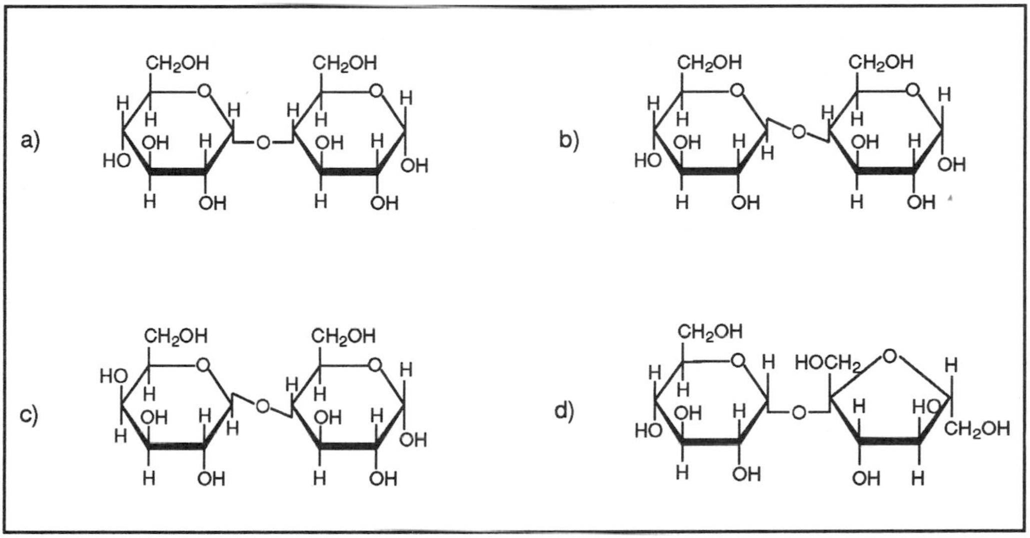

Figure 5.18 Disaccharides a) maltose b) cellobiose c) lactose d) sucrose.

hydrolysis of glycosidic bonds

Lactose is found in milk, and is thus of major importance in the nutrition of young mammals. To be utilised by humans, lactose must be hydrolysed by the enzyme lactase in the intestine (lactose cannot be absorbed); the glycosidic bond is broken by incorporation of a molecule of water. This hydrolysis is thus a reversal of the condensation reaction involved in synthesis of disaccharides. Although this enzyme is present in children, it is lost with age by many of the world's population, especially amongst Oriental races. This results in an inability to utilise lactose and may cause lactose intolerance. In such cases, ingestion of lactose causes intestinal pain. A biotechnological solution to this may be provided by enzymic hydrolysis of lactose by lactase.

sucrose

The most widespread naturally occurring disaccharide is sucrose, used as sugar in the home. This consists of one molecule each of α-D-glucose and β-D-fructose, linked by a 1,2 glycosidic bond (Figure 5.18d). This mean that both reducing moieties of the glucose and fructose are involved in the glycosidic bond and there is no potentially free reducing group. Sucrose is therefore a non-reducing sugar. This property is widely used as a means of distinguishing sucrose from monosaccharides and other disaccharides. Sucrose is one of the principal products of photosynthesis and is transported round the plant and to storage organs (roots, seeds, tubers). It is nutritionally important and is widely used as a sweetening agent. Like lactose and other disaccharides, sucrose must be broken down to its monosaccharide constituents before it can be absorbed by humans. This is accomplished by enzymatic hydrolysis.

non-reducing nature of sugar

SAQ 5.3

How many of the molecules drawn below should be called maltose? Identify the four disaccharides, looking carefully at the component monosaccharides and the type of glycosidic bond.

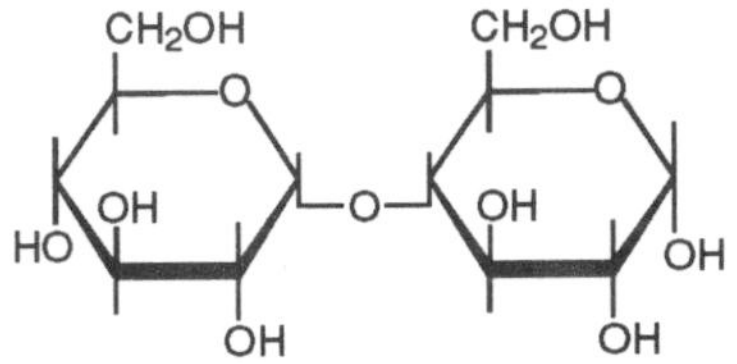

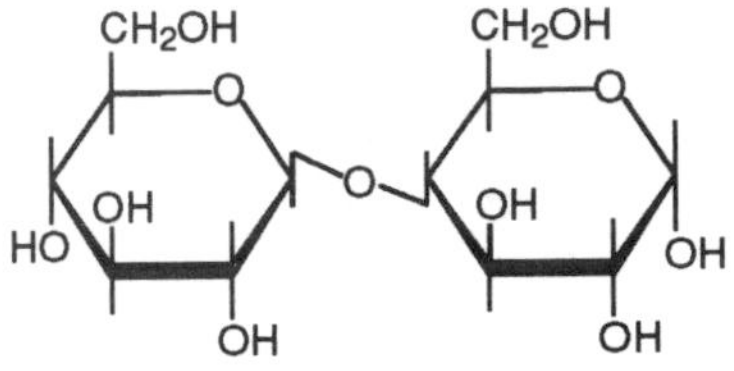

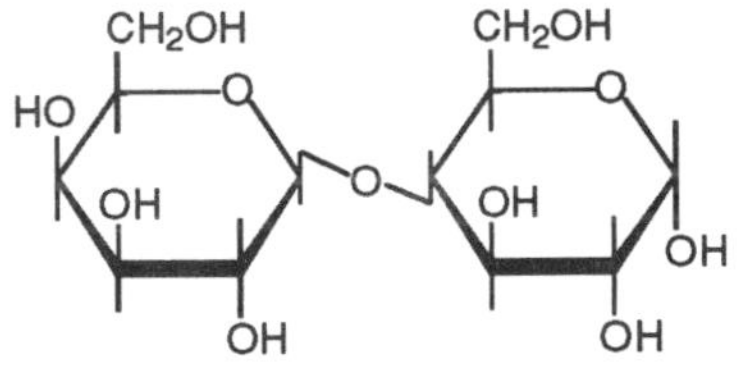

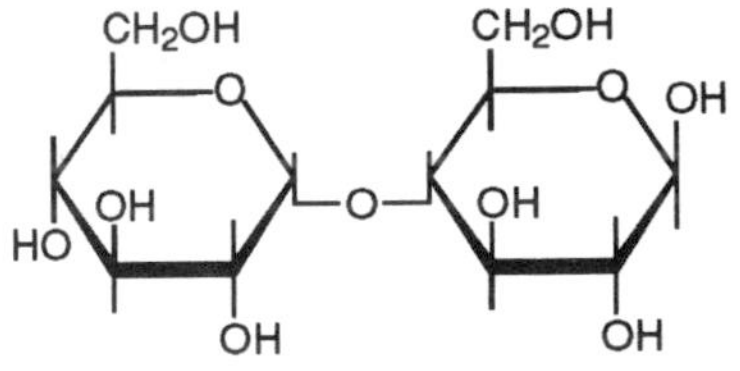

SAQ 5.4

Is cellobiose a reducing sugar? Give brief reasons for your answer.

5.6 Polysaccharides

Most of the carbohydrates found in nature occur as polysaccharides. Polysaccharides either have a structural role (eg cellulose) or serve as a stored form of energy (eg starch in plants and glycogen in animals). Polysaccharides have high molecular weights but, unlike proteins and nucleic acids, do not have a uniform structure or size. Because of their size, they are not readily soluble in water, although starch and glycogen are more soluble than cellulose. Starch and glycogen are deposited as granules in the cytoplasm of cells. To be utilised, these polymers must first be converted to monosaccharides for transport from cell to cell (in mammals, as glucose in the bloodstream; in plants, as previously mentioned, as the disaccharide sucrose).

Hydrolysis of polysaccharides, whether by acid or by enzyme, yields mono- and disaccharides. If hydrolysis results in only a single type of monosaccharide, the polysaccharide is termed a homopolysaccharide. When more than one type of monosaccharide is formed on hydrolysis, the polymer is described as a heteropolysaccharide. Although the more well-known polysaccharides are all homopolysaccharides, heteropolysaccharides have important roles and properties and are a valuable resource.

5.6.1 Storage polysaccharides - starch and glycogen

starch: glucan

amylose

In plant cells, carbohydrate reserves are stored in the form of starch. Starch is a homopolysaccharide consisting of glucose units (hence it is known as a glucan). It consists of two components, which are present in varying amounts. Amylose is an unbranched chain of D-glucose units linked by α-1,4- glycosidic bonds (Figure 5.19a); its molecular weight can vary from a few thousand to 150,000 daltons. Amylose gives a characteristic blue colour when treated with iodine. In solution it forms a gentle non-rigid helical structure (Figure 5.19b). Iodine can occupy the inside of the helix, resulting in the formation of a deep blue starch-iodine complex. The helical structure allows penetration by solvent. As a result of this, starch is more soluble and more flexible than cellulose. None-the-less, starch is deposited in plant cells as insoluble granules.

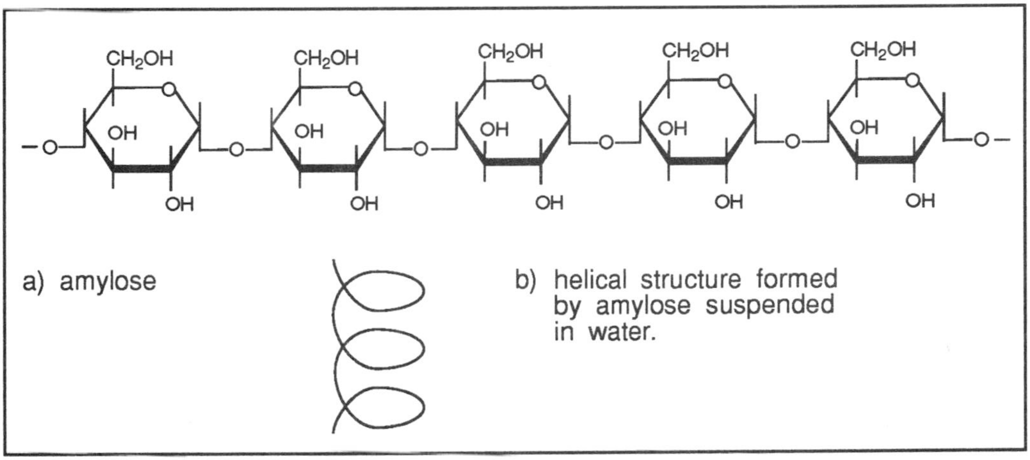

Figure 5.19 The structure of amylose. a) The polymer consists of linear chains of glucose units linked by α-1,4- glycosidic bonds. b) In solution, amylose forms a helix.

amylopectin

Amylopectin is the other component of starch. It is a branched polymer of glucose. Chains again consist of α-1,4- linked glucose but branches are made through occasional α-1-6- glycosidic bonds (Figure 5.20a). The branches typically occur about every 20-30 glucose units (Figure 5.20b). This branched structure means that there are numerous non-reducing ends (see Figure 5.20b) but very few reducing termini (ie those with a C - 1 carbon atom which is not involved in a glycosidic bond). In the amylopectin structure shown in Figure 5.20b, there are 17 non-reducing ends and only 1 reducing end. Thus starch tends to be a non-reducing polysaccharide. Amylopectin also reacts with iodine, but gives a purple-red colour.

glycogen

In animal cells, carbohydrate reserves are stored as glycogen, mainly in muscle and liver. Like amylopectin, glycogen consists of branched polymers of glucose. α-1,4-linkages result in chains, with highly frequent branches being given by α-1,6- glycosidic bonds. The degree of branching is greater in glycogen than in amylopectin. In glycogen, the branches occur every 8-12 units. Because of the branching, glycogen forms a very compact structure.

Several questions arise from consideration of the structure of glycogen and starch.

1)   Why are polymers used as a storage form of carbohydrate, since there is a cost (in energy) in constructing the polysaccharide?

2)   What advantage is provided by branched polysaccharides as an energy store?

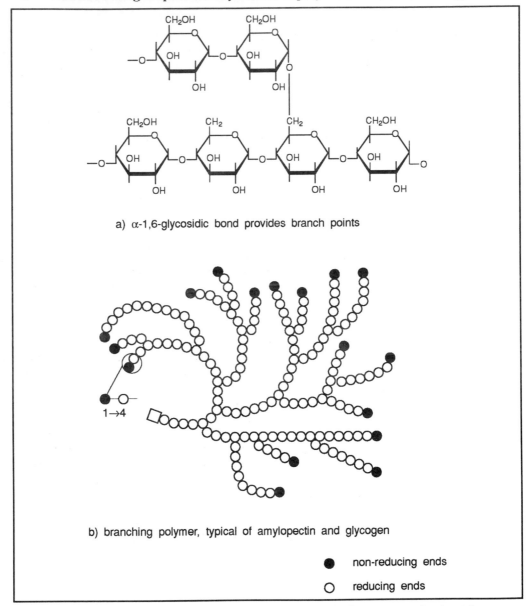

a) α-1,6-glycosidic bond provides branch points

b) branching polymer, typical of amylopectin and glycogen

● non-reducing ends

○ reducing ends

Figure 5.20 The structure of amylopectin. a) As well as α-1,4- bonds, there are occasional α -1,6- glycosidic linkages. This gives a highly branched structure b), in which there are numerous non-reducing ends and few reducing ends. In this model there are 17 non-reducing ends and only 1 reducing end.

reduction of osmotic potential

The following reasons can be put forward for the use of branched polysaccharides as energy stores. The use of a polymer means that the osmotic potential associated with the stored material is markedly reduced. Osmotic potential is related to the number of molecules present in solution. By combining several thousand glucose molecules into a single polymer, the osmotic potential is reduced from that generated by thousands of molecules to that of one. Further, starch and glycogen are deposited as insoluble granules, which further lowers any osmotic potential. Without the use of polymers, accumulation of carbohydrate would lead to uptake of water into cells and dramatic swelling. In addition, soluble low molecular weight sugars would tend to diffuse out of the storage cell.

relative insolubility of polymers

compact nature of polymers

Branching makes the molecule more compact, thereby facilitating economical use of storage 'space'. Probably of greatest significance is the fact that glycogen and amylopectin have numerous non-reducing ends, because it is onto these ends that more glucose units are added and, in some cases, from these ends that mobile low molecular weight sugars are released. When glycogen is being broken down by enzymes (eg glycogen phosphorylase and amylase), reaction occurs at the non-reducing ends. Thus the use of branches means that 'access' by the cell to its carbohydrate reserves is made easier and is likely to be more rapid than with unbranched chains with only one non-reducing end.

access by hydrolysing enzymes

∏ Write down the reasons why glycogen and starch are suitable for the storage of carbohydrate reserves but not for the transport of carbohydrates?

We have already explained that if the monosaccharide units in glycogen or starch were present in free (ie unpolymerised) form, a very high osmotic potential would occur: this would result in swelling and bursting of the cell through uptake of water. However, polysaccharides are unsuitable for transporting carbohydrates around the body because they are insoluble and cannot cross membranes. For this purpose, small soluble sugars are used. Storage polysaccharides in cells are deposited as granules; but this still allows access by solvent and enzymes at their surfaces.

uses of starch

Quorn

high fructose syrups

Starch is a plentiful resource which has formed a staple part of our diet for centuries, through cereals and potatoes. Biotechnology has already contributed to the further utilisation of starch by growth of micro-organisms as a high protein foodstuff (eg growth of the fungus *Fusarium* on starch by Ranks Hovis Macdougall, subsequently marketed as 'Quorn'). Isolated enzymes have also been used to convert 'low-value' starch to 'added-value' products such as high fructose syrup, which is extensively used as a sweetener. In this process, maize starch is hydrolysed by amylase and other enzymes to yield D-glucose, which is then partially converted to its structural isomer D-fructose. Although sucrose is reasonably effective as a sweetening agent, fructose is sweeter than sucrose. Starch thus provides an alternative source of a sweetening agent for the food and confectionary industries.

Other storage polysaccharides occur in nature. One, mentioned earlier, is inulin and consists of fructose units linked by β-2,1- glycosidic bonds.

SAQ 5.5

Consider the following highly hypothetical polymer of glucose. Identify on it the following features:

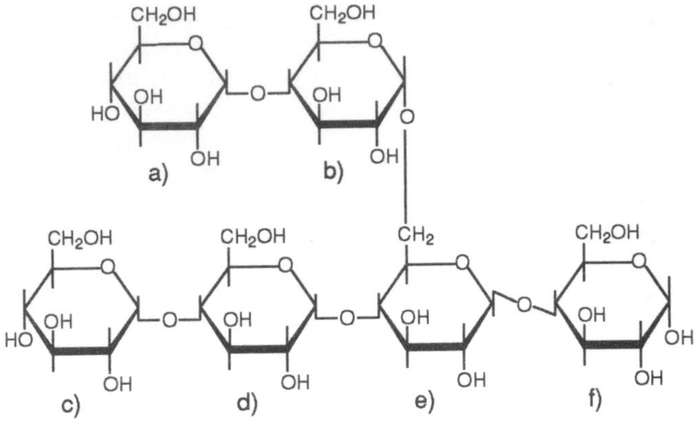

1) Non-reducing termini

2) Reducing terminus

3) β -1,4- glycosidic bond

4) α -1,6- glycosidic bond

5) α -1,4- glycosidic bond(s)

5.6.2 Structural polysaccharides - cellulose

hydrogen bonding between cellulose chains

The most important structural polysaccharide in plant cells is cellulose. Like amylose, cellulose is a polymer of glucose, in which each glucose unit is linked to the next by a 1,4- glycosidic bond. Unlike amylose, the linkage involves the carbon - 1 hydroxyl in the β- position, giving a β-1,4- bond (Figure 5.21). This apparently trivial difference has a dramatic influence on the structure of the polymer and is an excellent illustration of the influence of bonding on structure and hence properties of biological molecules. Instead of the flexible helical coil which amylose forms, cellulose forms rigid parallel chains which are highly insoluble. Approximately 100-200 individual chains combine to form fibrils in which there is extensive hydrogen bonding (which stabilises the fibril) and few hydroxyls are exposed to the surrounding solvent. This configuration is responsible for the rigidity of cellulose fibres and their insolubility. This contrasts with the helical α-1,4- linked polymers (starch, glycogen) where open spaces allow solvent penetration, promoting flexibility and making them relatively soluble.

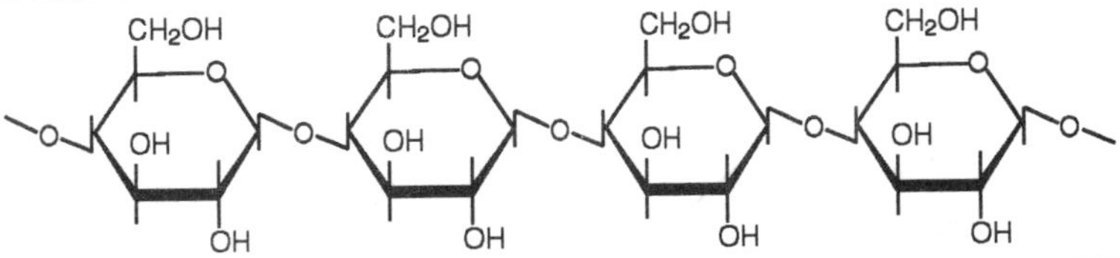

Figure 5.21 The β-1,4-glycosidic bonds in cellulose.

Fibres of cellulose are laid down in the cell wall of plants in a regular manner to form long fibres. This structure provides the strength to maintain the shape of the cell, and ultimately the whole plant. Cellulose is the most abundant carbon compound in the world, representing more than half of all the carbon in higher plants. Some plant components, for example cotton, are almost pure cellulose.

importance of gut micro-organisms in using cellulose

Another aspect of cellulose arising from the β-1,4- linkage is the nutritional availability of the glucose of which it is composed. The β-1,4- linkages are not hydrolysed by amylases, which are widespread in nature and found for example in the digestive system of humans. Humans lack an enzyme which can hydrolyse the β-1,4- linkage and hence we can obtain no useable carbohydrate from cellulose. However, some micro-organisms do contain the appropriate β-glucosidase (usually called cellulase). Ruminants contain these micro-organisms which secrete cellulase in their alimentary tracts and therefore hydrolyse cellulose. The micro-organisms grow on the sugars released by this hydrolysis. The grazing animal then benefits by degrading and utilising the constituents of the micro-organisms. Utilisation of cellulose by humans is at best third-hand (by eating the ruminants or dairy products). Biotechnological solutions to this problem of using cellulose (eg by the *in vitro* use of cellulase), have long been sought.

SAQ 5.6

Write a brief explanation of the significance of the following factors on the structure and role of a polysaccharide:

1) nature of the glycosidic bonds involved;

2) occurrence of branching.

5.6.3 Complex polysaccharides

Structural polysaccharides also play an important role in animal and microbial cells. They are found, for example, as components of the bacterial cell wall, in cell walls of fungi and on the outside of animal cells and in animal intercellular material. In most cases these polysaccharides are polymers of modified (or derived) monosaccharides, rather than of glucose. These derived polysaccharides contain various groups, including carboxylic acid groups and amino groups, often substituted in various ways (Figure 5.23). Monosaccharides containing amino groups are called amino sugars. We shall briefly consider 3 examples of these more complex polysaccharides.

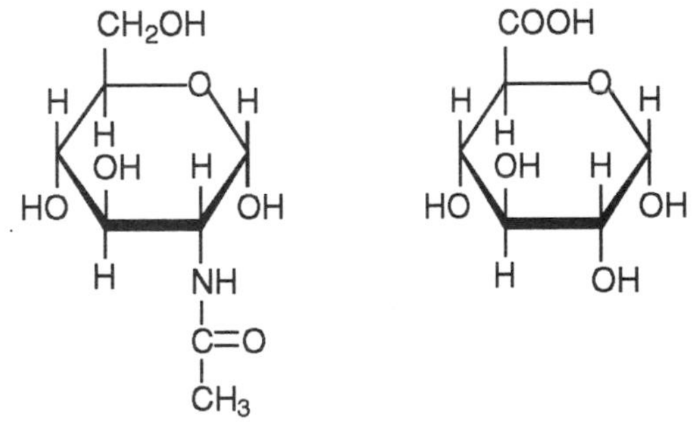

Figure 5.22 Two modified sugars found in polymers, N-acetyl glucosamine and D-glucuronic acid.

peptidoglycan

N-acetyl
glucosamine
N-acetyl
muramic acid

Bacterial cell walls contain a heteropolysaccharide whose chains are cross-linked by short peptides; the whole structure is called peptidoglycan. The polysaccharide consists of alternating units of N-acetyl muramic acid and N - acetyl glucosamine (Figure 5.23a), joined by β-1,4- linkages. When these are cross-linked by the peptide bridges, a 'net' is produced which encloses and gives shape to the bacterial cell (Figure 5.23b). Penicillin acts as an antibiotic through inhibition of the synthesis of peptidoglycan.

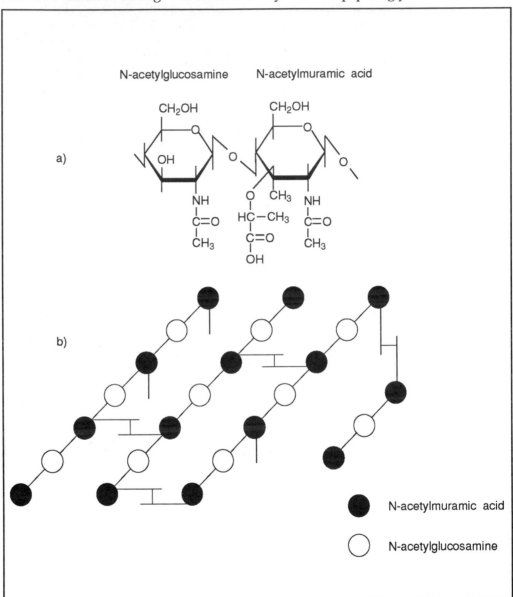

Figure 5.23 The structure of the peptidoglycan of bacterial cell walls. a) The repeating disaccharide unit of N - acetyl muramic acid and N - acetyl glucosamine. b) The complete peptidoglycan, in which individual polysaccharide chains are cross-linked by short peptides.

chitin | N-acetyl glucosamine is also used in the construction of the homopolysaccharide chitin, which forms the exoskeleton of insects and crustaceans and the cell wall of some fungi. The monosaccharide units are linked β-1,4-, giving a polysaccharide which is even more rigid and insoluble than cellulose. Thus amino sugar derived polysaccharides can have important structural roles.

muco-polysaccharides | The polysaccharide coat found on most animal cells and in the intercellular material of animal tissues contain mucopolysaccharides. These are gelatinous polysaccharides containing amino sugars which are gelatinous. They have high molecular weights and they act as lubricants and as adhesives. An example is hyaluronic acid (Figure 5.24), which consists of alternating units of D-glucuronic acid and N-acetyl glucosamine.

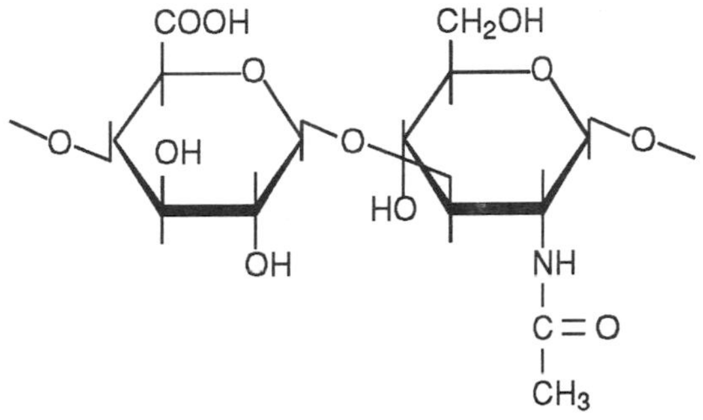

β-1,3- linkage in hyaluronic acid

Figure 5.24 One of the constituents of the mucopolysaccharides: hyaluronic acid.

agar

agarose

alginate | The third example of complex polysaccharides includes gelling polysaccharides. These are produced by micro-organisms and lower plants (especially seaweeds) and are characterised by a transition from a fluid (although viscous) state to a 'set' gel state. This transition may be controlled by temperature (eg agar, agarose) or by divalent cations such as calcium (eg alginate). These materials are of considerable use and interest in the food industry. Biotechnologists are continually seeking ways of using living systems to make more of these valuable materials. Note that gelatine used in confectionary and table jelly is not a polysaccharide but is derived from the protein collagen.

Summary and objectives

In this chapter we have discussed the structure and properties of the major groups of carbohydrates and related these to some of the functions they fulfil within cells. You will also have learned much about the importance of stereochemical configuration on the properties of biochemicals. Now that you have completed this chapter you should be able to:

- use appropriate terms to describe the structure of mono-, di- and polysaccharides;

- use the terms enantiomers, anomers and epimers, and identify D and L sugars;

- identify aldose and ketose monosaccharides and distinguish between reducing and non-reducing sugars;

- predict whether a disaccharide is a reducing sugar or not;

- distinguish between homopolysaccharides and heteropolysaccharides and give examples of each;

- describe a variety of carbohydrate derivatives, especially amino sugars and their occurence;

- use correct nomenclature to describe different glycosidic bonds and explain the consequences of the type of glycosidic bond for polysaccharide structure;

- describe roles of carbohydrates and show how their structure makes them suited to their roles.

Lipids

Lipids

Lipids are insoluble in water: this means that to extract lipids in order to study them, non-polar solvents such as chloroform or ether must be used. Although in molecular structure lipids are a heterogeneous group, their solubility properties represent a common property. As we shall see, precise properties of individual classes of lipids result from their structures. Their properties in turn influence the roles they perform. Their major roles are:

- storage (and transport) form of food reserves (ie metabolic fuel);

- major structural component of cell membranes;

- regulatory roles via hormones and vitamins;

- protective roles on the surface of organisms, for example waxes on leaves and insects.

In this chapter, we will examine each of the major groups of lipids in turn and consider their important properties.

6.1 Neutral fats

fatty acids
glycerol
triglycerides

The main intracellular storage lipids are esters formed between fatty acids and glycerol (Figure 6.1). Most frequently, all three hydroxyl groups of glycerol are esterified (as shown in Figure 6.1), resulting in triglycerides (also known as 'triacyl glycerols'). These are called neutral fats because the acidic carboxyl group of each fatty acid has been neutralised in forming an ester. Most naturally occurring neutral fats are triglycerides. Glycerides in which only one or two of the glycerol hydroxyl groups have been esterified are only found in small amounts.

simple and
mixed
triglycerides

How can triglycerides vary and why do they vary? Obviously, glycerol is present in all triglycerides, but many different triglycerides occur through variation in the fatty acid component. Furthermore, in a triglyceride the three fatty acids may all be of the same type. These are known as simple triglycerides. Those containing two or more different fatty acids are called mixed triglycerides. Most fats that are found naturally are mixtures of simple and mixed triglycerides.

The reason that triglycerides differ in their structures is that the structure of a fatty acid influences the properties of the fats containing it. Thus building fats from different fatty acids allows fats of differing properties to be formed. It is therefore important to have some understanding of the structure and properties of fatty acids.

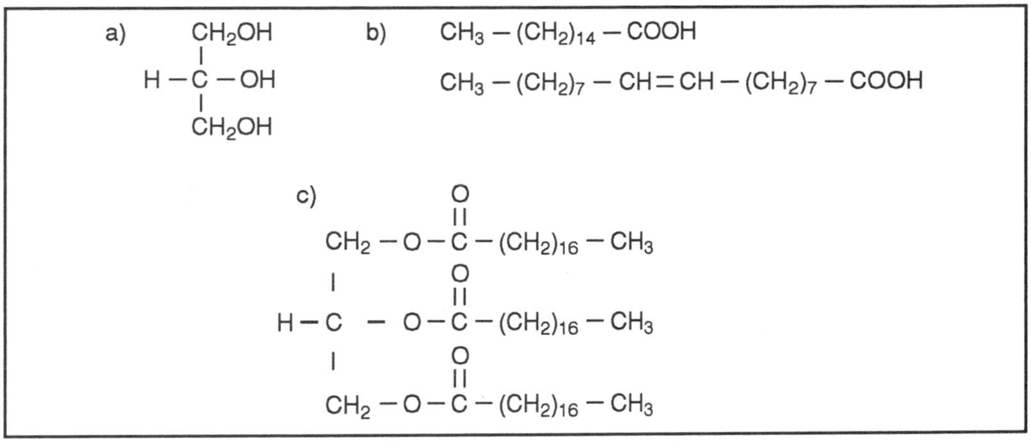

a)
$$CH_2OH$$
$$H-C-OH$$
$$CH_2OH$$

b)
$$CH_3-(CH_2)_{14}-COOH$$
$$CH_3-(CH_2)_7-CH=CH-(CH_2)_7-COOH$$

c)
$$CH_2-O-\overset{\overset{\displaystyle O}{\|}}{C}-(CH_2)_{16}-CH_3$$
$$H-C-O-\overset{\overset{\displaystyle O}{\|}}{C}-(CH_2)_{16}-CH_3$$
$$CH_2-O-\overset{\overset{\displaystyle O}{\|}}{C}-(CH_2)_{16}-CH_3$$

Figure 6.1 The components of neutral fats a) glycerol b) fatty acids c) neutral triglyceride.

saturated and unsaturated fatty acids

When naturally occurring neutral fats are analysed, different fatty acids may be isolated. All possess a long hydrocarbon chain (which is non-polar and hence hydrophobic) and a terminal carboxyl group (Figure 6.2). The carbon chain may be saturated or it may contain one or more (maximum 3) double bonds (ie unsaturated). Most of the naturally occurring fatty acids have an even number of carbon atoms with chains consisting of 4-22 carbon atoms. Chains of 16 and 18 carbon atoms are the most common. Fatty acids differ from each other in chain length and the number and position of their double bonds. Double bonds in naturally occurring fatty acids are in the cis configuration (Figure 6.2). Structures of typical fatty acids are shown in Figure 6.2: a saturated fatty acid will adopt a linear configuration; the presence of a double bond causes a rigid bend in the structure. A second double bond will produce a further bend. Fatty acids with more than one double bond are termed polyunsaturated.

polyunsaturated fatty acids

low solubility in water

Since fatty acids are largely non-polar hydrocarbon chains, they (and fats derived from them) are very poorly soluble in water. Although low-molecular weight fatty acids (such as the C4 butyric acid) are miscible with water, those with more than 6-8 carbon atoms are effectively insoluble in water. They are, however, soluble in non-polar solvents.

Table 6.1 lists some naturally occurring fatty acids, together with their melting points. This table gives the commonly used names of fatty acids, such as palmitic and stearic acids. There is also a systematic nomenclature which can be found in some biochemistry textbooks. For simplicity we have not included these here as they are not in common usage by biologists. Table 6.1 also contains convenient shorthand notations to describe important features of fatty acids. Thus a saturated fatty acid with 16 carbon atoms is designated '16:0', whilst an unsaturated fatty acid of 18 carbon atoms with one double bond is referred to as '18:1'.

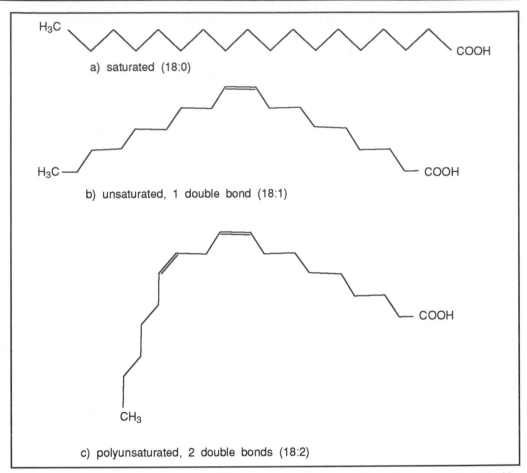

Figure 6.2 Common fatty acids a) saturated b) unsaturated, with one double bond c) polyunsaturated, with two double bonds. Note the simplification of the structures. Each 'angle' represents a carbon atom and the hydrogen atoms attached to these carbons are not shown. This is a common convention used in drawing hydrocarbon structures.

Π   As the size of the fatty acid increases what happens to its melting point? What effect does the presence of unsaturated bonds have on the melting point? Table 6.1 will provide the answers.

fats and oils   A striking consequence of change in length of the fatty acid is the increase in melting point with increase in size. Saturated fatty acids with 10 or more carbon atoms are solids at room temperature. Note also the effect of double bonds: for a given chain length, introduction of double bonds leads to lowering of the melting temperature. This is because the kink in the hydrocarbon chain of the unsaturated fatty acid prevents the close packing together of numerous hydrocarbon chains. This reduces the strength of interaction between fatty acids and hence less thermal energy is required to cause melting. The melting point of a neutral triglyceride is a consequence of its fatty acid composition. Thus the precise melting point, its sharpness and whether a lipid is liquid or solid at room temperature is controlled by the fatty acid composition. Triglycerides which are solid at room temperature are called fats, whereas those which are liquids are termed oils. Fatty acids are also components of most of the lipids present in membranes:

No. of carbon atoms		Name	Designation (*)	Melting point (°C)
Saturated fatty acids				
4		butyric	4:0	-4
8		caprylic	8.0	17
10		decanoic	10:0	32
12		lauric	12:0	44
14		myristic	14:0	52
16		palmitic	16:0	63
18		stearic	18:0	70
20		arachidic	20:0	75
Unsaturated fatty acids				
18	pne double bond	oleic	18:1	13
18	two double bonds	linoleic	18:2	-9
18	three double bonds	linolenic	18:3	-17
20	four double bonds	arachidonic	20:4	-40

Table 6.1 Some naturally occurring fatty acids and their melting points
* This is a convenient shorthand way to describe the number of carbon atoms and number of double bonds in fatty acids. Data obtained from Dawson R.M.C., Elliott, D.C., Elliott, W.H. and Jones K.M. Data for Biochemical Research, (3rd Edition), Oxford University Press, 1986.

hence differences in fatty acids lead to alterations in the fluidity of the membrane; this will be further discussed later.

Π Stearic acid, oleic acid and linoleic acid all have 18 carbon atoms; their melting points, however, differ greatly. Explain this divergence in melting points (Again use Table 6.1 to help you answer this).

Stearic acid is saturated, oleic acid has one double bond and linoleic acid has two double bonds. Double bonds produce kinks in the nonpolar fatty acid tails. As a consequence, fatty acids with double bonds will be less closely packed in the solid phase: this causes a decrease in intermolecular interaction and a lower melting point.

Given the influence of fatty acid structure on the properties of fats, it is not surprising that the food industry has spent many years studying lipids in order to obtain and design fats for particular properties.

Π Assume you have available large quantities of an edible oil, containing predominantly polyunsaturated fatty acids, which is liquid at room temperature. A potential client requires a neutral fat which is solid at room temperature, to form the basis of a margarine. Without specifying the precise approach, could your oil be made to be solid at room temperature? If so, how?

Careful thought may have enabled you to answer 'Yes'. The oil could be made solid at room temperature, providing you can reduce the number of double bonds. If this is done, the disruption to close packing (caused by the bends in the unsaturated fatty acids) will be eliminated and the oil will solidify to a fat. The double bonds can be removed by hydrogenation, as shown:

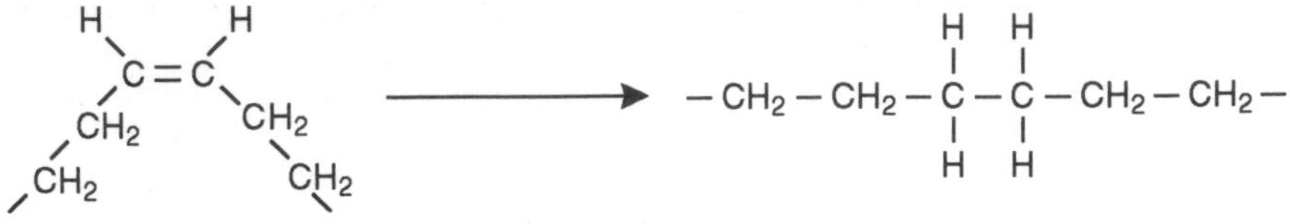

This is precisely what is done in the synthesis of both margarines and soaps.

SAQ 6.1

1) Which of the following neutral fats is likely to be liquid at room temperature?

2) Which is likely to still be a solid at 40°C? (You will find the data in Table 6.1 helpful).

Neutral fats containing:

a) two molecules of oleic acid (18:1) and one molecule of stearic acid (18:0);

b) three molecules of palmitic acid (16:0);

c) two molecules of linoleic acid (18:2) and one molecule of lauric acid (12:0).

NB Do not try to learn the names of all the fatty acids; they are used here solely for identification.

SAQ 6.2

Based solely on the properties of the fatty acids, write the structures of neutral fats which should be:

1) Solid at room temperature (20°C), whilst containing only saturated fatty acids.

2) As a), but also liquid at 37°C, it should 'melt in the mouth'!

3) An unsaturated fat which is an oil at 20°C.

essential fatty acids

Mammals can synthesise saturated fatty acids and fatty acids which contain one double bond. They are unable to synthesise fatty acids containing two or more double bonds: these must therefore be obtained in the diet. These fatty acids are termed essential fatty acids. In contrast, plants and microorganisms are able to synthesise all the fatty acids they require.

We have now examined the structure of neutral lipids and how their structure affects their properties. We have also probed, in outline terms, how the food industry can manipulate neutral fats to achieve fats with particular properties. We have not yet considered the benefits of neutral fats to the organism which makes them.

neutral fats storage of energy

Neutral fats provide an excellent store of energy, for both animals and plants, which can be rapidly mobilised to provide energy. Since we have already described storage forms of polysaccharides, why are neutral lipids used as well? Remember that neutral fats are highly hydrophobic and thus insoluble in water. They coalesce into droplets which are essentially anhydrous (ie no water present). This contrasts markedly with starch and glycogen which interact favourably with water and are hydrated. Neutral fats are also highly reduced when compared to polysaccharides. When completely oxidised to carbon dioxide and water, fatty acids yield about 38 kJ per gram, whereas carbohydrates and proteins only produce about 17 kJ per gram. These figures are for dried carbohydrates. Hydrated glycogen (as found in the body) contains substantial amounts of water. About 3g wet weight of glycogen contains only about 1g of glycogen. When this is included in the comparison, neutral fats are found to contain about 6 times more energy per gram than glycogen. This achieves a substantial weight saving, for the provision of a given energy store.

fat globule

In animals, neutral fats are stored in the cytoplasm of fat cells. Numerous triglycerides coalesce to form a globule, which may occupy the vast majority of the volume of the cell. These specialised cells are usually packed together in adipose tissue. Note also that

adipose tissue

neutral fat is an excellent heat insulator. Thus adipose tissue provides an energy store and also helps to lessen wastage of energy through heat loss.

∏   Given the behaviour of neutral fats to form globules in fat cells which are localised beneath the skin and their apparent metabolic role, write down explanations why they are useful in the following situations,
    1) to birds;
    2) to dolphins or whales;
    3) in seeds such as castor beans.

1)   The principal benefit of neutral fats for birds is the high energy content of neutral fats. Remember our comment that, per gram of stored material in its normal *in vivo* form, fat contains some 6 times more energy than glycogen. If glycogen replaced fat reserves in birds they would be a lot heavier: their ability to fly might disappear! Sustaining a bird on a long flight (eg during migration) will also be important.

2)   Whilst an energy store is clearly important, weight is much less of an issue for an animal in water, given the buoyancy effect of water. Here, insulation against heat loss to the surrounding water (which would be expected to be rapid) is vital: the layer of fatty tissue (blubber) ensures minimal loss of energy as heat through the skin.

3)   Oil storing seeds, such as castor bean and rape, are variations of the general theme of providing an energy store in seeds to ensure that, during germination, there is sufficient energy available to enable the seedling to survive until it becomes self sufficient. There is no clear survival advantage in using fats rather than starch. Equally, there is no particular disadvantage. Thus amongst seeds we also find those that store starch (eg cereals). Oil containing plants are part of the richly varied plant resources which man can utilise, and whose diversity we must be careful to preserve.

6.2 Phospholipids

membranes

Phospholipids are found almost exclusively in membranes and are the most important component of membranes. In these lipids, one of the hydroxyl groups of glycerol (the C -3) is esterified to phosphoric acid whilst the remaining hydroxyl groups are esterified to fatty acids; the product is called phosphatidic acid (Figure 6.3). Free phosphatidic acid only occurs in small amounts in cells, but is an important intermediate in the biosynthesis of phospholipids.

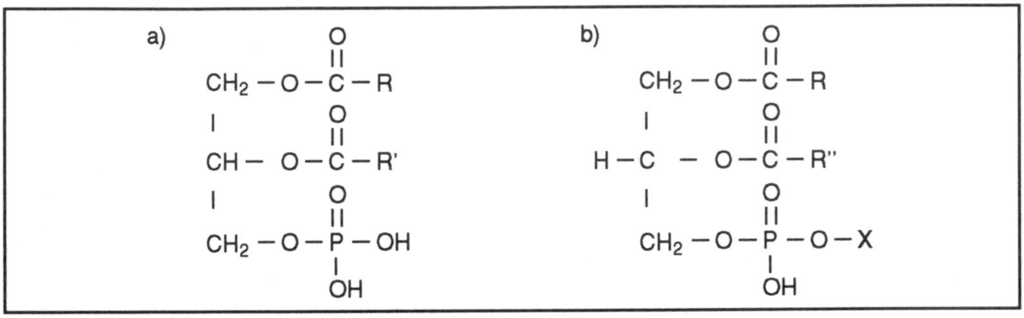

Figure 6.3 General structure of a phospholipid a) phosphatidic acid b) phospholipid, R' and R" indicate fatty acid hydrocarbon chains. The group identified as X is normally an ester resulting from reaction with an alcohol containing molecule.

phosphatidyl choline

Naturally occurring phospholipids are of the general structure shown in Figure 6.3b. The fatty acid which is esterified to the C -2 carbon atom of glycerol is often an unsaturated fatty acid. Normally there is an additional group present in a phospholipid by substitution on the phosphoric acid group (represented as X in Figure 6.3b) This involves esterification of the hydroxyl group of an alcohol to the phosphoric acid group. A typical phospholipid, phosphatidyl choline, is shown in Figure 6.4. Examples of the hydroxyl-containing compounds that are often found in these naturally occurring phosphoglycerides are choline, ethanolamine, serine and inositol. We have already met the structure of serine, it is an amino acid. Ethanolamine has the structure $HO-CH_2CH_2-NH_3^+$. Inositol is a polyhydroxy derivative of a monosaccharide rather like sorbitol. In each case the addition of these to the phosphate groups of phosphatidic acid strengthens the polar nature of the 'head' of the phospholipid.

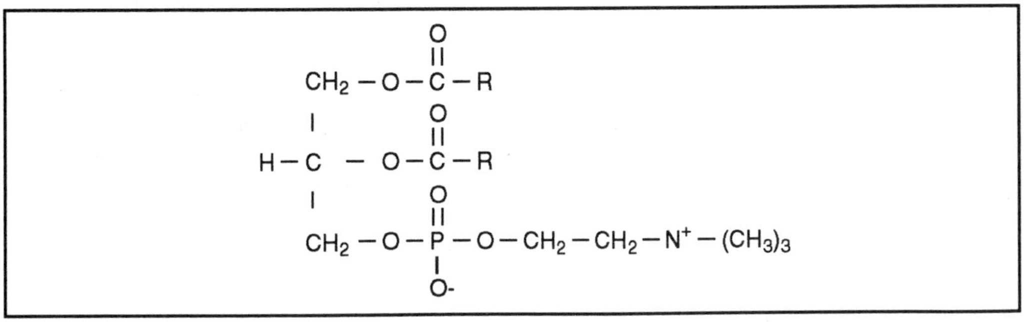

choline
ethanolamine
serine
inositol

Figure 6.4 A typical phospholipid, phosphopatidyl choline. Note that, at neutral pH, the phosphate group will have a negative charge.

polar groups

What is the significance of the phosphoric acid and alcohol containing groups? You will recall that fatty acid chains are extended and linear (although double bonds introduce bends into them) and very nonpolar (ie hydrophobic). The substituted phosphoric acid group, on the other hand, is highly polar: in every case there is a polar phosphate group containing, at neutral pH, a negative charge (as shown in Figure 6.4), substituted either by groups with ionised components (amine, carboxyl groups), or with hydroxyl groups. This produces a highly polar 'head', which reacts favourably with water. Phospholipids thus have two distinct components, a hydrophilic polar 'head' and a hydrophobic nonpolar 'tail'. Substances with both polar and nonpolar properties are termed

amphipathic

amphipathic: most membrane lipids are amphipathic.

Amphipathic molecules behave in a characteristic way when added to water. The polar head associates with water molecules, whilst the nonpolar chains 'prefer' a nonpolar environment devoid of water. This is because the nonpolar component disrupts the bonding between water molecules; this disruption is minimised if all the nonpolar

hydrophobic interaction

components are associated together, away from contact with water. This is the basis of what are often called 'hydrophobic interactions' or 'hydrophobic forces'. We described these type of interactions when we were discussing protein structure in Chapter 3. Phospholipids form characteristic structures when mixed with water. These aggregates may be spherical or bilayers (Figure 6.5) in which the polar heads are in contact with the aqueous phase and the nonpolar tails are hidden from the aqueous phase due to the favoured aggregation of hydrophobic groups. These structures are described in the next chapter when we discuss membranes.

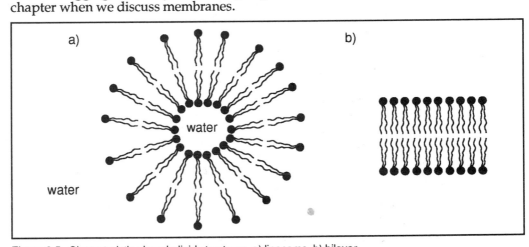

Figure 6.5 Characteristic phospholipid structures a) liposome b) bilayer.

plasmalogens

Other types of phospholipids are widespread and will be briefly mentioned. The plasmalogens (Figure 6.6) are based on glycerol and are similar to the phosphoglycerides described except that the hydrocarbon component attached to the glycerol C -1 carbon is linked by an ether (rather than an ester) linkage. The overall shape and amphipathic character is not affected by this. Plasmalogens are commonly found in the membranes of nerve cells.

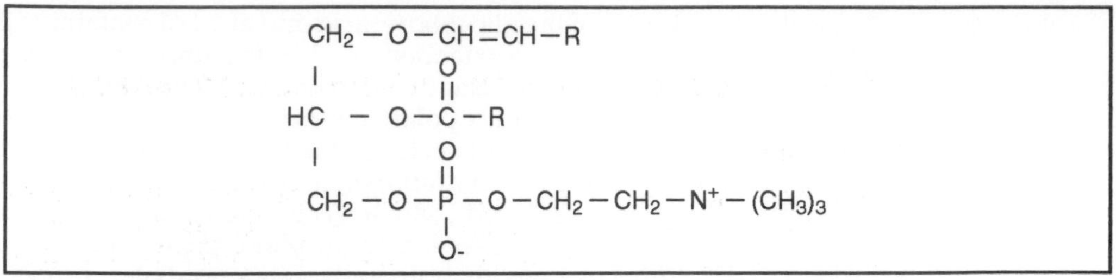

$$CH_2 - O - CH = CH - R$$
$$|$$
$$\qquad\qquad O$$
$$\qquad\qquad ||$$
$$HC \quad - \quad O - C - R$$
$$|$$
$$\qquad\qquad O$$
$$\qquad\qquad ||$$
$$CH_2 - O - P - O - CH_2 - CH_2 - N^+ - (CH_3)_3$$
$$|$$
$$O-$$

Figure 6.6 A typical plasmalogen. Note that the molecule is amphipathic, having polar and nonpolar components; it thus behaves in a similar manner to phospholipids.

sphingolipids
sphingosine

Sphingolipids are another important group of naturally occurring phospholipids. Sphingolipids do not contain glycerol but instead are based on sphingosine, which contains a long hydrocarbon chain (like a fatty acid). They also contain a hydroxyl and an amino group (Figure 6.7a), to which a fatty acid and, in some cases, a substituted phosphoric acid group may be attached. A typical sphingolipid is shown in Figure 6.7b.

glycolipids

Other sphingolipids contain carbohydrates at the polar end of the molecule and are known as glycolipids. They are especially abundant in plant cell membranes and in nerve and brain cells. The overall result is of an amphipathic molecule with a polar head group and two long chain hydrocarbon nonpolar tails. Although phosphoglycerides and sphingolipids may seem to differ considerably in their structure, they in fact have a high degree of similarity in so far as they have a hydrophobic portion and a hydrophilic (polar) end. They will also behave in a similar manner to that shown for phosphoglycerides when mixed with water (Figure 6.5). Thus these different individual molecules (and there are many variants of these 'typical' structures) are simply alternative ways of achieving molecules with the property of spontaneously forming bilayers. This is vital for the production of membranes.

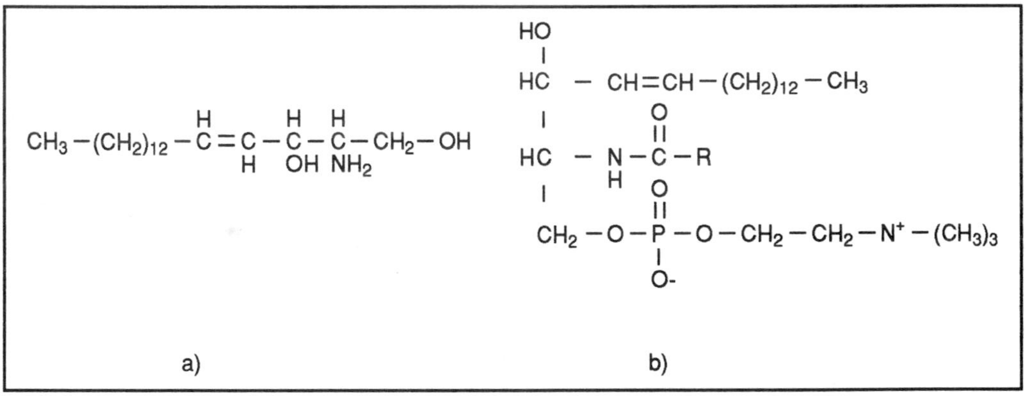

$$CH_3 - (CH_2)_{12} - \overset{H}{\underset{H}{C}} = \overset{H}{C} - \overset{H}{\underset{OH}{C}} - \overset{H}{\underset{NH_2}{C}} - CH_2 - OH$$

a)

$$HO$$
$$|$$
$$HC \quad - CH = CH - (CH_2)_{12} - CH_3$$
$$|$$
$$\qquad\qquad O$$
$$\qquad\qquad ||$$
$$HC \quad - N - C - R$$
$$|\quad\;\; H \;\; O$$
$$\qquad\qquad\qquad ||$$
$$CH_2 - O - P - O - CH_2 - CH_2 - N^+ - (CH_3)_3$$
$$|$$
$$O-$$

b)

Figure 6.7 Sphingosine and sphingolipids a) sphingosine b) a typical sphingolipid (sphingomyelin).

6.3 Steroids, carotenoids, vitamins and waxes

6.3.1 Steroids

Steroids are included amongst lipids because their solubility properties are similar to those of neutral fats and phospholipids. Their structures are, however, very different. They are derived from the condensation of 5 carbon isoprene units (Figure 6.8a). All such compounds are referred to as isoprenoids. Steroids have a complex structure which consists of four fused rings of carbon atoms (Figure 6.8b). This basic skeleton of carbon rings may contain double bonds and various side chains. Those steroids which have an 8-10 carbon atom side chain at position 17 and also a hydroxyl group at position cholesterol 3 are classified as sterols. Cholesterol (Figure 6.9) is the most abundant animal steroid, being found in high concentration in nervous tissue and most membranes. It has a structural role in membranes, influencing membrane fluidity and acting as a kind of 'antifreeze' (See also Chapter 7).

a)

$$CH_2 = C - CH = CH_2$$
$$|$$
$$CH_2$$

b)

Figure 6.8 a) Isoprene b) The steroid 'nucleus'.

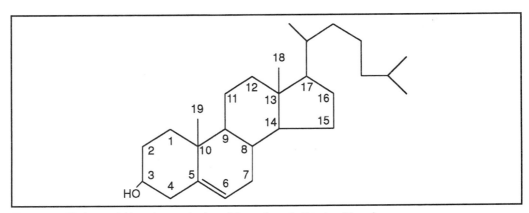

Figure 6.9 Cholesterol. Note the numbering of the carbons in the steroid nucleus.

animal sex
hormones

bile acids

nonpolar
nature of
sterols

Many animal hormones are also steroids. For example, the female and male sex hormones oestradiol and testosterone (Figure 6.10). Bile acids, such as cholic acid (Figure 6.10) and related molecules also contain the sterol nucleus. These molecules are responsible for the detergent action of bile. Steroids and sterols are predominantly nonpolar (with the exception of the bile acids), polar groups being confined to the solitary hydroxyl group. Thus they are only weakly amphipathic.

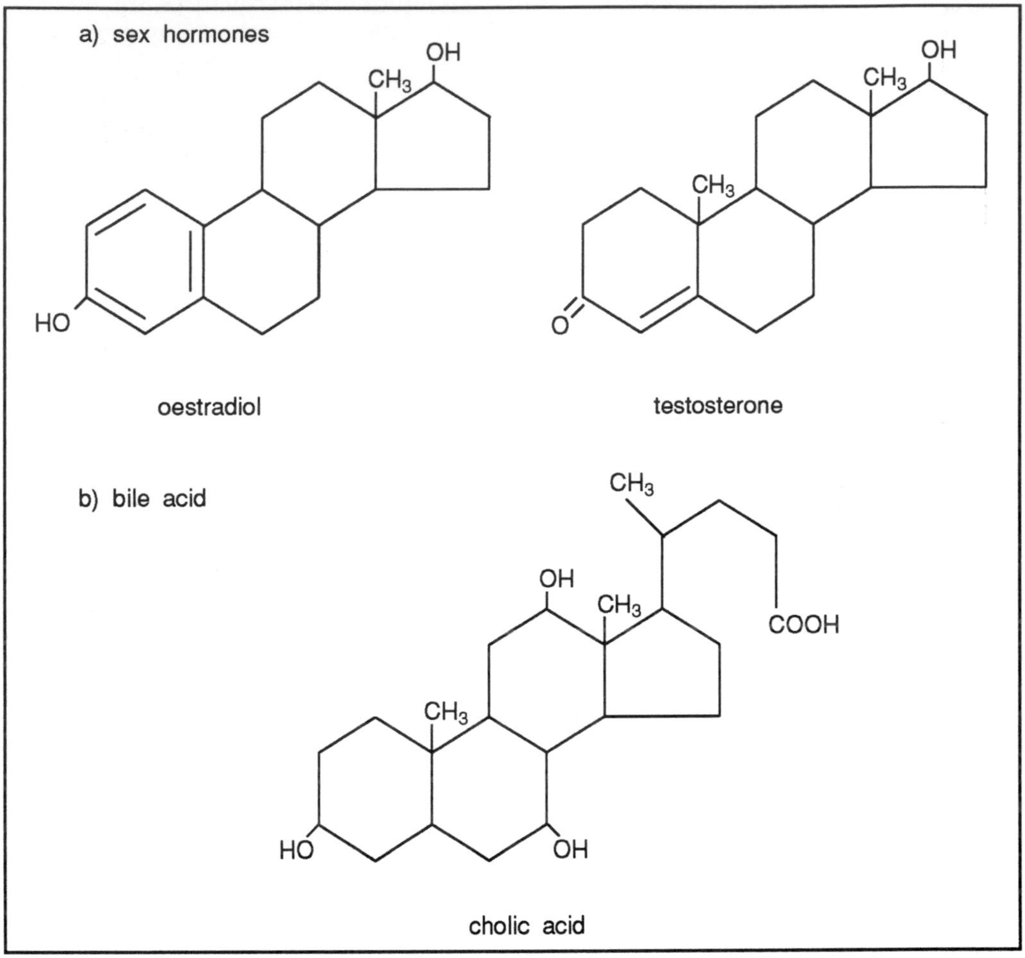

a) sex hormones

oestradiol

testosterone

b) bile acid

cholic acid

Figure 6.10 Steroids. a) The female and male sex hormones oestradiol and testosterone. b) Cholic acid, responsible for the detergent action of the bile.

SAQ 6.3

Match the following names of lipid groups (1-5) to the appropriate structures a-e:

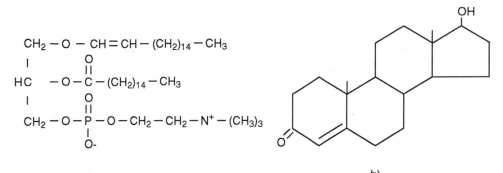

$$CH_2 - O - CH = CH - (CH_2)_{14} - CH_3$$
$$| \qquad\qquad O$$
$$\qquad\qquad ||$$
$$HC \quad - O - C - (CH_2)_{14} - CH_3$$
$$| \qquad\qquad O$$
$$\qquad\qquad ||$$
$$CH_2 - O - P - O - CH_2 - CH_2 - N^+ - (CH_3)_3$$
$$| \\ O\text{-}$$

a)

b)

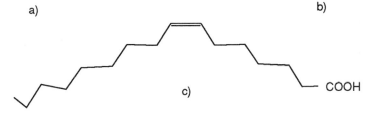

c)

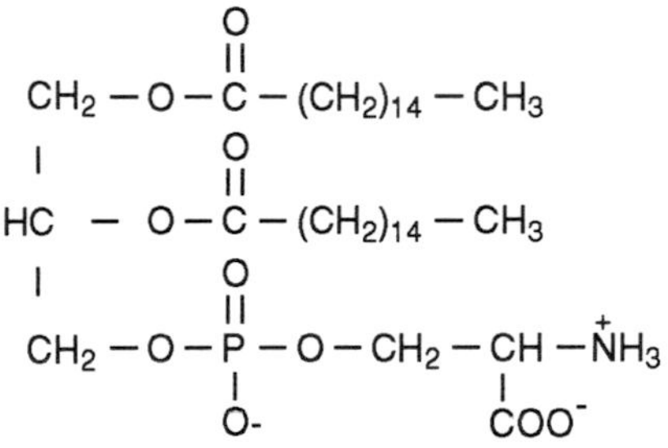

$$CH_2 - O - \overset{\overset{\textstyle O}{||}}{C} - (CH_2)_{14} - CH_3$$
$$| \qquad\qquad O$$
$$\qquad\qquad ||$$
$$HC \quad - O - C - (CH_2)_{14} - CH_3$$
$$| \qquad\qquad O$$
$$\qquad\qquad ||$$
$$CH_2 - O - P - O - CH_2 - CH - \overset{+}{N}H_3$$
$$| \qquad\qquad\qquad |$$
$$O\text{-} \qquad\qquad\quad COO^-$$

d)

$$CH_2 - O - \overset{\overset{\textstyle O}{||}}{C} - (CH_2)_{14} - CH_3$$
$$| \qquad\qquad O$$
$$\qquad\qquad ||$$
$$HC \quad - O - C - (CH_2)_{14} - CH_3$$
$$| \qquad\qquad O$$
$$\qquad\qquad ||$$
$$CH_2 - O - C - (CH_2)_{14} - CH_3$$

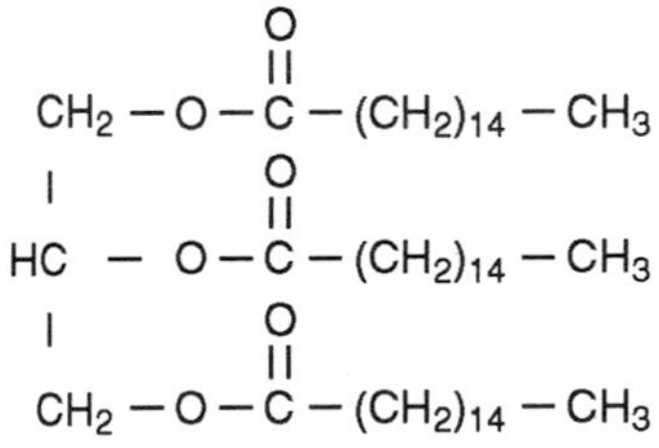

e)

1) triglyceride;

2) phospholipid;

3) plasmalogen;

4) unsaturated fatty acid;

5) steroid.

6.3.2 Carotenoids and vitamins

retinol
β carotene

Vitamins are substances which animals require in trace amounts for growth and normal health. Vitamins are classified as either water - soluble or fat - soluble. Vitamin C and those of group B are water soluble. The fat - soluble vitamins includes vitamins A, D, E and K. Vitamin A (retinol), can be derived from β carotene, an example of a carotenoid (Figure 6.11).

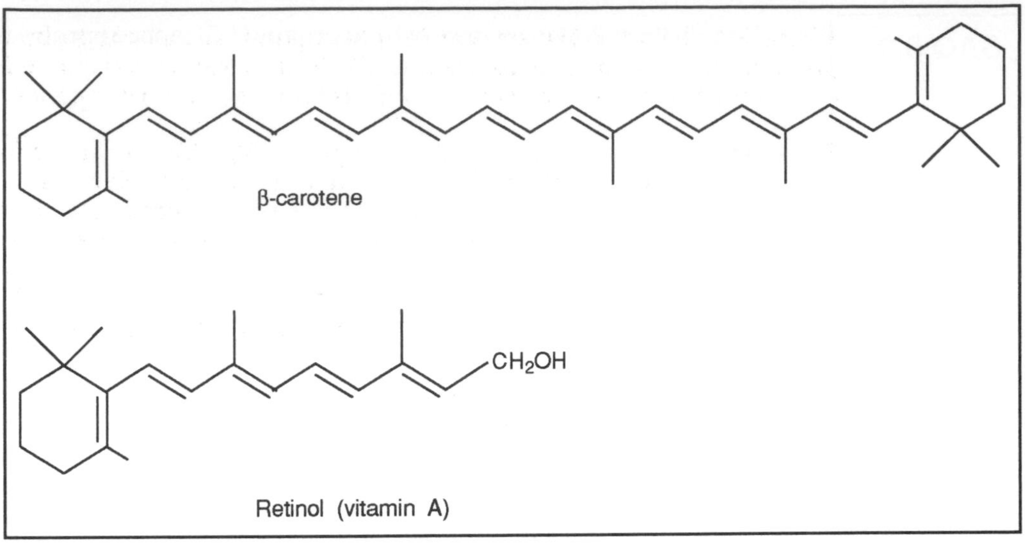

Figure 6.11 The structures of β carotene and retinol (like the steroids, they are based on isoprene units).

damage to
epithelial and
night blindness
Carotenoids are plant pigments involved in photosynthesis. β carotene is found in substantial amounts in most green and yellow/orange vegetables. Vitamin A is also abundant in fish oils. Vitamin A is essential for good vision and also maintenance of epithelial tissues. Night blindness is an indication of vitamin A deficiency.

Vitamin D is another fat-soluble vitamin. Active forms of vitamin D are produced by irradiation (by sunlight) of naturally occurring sterols. Its structure is given in Figure 6.12. Deficiency of vitamin D causes rickets which is characterised by brittleness of bones. The underlying cause of this is the influence of vitamin D on serum calcium and phosphorus levels. Vitamin D has a major effect on the absorption of calcium in the intestine. Deficiency of the vitamin means that bone calcification does not occur. Fish liver oils are an excellent source of vitamin D.

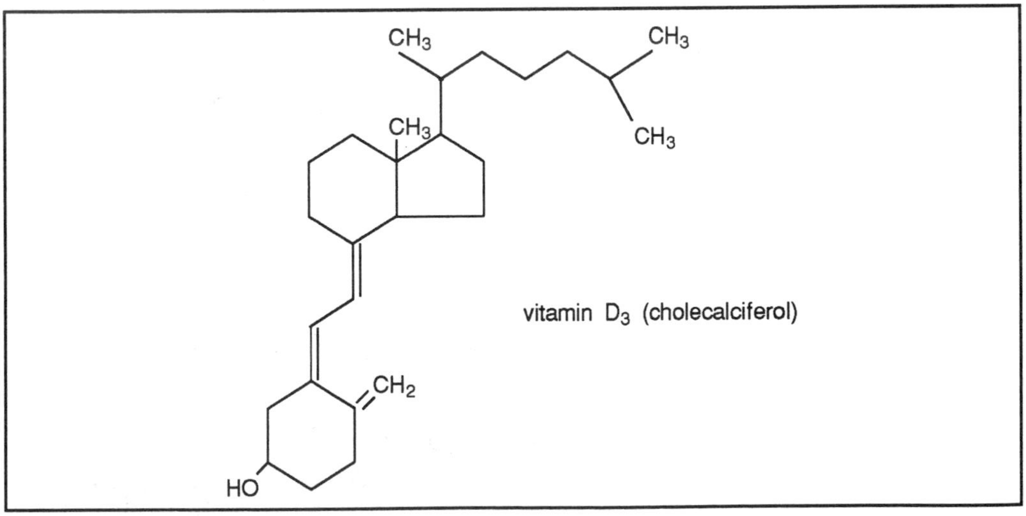

rickets

Figure 6.12 The structure of vitamin D.

Vitamin E (Figure 6.13), present in wheat germ, is thought to be involved in the prevention of damaging oxidations. Deficiency causes various effects, including erythocyte haemolysis and sterility. Effects however vary with species.

Plants are a good source of vitamin K (Figure 6.13), which is needed for normal blood clotting. Deficiency of vitamin K leads to defective prothrombin, resulting in its failure to be activated to thrombin. This is an essential step in normal blood clotting.

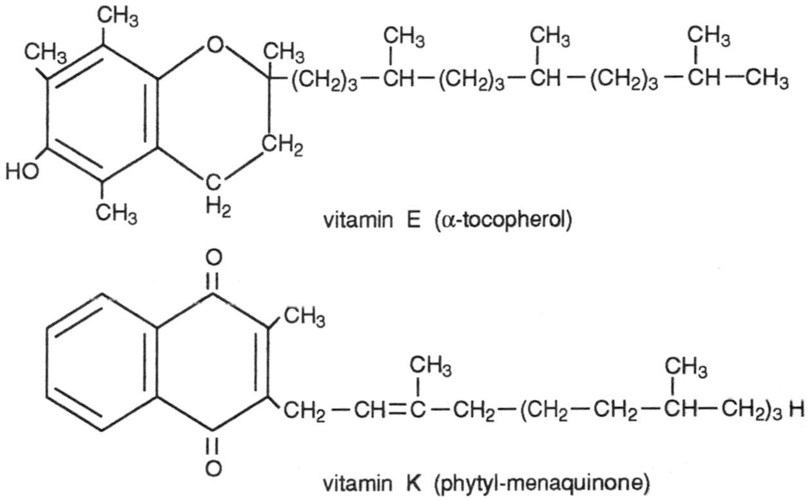

Figure 6.13 Structure of vitamin E and vitamin K. Note that the sidechains of both these vitamins are derived from isoprene.

prostaglandins

Lipids can also act as hormones and hence exert a regulatory role in animals. We have already met with some of the steroid hormones. Other important lipid hormones are the prostaglandins (Figure 6.14). At very low concentrations these influence a variety of metabolic processes, including both vasodilation and vasoconstricting activity.

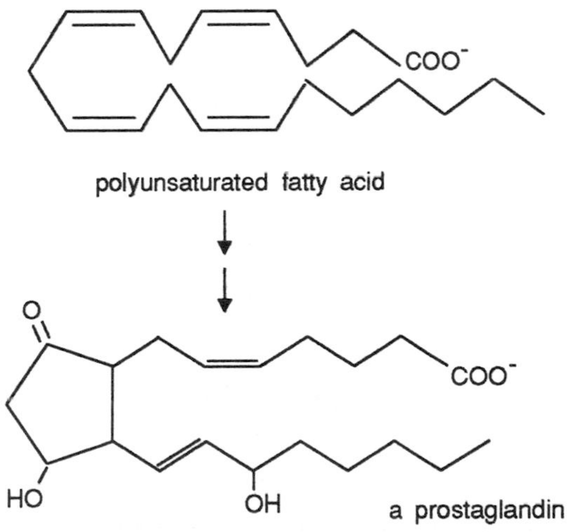

Figure 6.14 A typical prostaglandin. (Note that they are produced by modification of polyunsaturated fatty acids).

6.3.3 Waxes

waxes   Waxes are esters of fatty acids and hydroxyl containing compounds other than glycerol. They are highly nonpolar. Myricyl palmitate occurs in beeswax. Waxes are used to provide a water repelling layer on insects. On leaves they form a shiny layer (cuticle) which helps to minimise water loss.

6.4 Lipoproteins

consistent proportion of protein and lipid

hydrophillic interaction

Lipoproteins consist of proteins coupled to a lipid component, which may be fatty acid, neutral fat, phospholipid or cholesterol. Within each category of lipoprotein, they tend to have consistent proportions of components. This suggests that binding between lipid and protein is not random or haphazard. The protein constituent typically has a high proportion of nonpolar amino acid residues. Since neither ionic nor covalent bonds appear to be involved in the linkage between lipid and protein, it is likely that these nonpolar amino acid residues are responsible for the association through hydrophobic interactions.

chylomicrons
VLDL
LDL
HDL
serum albumin

Lipoproteins have two main roles. The first is the transport of lipids. In the bloodstream, lipids are transported as chylomicrons, very low density lipoproteins (VLDL), low density lipoproteins (LDL), high density lipoproteins (HDL) or as complexes of albumin (a highly abundant serum protein) and free fatty acids. Chylomicrons are stable particles of about 200 nm diameter, containing mainly neutral triglycerides, together with some phospholipid, cholesterol and protein. High density lipoproteins also occur in blood.

Lipoproteins are also components of membranes, such as those of the mitochondria, endoplasmic reticulum and nuclei. The inner mitochondria membrane, containing the electron transport system, contains large amounts of lipoproteins.

| SAQ 6.4 | Which of the following are derived from isoprene units? |

1) sphingosine.

2) cholesterol.

3) prostaglandin.

4) vitamin A.

5) palmitic acid.

6) vitamin D.

7) vitamin E.

8) vitamin K.

9) testosterone.

Summary and objectives

In this chapter we have examined the structure of the major groups of lipids and have outlined the metabolic and structural roles played by lipids. We paid particular attention to the importance of the nature of the fatty acids in neutral fats and phospholipids on the overall properties of these molecules. We also showed that as a result of the amphipathic properties of the phospho- and sphingolipids, they have a tendency to form bimolecular layers or aggregated structures in aqueous environments.

Now that you have completed this chapter, you should be able to:

* identify the major groups of lipids including the neutral lipids, phospholipids, sphingolipids, sterols and carotenoids;

* explain the effects of chain length and degree of unsaturation on the properties of fatty acids and neutral lipids;

* describe the formation of lipid bilayers from amphipathic lipids;

* describe in general terms the roles of lipids as energy stores, components of membranes, vitamins and hormones and to give examples of each.

Biological membranes

Biological membranes

Membranes are vitally important components of living systems. They are made largely of lipids and proteins although some membranes contain significant quantities of carbohydrates. In this chapter we will consider the roles of membranes and examine their structure. You will learn that membranes are examples of structures which are made of a variety of biochemicals and that they have important roles to play in organising and compartmentalising the activities of cells. Because of this fundamental importance, it is essential that you understand the structure and properties of membranes.

7.1 Roles of membranes

special
property of
lipids

The role fulfilled by lipids in forming membranes is a vital function for cells. Whilst some of the other functions of lipids (such as energy storage discussed in Chapter 6) are, or could be, performed by other molecules, no other groups of naturally occurring molecules can form membranes. Without membranes, there would be no cells. Membranes, of course, allow the inside of a cell to be kept distinct from the outside. It also allows internal compartmentalisation to take place. Membranes represent the walls of the cellular factory. We shall begin considering membranes by summarising the roles of membranes and then examine how these roles are achieved.

Permeability barrier

influx

efflux

The cell membrane (also known as the plasma membrane or plasmalemma) is a permeability barrier, which can regulate the influx and efflux of most materials into and out of cells. The outer surface of the membrane is the area of contact of the cell with its surrounding medium. Everything the cell requires must go through this membrane.

Regulation of transport across membranes

gates

channels

pumps

Whilst some materials cross membranes easily and will diffuse according to concentration gradients, many compounds cannot do so. Thus membranes have intrinsic permeability properties. Membranes regulate transport of materials by a variety of systems such as gates, channels and pumps. These are frequently specific for particular compounds. In addition, membrane bound vesicles may be formed to allow ingestion of material or the reverse process (elimination of material) to occur.

Subcellular compartments

Intracellular membranes allow the internal space of the cell to be subdivided or compartmentalised. This allows particular processes (eg generation of ATP) to take place in a confined place in specialised organelles (eg mitochondria).

Signal recognition and transduction

Since the plasma membrane is the area of the cell in contact with the external environment, it is here that reception of chemical signals must take place. The 'signal', such as a hormone, must be recognised and a suitable biochemical response triggered within the cell. This is accomplished through specific proteins known as receptors, which are located on the outer surface of the cell membrane. An example is the receptor for the hormone insulin. Thus membranes may have particular proteins within, or on, them, orientated in a particular manner. Some of these proteins may be enzymes.

receptors

Metabolic processes may be structurally organised

One attraction of locating enzymes in membranes is that it then becomes possible to position them optimally with respect to other enzymes. Take a situation where there is a metabolic pathway in which the product of enzyme 1 is the substrate of enzyme 2; the product of enzyme 2 is the substrate of enzyme 3, and so on. The probability of material passing rapidly through this pathway may be improved by fixing the enzymes in a membrane (Figure 7.1a) as compared to the sequence of reactions occurring in the cytoplasm of the cell (Figure 7.1b). This approach is exemplified by the careful arrangement of electron transport components of the respiratory chain, within the inner mitochondrial membrane.

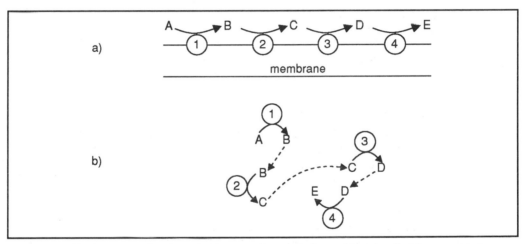

Figure 7.1 a) Enzymes associated together in a membrane may be positioned so that they operate in a sequential, organised manner, perhaps eliminating free diffusion of substrates. b) If the same reactions take place in solution in the cytoplasm, diffusion from one enzyme to the next is required.

Cell - cell interaction

We have already said that the outside of the plasma membrane is the part of the cell which is in contact with the 'outside world'. Thus interaction between cells, such as aggregation to form tissues, must involve the cell membrane. There is ample evidence that particular cell types can be recognised by individual cells.

All of these aspects and properties of membranes are further developed in the companion BIOTOL text 'The Infrastructure and Activities of Cells'. In this section, we will show, in general terms, how this multitude of roles can be provided by membranes. It should be clear however that membranes are very important in providing the basis for many of the properties of cells. In our chemical factory analogy, membranes control what comes into the factory and what leaves. They divide the factory into departments and control the transfer of material between departments. They also enable the machines of the factory (enzymes) to be organised into highly efficient production lines.

7.2 Composition of membranes

Most biological membranes contain about 40% lipid and 60% protein, although marked variations can occur. Table 7.1 shows the composition of some typical membranes. The high protein content may seems surprising.

	Lipid	Protein	Carbohydrate
	%	%	%
Red blood cell membrane (human)	43	49	8
Myelin	79	18	3
Gram +ve bacteria	25	75	10
Inner mitochondrial membrane	24	76	1-2
Outer mitochondrial membrane	48	52	2-4
Liver plasma membrane	42	53	5-10

Table 7.1 Composition of some typical membranes. Data taken from Guidotti, G (1972) Annual Reviews of Biochemistry 41, 731.

Proteins are responsible for many of the more specialist functions identified earlier. The lipid fraction of the membrane consists mainly of phospholipids, although other lipids are frequently present (Table 7.2). The plasma membranes of animal cells contain considerable amounts of cholesterol. Sterols are completely absent from bacterial membranes. The ratio of the different components in the membrane is characteristic of:

*   the type of organism (ie the source);

*   the type of cell (eg liver cell, red blood cell);

*   the type of membrane (plasma membrane, mitochondrial inner membrane).

However, for a given type of membrane, the ratio between various phospholipids (several are usually present) and cholesterol is approximately constant.

∏ Write down the reasons why you might think that the composition of biological membranes is likely to be genetically determined?

The fact that the ratio of different types of membrane lipids (phosphatidyl choline, phosphatidyl ethanolamine, sphingolipids etc) is constant for each type of membrane strongly suggests that membrane composition/structure are under genetic control. The same argument would also apply to the protein content of membranes. It is clearly

	Phospholipid	Glycolipid	Cholesterol
	%	%	%
Red blood cell (rat)	61	11	28
Myelin (rat)	41	42	17
Spinach chloroplast (+ 8% pigments)	12	80	trace

Table 7.2 Lipid component of membranes

advantageous to ensure, through genetic control during synthesis, that the 'right' (ie most appropriate) molecules are incorporated into a particular membrane. Note that other factors, such as diet, can influence the lipid composition, but only in terms of the fatty acid composition of the various membrane lipids.

7.3 Membrane lipids and aggregates

amphipathic nature of some lipids

Membrane lipids, especially phospholipids and sphingolipids, possess a polar, hydrophilic head and a long nonpolar, hydrophobic tail. As we described in Chapter 6, they are said to be amphipathic. Because of these two contrasting regions, they will behave in particular ways when added to water. They will associate into aggregates in which the polar heads are exposed on the surface and the nonpolar tails are orientated inwards and are thus 'hidden' from the aqueous environment. Various structures can be formed.

7.3.1 Monolayers

If amphipathic lipids are added to water, they will not dissolve but will tend to form a monolayer at the surface of the water. The polar heads are in contact with the water, the nonpolar tails protruding into the air phase. They tend to spread extensively (hence the appearance of oil on the surface of ponds and rivers; 'spreading oil on troubled waters' serves to lessen waves by lowering the surface tension of the water).

7.3.2 Micelles

sonication

Micelles (Figure 7.2a) are colloid - sized particles in which the polar groups interact with the surrounding aqueous phase, with the hydrocarbon on the inside. They will form if amphipathic lipids are sonicated or blended with water. Stable micelles are particularly formed by fatty acids, which, because of the nonpolar hydrocarbon chain, are very poorly soluble in water. Micelles are still in equilibrium with the free fatty acids from which they form. Phospholipids do not tend to form stable micelles, probably because there are two fatty acid chains to be shielded by the polar head group: the hydrocarbon tails of phospholipids are thus too bulky. Soaps work on the micelle principle, they are salts of fatty acids which can enclose a droplet of nonpolar fat. They thereby give it a high surface charge (Figure 7.2b), so that similar particles are repelled. The fat droplets therefore do not coalesce into large drops. All detergents (of which soaps are one type) work on this principle.

soaps

Rdetergents

We used the word sonicated - do you know what it means?

In the biological sciences we can use the energy of ultra high sound to break up large biological structures (eg cells, organelles). If we apply ultra high sound to a mixture of lipids and water, the sound agitates the mixture and the lipid molecules are dispersed in the water. This process is called sonication.

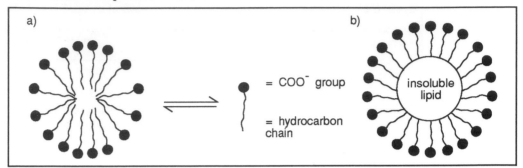

Figure 7.2 (a) Micelles, formed by fatty acids in water. Note that the hydrophobic tails are shielded from water by the polar carboxyl heads. Micelles of fatty acids are in equilibrium with the (poorly soluble) fatty acids from which they form. (b) Emulsification of insoluble lipid by a soap; the fatty acids of the soap coat the lipid and aid its solubilisation.

7.3.3 Liposomes

study of lipid bilayer permeability

If water were to be enclosed in a micelle. it would tend to destabilise it. If, however, a double layer of amphipathic lipid is formed (Figure 7.3), enclosed water does not destabilise the structure and it remains stable. Such aggregates, formed when water/phospholipid mixtures are sonicated or blended, are called liposomes. Both entrapped and surrounding water is bordered by the polar parts of amphipathic lipids. These liposomes can be formed by phospholipids. Liposomes are very useful for studying membrane permeability. If liposomes are generated in an aqueous phase containing dissolved compounds (eg amino acids), some of the dissolved amino acid will be entrapped within the liposomes. It is possible to transfer these to medium lacking the amino acid (ie change the external medium). The rate of leakage of the amino acid can then be studied. Studies of this type enable the permeability of lipid bilayers to different compounds to be determined.

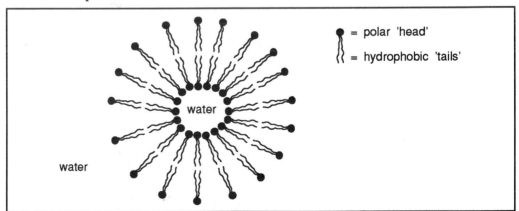

Figure 7.3 A liposome, as formed by phospholipids in water. Note the double layer of phospholipid so that all hydrophobic tails are together and protected from water by the polar heads. Liposomes provide an excellent experimental system for studying membrane permeability.

The permeability studies show that:

- lipid bilayers are highly impermeable to polar or ionic compounds;

- permeability can be correlated with solubility in nonpolar solvents.

Molecule	Permeability Coefficient $\mu m.sec^{-1}$
Water	40
Tryptophan	1×10^{-3}
Glucose	6×10^{-4}
Chloride (Cl⁻)	8×10^{-5}
Sodium ion (Na⁺)	1×10^{-8}

Table 7.3 Permeability coefficients of some low molecular compounds through lipid bilayers
After Stryer, L. Biochemistry 3rd Edition, Freeman, 1988

Water is a notable exception, in that it readily permeates lipid bilayers (perhaps because of its size). Table 7.3 contains permeability coefficients for a number of ions and molecules. These studies indicate that transport of polar compounds across membranes must involve some form of assistance: this is clearly one role of membrane bound proteins.

clinical use of liposomes

Liposomes may have clinical uses. Since they are impermeable to many compounds, yet can fuse with plasma membranes, it is possible to use them to transport materials into cells. If liposomes could be targeted to particular cells (perhaps by making use of the ability of cell membranes to recognise particular membranes), it would be possible to use them to deliver drugs or enzymes in a highly specific manner.

7.3.4 Lipid bilayers

spontaneous assembly of bilayers

As we indicated, although fatty acids will spontaneously form stable micelles, this does not occur with phospholipids, because of the bulk of the hydrophobic tails. Liposomes can be generated experimentally but do not form unaided. So what happens when phospholipids and glycolipids encounter water? They rapidly and spontaneously form lipid bilayers (Figure 7.4). This spontaneous self assembly of a sheetlike membrane is of vital importance for cells. It is also important to appreciate that it occurs as a direct consequence of the amphipathic character and particular dimensions of phospholipid molecules.

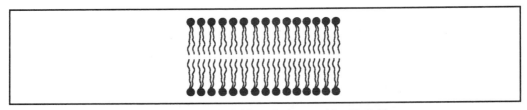

Figure 7.4 Lipid bilayer, as formed by phospholipids and other amphipathic lipids.

These lipid bilayers can be highly extensive (millimetres) and, in an aqueous environment, they represent a particularly stable configuration. The polar head groups make ionic and hydrogen bonds with water molecules. The nonpolar tails are protected from water and hence do not disrupt the hydrogen bonding which occurs between water molecules. Hydrophobic interactions are the driving force for the formation of lipid bilayers. If any gaps or tears in the bilayer were to occur, this would expose hydrophobic sections to water, which would be energetically unfavourable: thus the membrane will rearrange to 'seal up' such gaps.

SAQ 7.1

1) What are the important structural features of an amphipathic lipid?

2) Which of the following is an amphipathic lipid? Identify important regions of the amphipathic lipid(s).

a)
$$CH_2-O-\overset{\overset{\displaystyle O}{\|}}{C}-(CH_2)_{14}-CH_3$$
$$HC\ -\ O-\overset{\overset{\displaystyle O}{\|}}{C}-(CH_2)_{16}-CH_3$$
$$CH_2-O-\underset{\underset{\displaystyle O^-}{|}}{P}-O-CH_2-CH_2-{}^+NH_3$$

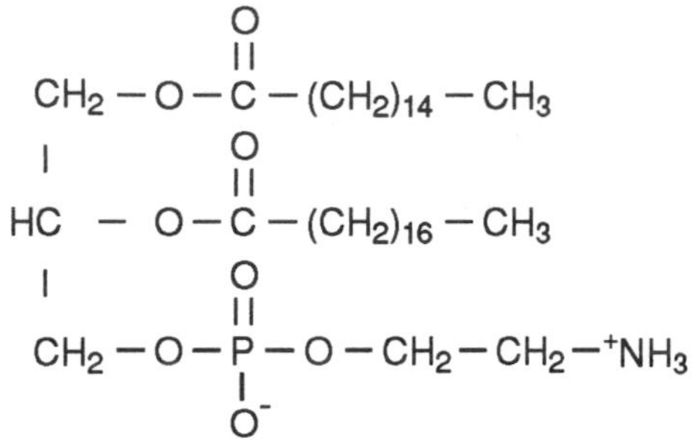

b)
$$CH_2-O-\overset{\overset{\displaystyle O}{\|}}{C}-(CH_2)_{14}-CH_3$$
$$HC\ -\ O-\overset{\overset{\displaystyle O}{\|}}{C}-(CH_2)_{16}-CH_3$$
$$CH_2-O-\overset{\overset{\displaystyle O}{\|}}{C}-(CH_2)_{14}-CH_3$$

c)
$$CH_2-O-\overset{\overset{\displaystyle O}{\|}}{C}-(CH_2)_{14}-CH_3$$
$$HC\ -\ O-\overset{\overset{\displaystyle O}{\|}}{C}-(CH_2)_{16}-CH_3$$
$$CH_2-O$$

with sugar ring bearing CH$_2$OH, OH, HO, and OH groups.

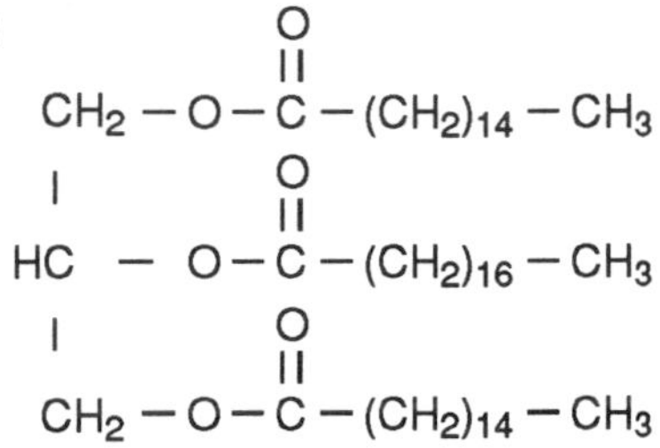

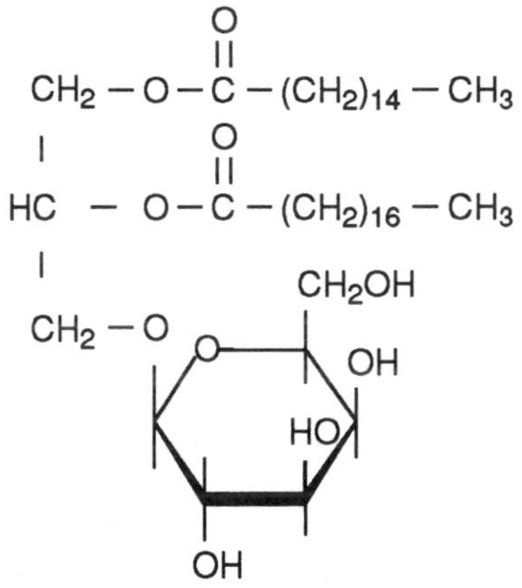

3) If an amphipathic lipid is mixed with water, what can happen? In what ways is this different to the outcome with non-amphipathic lipids?

7.4 Membrane proteins

specialist
function of
membrane
protein

peripheral and
integral proteins

Table 7.1 showed that membranes can contain large amounts of protein. These are responsible for the specialist functions associated with membranes, such as detection of hormonal signals or energy transduction. Early studies showed that whilst solutions of high salt concentration (eg 1 mol l^{-1} NaCl) could remove some protein from membranes, substantial amounts could not be removed in this way. These could only be removed with a detergent or an organic solvent. Membrane proteins are therefore categorised as either peripheral or integral (Figure 7.5). Integral proteins interact extensively with the inner parts (ie hydrophobic) of the lipid bilayer. This is why compounds that disrupt membranes and compete for nonpolar interactions are needed to release them. They may be thought of as being embedded to varying extents in the membrane, sometimes spanning the membrane (Figure 7.5). Peripheral proteins are bound to the outer surface of the bilayer by electrostatic and hydrogen bonds (remember that the polar head of phospholipids is negatively charged at neutral pH and is also well suited for hydrogen bonding). These bonds will be disrupted by added salt.

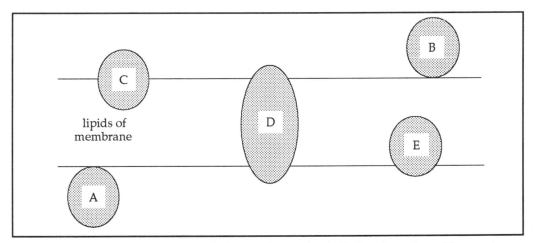

Figure 7.5 Membrane proteins. some (eg A, B) are only associated with the polar surface of the membrane and are called peripheral proteins. Others (C, D, E) are embedded (possibly to varying extents) and are called integral proteins.

Peripheral proteins will be broadly similar to those found in solution in the cell. Thus their surfaces predominantly contain polar amino acid residues which interact favourably with water. Integral membrane proteins are likely to be quite different: they tend to have hydrophobic amino acid residues exposed on their surfaces. These then associate with the hydrophobic interior of the membrane. This association of hydrophobic regions of a protein with the hydrophobic component of phospholipids will ensure that the protein remains positioned in the membrane.

∏ Study the following diagram and then shade: 1) areas of the surface of protein A which are likely to be hydrophobic and 2) areas of protein B which are likely to bear polar amino acid residues on their surface.

Its almost impossible to do this in text activity without glancing down the page, so try not to cheat! Your shading should look something like that shown below. The hydrophobic areas of protein A will be immersed in the lipid membrane, whilst the areas of protein B with significant amounts of polar amino acid residues will protrude from the lipid membrane, in contact with the aqueous surroundings.

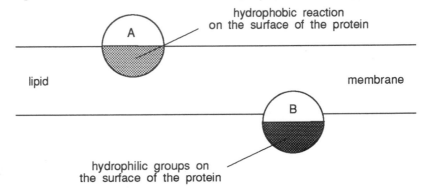

The technique of freeze fracture electron microscopy (which is described in the BIOTOL text 'The Infrastructure and Activities of Cells') allows the structure of membranes to be studied. This technique results in the peeling off of one half of the membrane, exposing the 'inside' of the membrane. This method provided the first direct experimental evidence for the occurrence of integral proteins in biological membranes. Those membranes with high protein content, such as those of red blood bells, have numerous globular particles, whilst synthetic bilayers composed only of phospholipid are smooth. The globular particles are removed by treatment with proteolytic (ie protein digesting) enzymes, strongly suggesting that they are proteins. Myelin membranes which surround neurons are predominantly lipid. Their role is primarily as an electrical insulator, and they are also largely free of these globular particles.

7.5 Properties of membranes

We have already mentioned the permeability properties of membranes. Two other general properties are important. These are the asymmetry and fluidity of the membrane.

SAQ 7.2

Four membrane preparations (A-D) were found to have the following lipid and protein contents (expressed as relative abundance): 1) upon isolation and 2) after treatment with 1 mol. l^{-1} NaCl. Comment on the possible origin of the membranes and the behaviour of the protein, particularly regarding its likely position in the membrane.

	Membrane Preparations			
1) Upon isolation	A	B	C	D
Phospholipid	24	32	23	40
Glycolipid	5	8	5	22
Cholesterol	13	0	12	20
Protein	58	60	60	18
2) After salt treatment				
Phospholipid	24	32	23	40
Glycolipid	5	8	5	22
Cholesterol	13	0	12	20
Protein	18	30	48	2

7.5.1 Asymmetry

different proteins on different sides of the membrane

It is now possible to appreciate how membranes can be made different and hence fulfil different roles. It is the protein complement which gives a membrane particular properties. The lipid bilayer is primarily to provide the sheetlike structure and permeability barrier. Thus transport of a particular polar compound is accomplished by the action of an appropriate carrier protein. Membranes are asymmetric and this too is achieved by protein distribution. The two sides of biological membranes have different proteins associated with them and different enzyme activities. When integral proteins span the membrane, a particular orientation is maintained. One well characterised case is the pump involved in regulating Na$^+$ and K$^+$ concentrations in cells. This is located in the plasma membrane of most cells (Figure 7.6). It also requires ATP. ATP must be on the inside of the membrane for the pump to work. Notice also that Na$^+$ is always pumped out and K$^+$ in. This can, of course, only be achieved if the membrane is asymmetric. Eukaryote plasma membranes usually have carbohydrate present as glycolipids and glycoproteins. Specific labelling techniques (such as the use of the plant proteins known as lectins which bind to specific carbohydrates) show that these are invariably on the outer surface of the plasma membrane. These properties clearly indicate that this system has a precise orientation.

carbohydrates on outer surface of plasma membranes

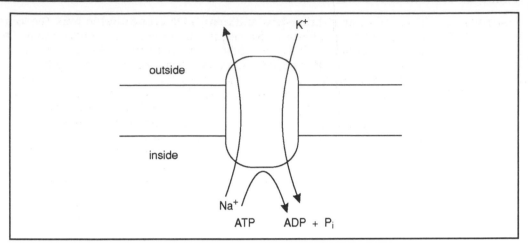

Figure 7.6 Orientation in membranes. In the plasma membrane, the sodium potassium pump has a distinct orientation. Na^+ is pumped out, K^+ is pumped in and ATP must be available on the inside.

7.5.2 Fluidity

membrane fluidity

Biological membranes are now known to be fluid, rather than rigid, structures. What this means is that both lipids and proteins can migrate or move about within the membrane. An excellent illustration of this was provided by the experiments of Frye and Edidin. They used fluorescent microscopy to test what happened when two different cell types were fused to each other. When this was done with cultured mouse and human cells, a heterokaryon (mixed cell) was formed in which part of the plasma membrane came from the mouse cell, part from the human cell. The two cell types have different membrane proteins. These can be identified by their ability to bind fluorescent labelled antibodies. The experiment showed that proteins from the two plasma membranes (mouse and human), whilst initially occupying distinct zones in the heterokaryon (Figure 7.7), rapidly became intermixed. Within an hour at 37°C, the two sets of marker proteins had completely mixed.

heterokaryon

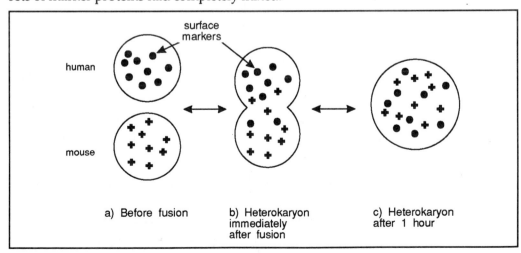

Figure 7.7 Membrane fluidity: the influential experiment of Frye and Edidin. Fluorescent markers on mouse and human cells are initially separate. Stage a), before fusion; Stage b), immediately after fusion, but rapidly intermingle. After one hour at 37°C, complete intermixing had occurred, Stage c).

This and other studies show that many membrane constituents can move very rapidly. Lateral diffusion (Figure 7.8a) of phospholipids is extremely rapid. They can diffuse 2μm in 1 second! This is a typical size of a bacterium! Some proteins, as shown above, are nearly as mobile as lipids but others seem to be more or less fixed. These comments apply only to lateral movement where the lipid or protein moves sideways but

maintains its orientation. Transverse diffusion (Figure 7.8b), in which the molecule moves from one side of the membrane to the other, is very slow and hence much less frequent. For phospholipids it does occur. For proteins, which have more extensive polar regions in contact with the aqueous phase, it has not been seen to occur. This is entirely consistent with the asymmetry of membranes and explains why asymmetry is maintained.

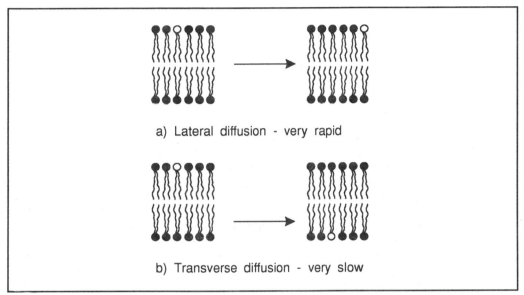

a) Lateral diffusion - very rapid

b) Transverse diffusion - very slow

Figure 7.8 Diffusion in membranes a) Lateral diffusion is rapid. b) Transverse diffusion is very slow, and has never been observed with proteins.

The fluidity of the membrane is affected by several factors including chemical composition and temperature. Lipid bilayers can undergo phase transitions between a fluid and a rigid state as temperature is lowered. The temperature at which this occurs depends on the length of the fatty acid chains and their level of unsaturation. The straight 'tails' of saturated fatty acids pack together well and favour the rigid state. Longer chains promote this effect. Unsaturated fatty acids, because of the bend in the fatty acid, do not pack together so tightly: these favour the fluid state. Two patterns by which membrane fluidity is controlled are apparent:

- in prokaryotes: the % unsaturated fatty acids determines, and is used to regulate, membrane fluidity;
- in eukaryotes: the cholesterol content affects membrane fluidity.

In eukaryotes, cholesterol fits into the membrane, between fatty acid chains, with its single hydroxyl group at the surface of the membrane (Figure 7.9). This tends to prevent the fatty acid chains adopting the rigid configuration. However, cholesterol can also physically block movement of phospholipids, which makes them less fluid. Thus inclusion of cholesterol into membranes is a way of regulating the properties of the lipid bilayer.

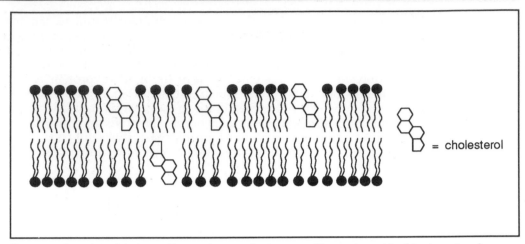

Figure 7.9 Location of cholesterol in membranes of eukaryotes. The weak amphipathic character of cholesterol results in its orientation as shown.

7.6 Models of membrane structure

model
inconsistent
with
experimental
evidence

The early models proposed for membrane structure consisted of a lipid bilayer. This failed to account for the presence of proteins in membranes and was thus extended to include protein on each surface of the membrane. These proteins were considered to be held in place solely by electrostatic bonds, with no association between the protein and the nonpolar parts of the lipids. Although this model was consistent with the universally similar appearance of membranes in the electron microscope (irrespective of biological source), this model is not compatible with all experimental observations. In particular, if proteins are on the surface, held by electrostatic bonds, they should be removed by treatment with high salt concentrations. This does not occur. Similarly, the globular appearance of the inside of membranes found after freeze fracture electron microscopy cannot be explained according to this model.

Fluid Mosaic
model

The currently accepted model for membrane structure is that proposed in 1972 by Singer and Nicholson, and is called the Fluid Mosaic model. The basic structure is a lipid bilayer, consisting of amphipathic lipids arranged as we have seen them do spontaneously. Into this sheet of lipids, proteins can be incorporated such that they may be completely peripheral, penetrate partially, span the membrane or be completely submerged within the hydrophobic centre of the membrane (Figure 7.10). The precise position adopted by a given protein depends on its structure. Polar proteins interact more strongly with the aqueous phase and the polar heads of the lipid molecules and are therefore located on either side of the membrane. For a protein which has both polar as well as nonpolar regions, the polar region protrudes from the membrane surface whilst its nonpolar region is embedded in the interior of the lipid bilayer. In this way the protein is 'anchored' in the membrane. One should think of these integral proteins as floating in a sea of lipid. The depth at which any protein floats depends on its surface residues: only those proteins which are completely surrounded by hydrophobic residues (ie are nonpolar) would be expected to be completely buried in the middle of the membrane. Both proteins and lipids are in principle free to move laterally, but neither can easily flip from one surface to the other.

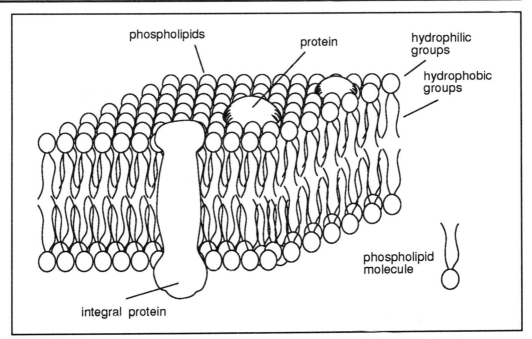

Figure 7.10 Fluid Mosaic model of membrane structure, proposed in 1972 by Singer and Nicholson. Proteins 'float' in a 'sea' of lipid.

Summary and objectives

This chapter has discussed the properties and composition of membranes. Membranes are very important structures that enable cells to compartmentalise their activities and provide an environment in which biochemical reactions may be organised. Now that you have completed this chapter, you should be able to:

- describe how lipids form the bilayer core of membrane sheets;

- distinguish between integral and peripheral proteins and explain how the properties of proteins may enable them to associate in particular orientations;

- describe the permeability properties of membranes especially with respect to polar molecules;

- describe the fluid mosaic model of membrane structure.

Enzymes

Enzymes

8.1 Introduction

In Chapter 1 we described cells as behaving like superb chemical factories, converting one set of chemicals (nutrients or substrates) into a vast number of complex products. We also described cells as being able to bring about these changes quickly and efficiently at a cool temperature. In subsequent chapters we examined many of the products of these chemical factories. We have learnt that many of the these are chemically very complex and are of specified stereochemical configuration. For example we have learnt that amino acids are produced in the L-configuration.

The ability of cells to carry out these chemical processes in such an efficient and specific manner undoubtedly depends upon the structural organisation of cells and on their ability to produce suitable catalysts for speeding up the desired reactions. We met with one aspect of the structural organisation of cells at a molecular level in Chapter 7, when we discussed the structure and properties of membranes. In this chapter, we examine the catalysts used by cells. These catalysts are all proteins and are called enzymes.

This chapter is a long one, reflecting the importance that must be attached to enzymes if we are to understand how living systems can bring about the vast array of chemical changes in a controlled and predictable manner. Do not attempt to study all of this chapter in one sitting. We will begin by discussing the general properties and measurement of enzymes and then move on to examine the factors which may influence their activities. Finally we will briefly explore the practical use we can make of enzymes other than their natural role within cells.

8.2 Introduction to enzymatic catalysis

Enzymes are the catalysts of cells and organisms. In Chapter 3 we examined the structure of proteins and reviewed some of their properties and roles. We are now going to assess enzymes to find out how they achieve catalysis and what factors affect their activity. A thorough understanding of these aspects is vital if we are to appreciate how cells work at a molecular level. Understanding enzymes also enables us to think of

enzymes as a resource

enzymes as a resource. What could we use enzymes for? How should we use them? What conditions should be chosen and, perhaps as importantly, what conditions must be avoided? These issues are developed in another BIOTOL text 'Technological Applications of Biocatalysis', but it is worth bearing them in mind as you think about enzymes working within the cell.

A catalyst is something which speeds up the rate of a chemical reaction, but remains unaltered itself. Enzymes are typical catalysts in this respect. All of them are proteins and enzymes are responsible for the vast majority of chemical reactions which take place in living organisms (ie very few reactions in cells occur spontaneously without an enzyme as catalyst).

8.2.1 General basis of catalysis

activation energy of the reaction

Later in this chapter we will examine in detail the ways in which enzymes achieve increased reaction rates. Essentially what an enzyme does is to lower the activation energy of a reaction. The following treatment is not a strict thermodynamic description but should enable the significance of the change in activation energy to be grasped.

free energy changes in a reaction

For a chemical reaction to proceed, there must be a change in free energy such that the products contain less free energy than the reactants. Free energy may be thought of as the capacity to do useful work. This implies that the difference in free energy content has been released in the reaction. This is known as the Free Energy Change, ΔG; under standardised conditions this becomes $\Delta G°$, the standard free energy change. An example may help: glucose may be burnt to give carbon dioxide and water, according to the reaction:

$$C_6H_{12}O_6 + 6O_2 \longrightarrow 6CO_2 + 6H_2O$$
$$\Delta G° = 2880 \text{ kJ.mole}^{-1}$$
$$= \text{standard free energy change}$$

If this is carried out in a device which measures how much energy is released in this reaction (a bomb calorimeter), you would find that 2880 kJ of energy per mole of glucose was evolved. Since this is release of energy, this is spoken of as a spontaneous reaction. However, at physiological temperatures, this reaction proceeds incredibly slowly: probably undetectably! Thus the fact that a large amount of energy can be released is irrelevant to the question of does the reaction take place at physiological temperatures. Why is this so?

Let us depict the reaction on a graph in which the vertical axis represents the energy content and the horizontal axis represents the progress of the reaction (Figure 8.1).

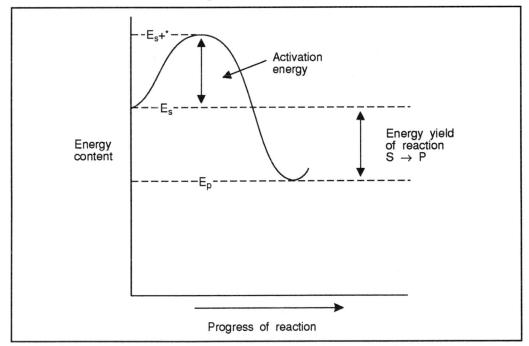

Figure 8.1 Energy relationship of substrate and product. For reaction to actually occur, an activation energy barrier must be overcome. E_s = energy of substrate E_p = energy of product.

energy barrier

activation
energy

We can see that the products are at a lower energy level than the reactants. If reactants (substrate) are converted to product, energy will be released. This means that the reaction can proceed. Whether it does proceed depends on an energy 'barrier' between substrate(s) and product(p). This energy barrier is known as the activation energy. For any molecule to actually react, it must possess energy corresponding to the top of this barrier (ie E_s + *). We will illustrate this in another way. Take the task of getting a rock (or a bicycle) from points A to B which are separated by a hill as depicted in Figure 8.2.

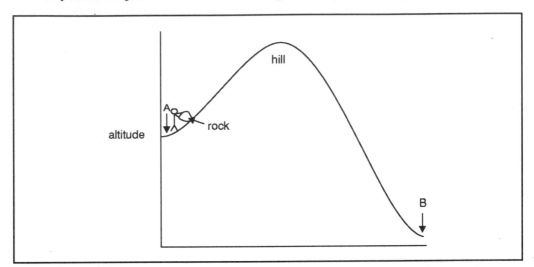

Figure 8.2 The energy requirement in getting a rock over a hill from points A to B is analogous to the situation faced by enzymes. Some energy must be supplied ('activation energy') but once over the hill, no further energy is needed (progress is spontaneous).

To do this, the rock must be pushed up the hill (ie some energy must be expended). Once at the top, it will get to B without further energy input (ie progress is spontaneous). If, instead of the rock, it was a bicycle fitted with a dynamo which generates electricity, travelling from the summit of the hill to B would release energy which could be used to generate electricity. (In other words when energy is given out there is a 'capacity to do work').

So how do molecules achieve the necessary energy to overcome the activation energy barrier? One way is to give them more energy in the form of heat; that is what is done with the bomb calorimeter experiment involving glucose, by first heating the glucose. In other cases, some molecules happen to possess enough energy to react. This is because the energy content of a population of molecules varies (Figure 8.3) such that, whilst the mean energy content corresponds to E_s, some molecules have more energy than E_s + * (shown shaded in Figure 8.3).

These molecules can undergo reaction. With glucose at room temperature, however, an insignificantly small number of molecules has the required energy. None-the-less, this example is important because enzymes can only accelerate reactions where the overall energy change is favourable. Whether the reaction is detectable in the absence of an enzyme is determined by the size of the activation energy barrier.

What do enzymes do to solve this problem? They do not provide the missing energy but instead provide an alternative route, for which less energy is needed. This is displayed in Figure 8.4.

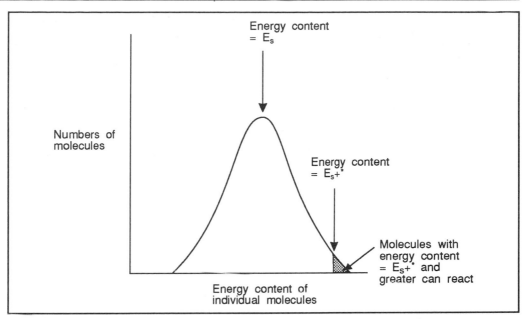

Figure 8.3 The energy content of different molecules varies. Whilst the mean corresponds to that of the starting energy level of E_s in Figure 8.1, some molecules (shown shaded) possess energy corresponding to $(E_s + {}^*)$ and can react.

enzymes work by lowering the activation energy barrier

The overall effect of this is to lower the activation energy barrier. In our boulder and hill analogy we have now found an alternative route round the side of the hill, for which much less energy needs to be supplied.

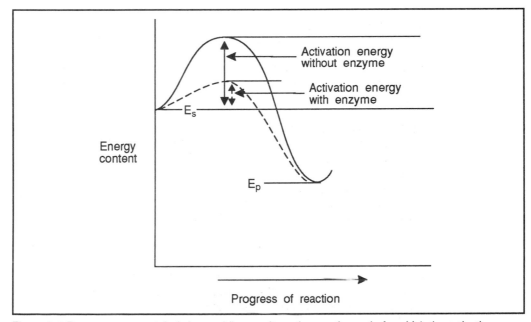

Figure 8.4 Enzymes cause catalysis by providing an alternative reaction path, for which the activation energy requirement is lower.

The effect of this lowering of the required activation energy can be seen most easily by examining the energy contents of molecules (Figure 8.5).

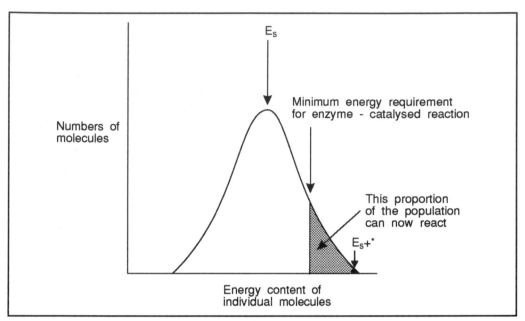

Figure 8.5 Energy profile of the substrate molecules: the enzymic reaction route means that a greater proportion of the molecules now possess the necessary energy (shown hatched).

∏ Examine Figure 8.5 carefully and see if you can explain why an enzyme can speed up a reaction.

the energy released during a reaction is unaffected by enzymes

In the absence of enzyme, very few molecules had sufficient energy to react (only those with energy greater than E_s + *). With the enzyme, the required energy is lowered (by the enzyme providing an alternative route) so that substantially more molecules have sufficient energy and hence can react. This is how enzymes achieve their catalytic effect. Note that the enzyme only affects the activation energy. The energy given out in the reaction is unaltered (this is the difference in energy levels of S and P). When glucose is converted to carbon dioxide and water by organisms, using enzymes, precisely the same amount of energy is released (2880 KJ. $mole^{-1}$) as in the bomb calorimeter.

catalase

The catalytic power of enzymes can be enormous. One of the most remarkable, in terms of rate enhancement, is catalase. This enzyme is very widely distributed in nature (plants, microbes, most animal tissues) and catalyses the breakdown of hydrogen peroxide

$$2H_2O_2 \rightarrow 2H_2O + O_2$$

enzymes work under mild conditions

One molecule of catalase can react with about 5 million molecules of hydrogen peroxide per minute. This means that catalase is accelerating the reaction rate by at least 10^{14} times (yes, that's right, 100,000,000,000,000 times!!) compared to the uncatalysed reaction.

One of the attractions of using enzymes industrially is that they frequently operate under mild conditions. A second example illustrates this as well as their catalytic power.

If we wish to hydrolyse all the peptide bonds in a protein, a common strategy is to use acid hydrolysis. Typical conditions are:

6 mol l^{-1} HCl (approximately 1/2 concentrated acid)

110-120°C

24-72 hours

These are very aggressive conditions which have to be maintained for an extended period. Yet the proteolytic enzymes present in our digestive systems will accomplish the same task at near neutral pH, at 37°C and in a few minutes! Precisely how enzymes achieve catalysis in chemical terms is discussed in a later section.

SAQ 8.1

Inspect the diagram below which shows the energy profile for a hypothetical reaction involving possible conversion of A to B. The solid line shows the uncatalysed reaction; dashed lines depict that in the presence of enzyme X; dotted lines in the presence of enzyme Y. Identify which of the following statements are true and which are false, giving brief reasons.

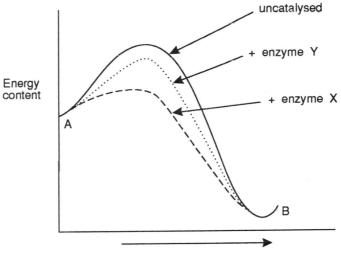

1) Conversion of A to B is rapid in the absence of any enzyme.

2) Addition of enzyme X accelerates the reaction by lowering the activation energy; it also lowers the energy yield of the reaction.

3) The activation energy barrier precludes conversion of A to B under any circumstances.

4) The reaction of A to B is spontaneous but, in the absence of enzyme X or Y, occurs to a negligible extent, because of the activation energy barrier.

5) Both enzymes accelerate the reaction by lowering the activation energy; neither enzyme affects the overall energy yield.

6) Enzyme X will produce (on the basis of energy considerations alone) a lower rate enhancement than enzyme Y.

In summary, enzymes only accelerate reactions which in principle are proceeding spontaneously although perhaps so slowly as to be undetectable. They do not change equilibria nor the amount of energy released. They only alter the speed with which equilibrium is achieved.

8.2.2 Specificity

We have already seen the remarkable rate increases which enzymes can achieve. Rate enhancement alone, however, is not the key to the crucial properties and functions of enzymes. A second vital property is that of specificity, in terms of what compound(s) an enzyme acts on, and what product(s) it produces. This is of enormous importance in cells because it allows the fate of compounds to be controlled and directed. Material is accurately and precisely passed along a pathway via a series of enzymes. Routinely, no side-products are formed and all the material will be converted to the product of the pathway. This high specificity is also vital for many of the commercial, medical, and analytical areas in which enzymes are used.

Enzyme specificity thus has two separate but important aspects. The first concerns which molecules any particular enzyme acts on. Enzyme's have an ability to discriminate between different molecules. Different enzymes show different degrees of discrimination. Urease, which hydrolyses urea according to the equation:

$$H_2N - \overset{\overset{\displaystyle O}{\|}}{C} - NH_2 \ + \ H_2O \longrightarrow CO_2 \ + \ 2 \ NH_3$$

high specificity for substrates is highly specific, failing to act even on other molecules in which minor modifications have been made. For example it will not act on:

$$H_2N - \overset{\overset{\displaystyle O}{\|}}{C} - NHCH_3$$

In other cases, specificity is less absolute, permitting one enzyme to act on closely related substances. Thus the proteolytic enzyme trypsin, which hydrolyses peptide bonds, acts only on the peptide bond to the C- terminal side of either arginine or lysine.

∏ If you can recall the structure of arginine and lysine, see if you can find a feature which the side chains of these two amino acids have in common. If not, look back to Chapter 2 to refresh your memory of their structures.

At the pH at which trypsin is most active (pH 8.0), the notable feature here is that these are the only amino acids whose side-chains are positively charged. Thus trypsin will tolerate minor differences in the precise shape of its substrate (Figure 8.6), providing the amino acid side-chain possesses a positive charge.

restricted specificity for substrates There are numerous cases of restricted specificity, as exemplified by trypsin. Note that for some proteolytic enzymes, specificity is much tighter. Thrombin, which is involved in blood-clotting, only cleaves the peptide bond between arginine and glycine, and only then when this dipeptide occurs in particular amino acid sequences. This ensures that thrombin only catalyses this beneficial step in blood clotting. Non-regulated protein hydrolysis in the blood system would be highly damaging!

There are also enzymes which display a relative lack of specificity. These are often hydrolytic enzymes, whose role is to break down proteins (or esters), perhaps to enable their constituent amino acids (or carboxylic acids/alcohols) to be used as a nutrient. In

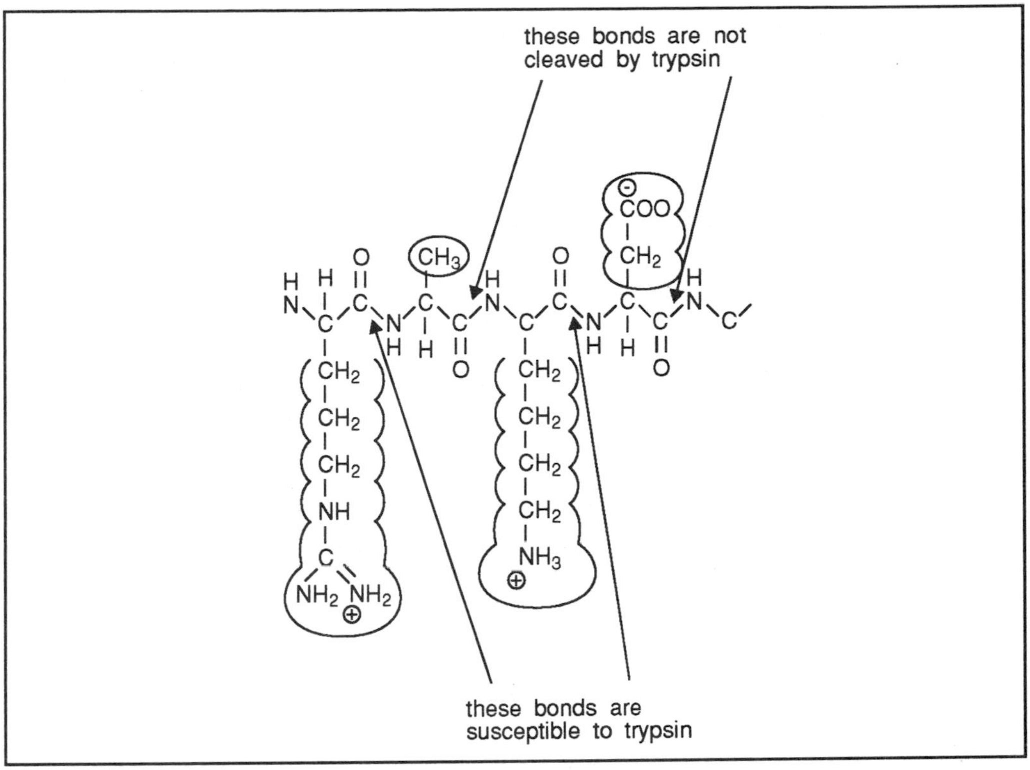

Figure 8.6 Trypsin cleaves the peptide bond after lysine and arginine amino acid residues. The difference in precise shape is less important than the presence of a positive charge at the end of the sidechain.

the same way that there are clear circumstances in which absolute specificity is desirable, so too are there circumstances (eg release of amino acids for nutrition) in which looser specificity has advantages.

specificity of reaction

The second important aspect of enzyme specificity concerns what reaction is performed. Enzymes tend to be highly specific in terms of which groups and bonds are acted upon and what products are formed. This is clearly vital for metabolism. The enzyme hexokinase phosphorylates glucose as the first step in glycolysis. Glycolysis is a metabolic pathway in which glucose (and other carbohydrates) are broken down. The reaction is as follows:

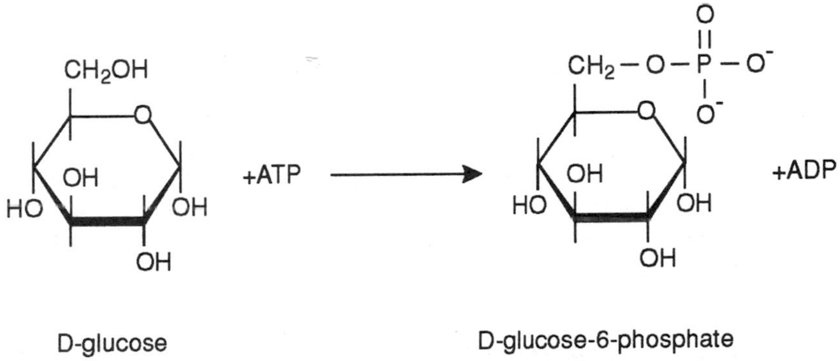

D-glucose   D-glucose-6-phosphate

There are 5 hydroxyl groups which could, in principle, be phosphorylated, yet only one, (at C-6) actually is. Thus only glucose-6-phosphate is formed. This shows the absolute positional specificity as far as the reaction is concerned. In passing, it may be noted that hexokinase displays loose specificity in terms of which sugars it will phosphorylate. It will also act on fructose, producing fructose-6-phosphate, which can also be used in glycolysis.

8.2.3 Stereospecificity

specificity for
particular
stereoisomers

Enzymes are highly discriminative of stereoisomers. Thus proteolytic enzymes such as trypsin only act on peptides made of L-amino acids. Peptide bonds in synthetic peptides made of D-amino acids are not hydrolysed. Similarly, hexokinase is inactive on L-glucose. This absolute specificity for either D- or L-isomers is normally observed with enzymes which act on optically active compounds.

Absolute positional and stereospecificity is displayed by our final example. Cortisone, a steroid hormone, may be synthesised from several precursors, including diosgenin and stigmasterol. One crucial step involves the introduction of a hydroxyl group at a particular position (position 11) on the steroid nucleus (Figure 8.7).

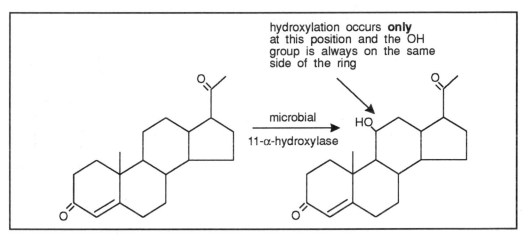

Figure 8.7 Enzyme specificity in the production of cortisone is vital in hydroxylation of progesterone to 11α-hydroxy progesterone.

The extra hydroxyl group can be in front of the plane of the steroid (designated α- and protruding forward out of the plane of the paper) or behind (configuration is β- and the hydroxyl is behind the plane of the paper). The α-configuration is the desired position. Chemically, this is extremely difficult to produce because when the molecule is hydroxylated in this way both α and β forms are produced. Thus a mixture of molecules is produced. Fortunately, a hydroxylating enzyme (hydroxylase) from *Rhizopus nigricans* has been obtained which hydroxylates specifically at position 11, and only gives 11-α hydroxyprogesterone, ie only one of the two possible isomers.

Π Write down a reason why stereospecificity might be achieved when a reaction is catalysed by an enzyme.

substrate
binding

The key to it is that catalysis involves the substrates binding transiently to the surface of the enzyme. This contrasts with a purely chemical reaction, which would take place in free solution and would not discriminate stereochemically. Whereas in free solution both stereoisomers could react (eg both D- and L-glucose; D- and L-amino acids), the requirement to bind to an enzyme enables the enzyme to discriminate between them. This is easily seen with simple models, which you could make (with plasticine and matchsticks if atomic models are not available).

Stereoisomers arise when an asymmetric centre occurs. These occur when a carbon atom has 4 different groups attached to it. Typical examples, described in detail elsewhere in this text (see Chapters 2 and 5), are D- and L-glyceraldehyde and D- and L-amino acids. Take the case of alanine. This amino acid can exist in two alternative configurations (Figure 8.8) which are mirror images of each other.

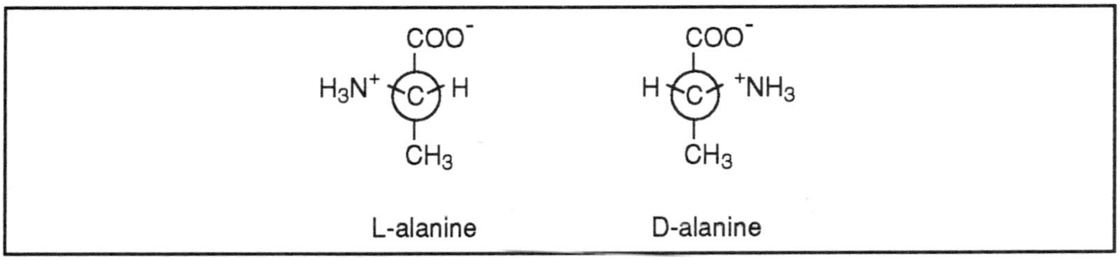

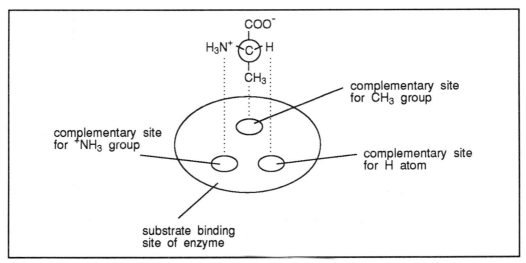

Figure 8.8 Alternative 3-dimensional configurations, D- and L- alanine.

The only difference lies in the distribution of the substituent groups around the central carbon atom. In free solution, these isomers will react in a similar manner. However, if they have to bind to the surface of an enzyme, and there are complementary sites for 3 of the substituent groups, (eg for $-CH_3$, $-COO^-$, and $-NH_3^+$ groups) then the position of these complementary sites dictates which isomer will bind and hence may react. In the example shown (Figure 8.9), only L-alanine has correctly positioned groups for binding to the enzyme.

Figure 8.9 Enzymes discriminate between stereoisomers by providing complementary binding sites to several groups of the substrate. If 3 such sites are used, only one of the stereoisomers can bind.

D-alanine will show no tendency to bind. Thus the prerequisite for a molecule to bind to an enzyme before catalysis can occur also underpins the absolute specificity for the 'correct' stereoisomer.

We have now seen that enzymes can readily distinguish between stereoisomers. Note that this ability does not have to be used: if only 2 binding sites had to be complementary, instead of 3, then the isomers would not be distinguished.

active site

How do enzymes distinguish between different compounds. For example, how does trypsin distinguish between alanine and arginine? This is accomplished by the particular shape and charge distribution of the active site of the enzymes. The active site of the enzyme is the part of the enzyme to which the substrate binds. Let us consider this in more detail.

8.2.4 Active sites

Only a small part of an enzyme directly interacts with the compounds it acts on (the substrate(s)). This part is known as the active site. If you think about the relative size of substrates and enzymes, this is not very surprising. Molecules of intermediary metabolism are typically small, with molecular masses of up to several hundred daltons. Enzymes usually have molecular masses in excess of 20,000 daltons. Thus catalytic activity is likely to be confined to only part of the enzyme molecule. The remainder of the enzyme none-the-less has vital roles, including maintaining the shape of the active site and providing binding sites for regulatory molecules (molecules which modify the activity of an enzyme). It may also ensure that the enzyme sticks to a membrane or associates with other proteins.

two properties
of an active site

The active site has to accomplish two objectives. Firstly, it must enable the correct substrate(s) to bind, whilst 'rejecting' other molecules, and ensure correct positioning of the substrate. This ensures that the second task (catalysis) can take place. The active site may conveniently be thought of as having two sub-sites: a recognition and binding site, and a reactive site. The former ensures only appropriate substrates bind and are correctly positioned. This is achieved by the precise shape and charge distribution of the active site. The classical depiction of enzyme substrate interaction involves a 'Lock-and-Key' analogy (Figure 8.10), in which only the 'correct' substrate binds. As a simple model this is fine, since we can readily see how some molecules are rejected and cannot bind (Figure 8.10b) whereas others possess the required shape (Figure 8.10a). Further, we can see how minor alterations in substrate structure may be acceptable (Figure 8.10c), although there may be differences in the strength of binding of different substrates.

lock and key
model

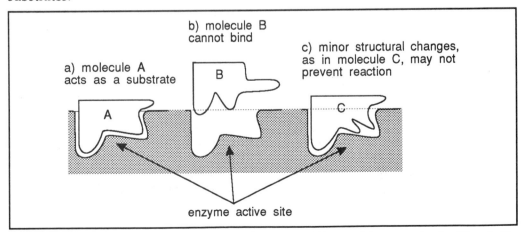

Figure 8.10 Models of the active site. According to the 'Lock-and-Key' model, some molecules 'fit' and will bind and react (Figure 8.10a), whereas other molecules (Figure 8.10b) do not possess the required shape. Molecules closely resembling the 'optimal' shape may also be acted on (Figure 8.10c).

The tendency to bind results from the complementarity of shape and charge of the substrate and active site. Particularly good binding results when there are a lot of close contacts but no cases where atoms are being 'forced' too close together. The strength of interaction is a function of the distance apart of atoms (Figure 8.11); if too close, repulsion occurs (ie within what are termed the van der Waals radii).

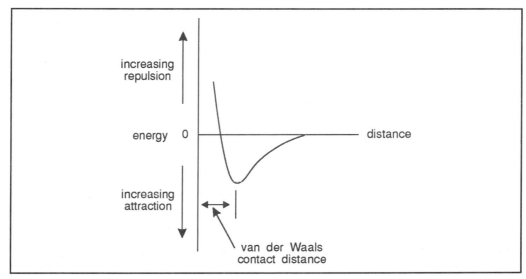

Figure 8.11 Effect of distance on van der Waals interaction. When atoms are too close (electron shells overlap) strong repulsion occurs. Attraction is maximised at the van der Waals contact distance (each atom has a characteristic van der Waals contact radius) and weakens as the distance between the atoms increases.

Increasing distance results in a progressively weaker interaction. Substrates bind to enzymes through numerous relatively weak bonds, including ionic and hydrogen bonds, and hydrophobic and van der Waals interactions. Only occasionally are covalent bonds formed and this is usually as part of the catalysis.

∏ The lock and key hypothesis of the active site implies that the enzyme is rigid and that substrate binds to it as a key fits a lock. From what you know about proteins, are these rigid molecules?

induced fit model

There is extensive evidence that enzymes are flexible. Koshland proposed an alternative model for enzyme action, based on Induced Fit (Figure 8.12a). In this hypothesis, the enzyme itself undergoes conformation changes as a result of substrate binding. The active site thus changes shape as the substrate binds and the 'active' form of the enzyme only exists when substrate is bound. Thus the substrate 'induces' the correct fit. This idea may be usefully extended in terms of the enzyme 'deforming' the substrate (Figure 8.12b). Some enzymes can only bind their substrate if it is strained or forced into an abnormal configuration. This provides one of the mechanisms by which rate enhancement is achieved and will be further discussed later.

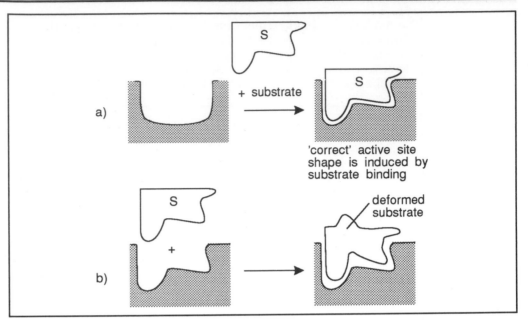

Figure 8.12 a) The 'Induced Fit' model for the active site. The active site changes shape when substrate binds: the 'active' conformation, which is responsible for catalysis, is 'induced' by the substrate and only occurs when substrate is bound. b) The substrate may be deformed or strained during binding.

SAQ 8.2

Comment on the truth, or otherwise, of the following statements. In each case write no more than 2 sentences to support your conclusion.

1) Enzymes are not true catalysts because they alter the equilibrium position of the reaction they speed up.

2) An enzyme-substrate (ES) complex is a pre-requisite for enzymatic catalysis.

3) When an enzyme binds substrate, binding sites for 2 groups on the substrate confer stereospecificity on the enzyme (ie the ability to distinguish stereoisomers).

4) Enzymes are rigid molecules.

5) Substrate specificity is a consequence of the particular shape and charge distribution of the active site.

substrate interaction with trypsin

Let us return to the example of trypsin cited earlier and ask how does trypsin distinguish between peptide bonds involving lysine and arginine, and all other amino acids? We will firstly examine how alanine and arginine are distinguished. We must look at the side- chains of the amino acids and at the active site of trypsin, as revealed by X-ray diffraction analysis. The binding site is a depression into which the side chain of the amino acid fits (Figure 8.13). The side chain of alanine is small (Figure 8.13a) and makes little contact with the amino acid residues present in the binding site.

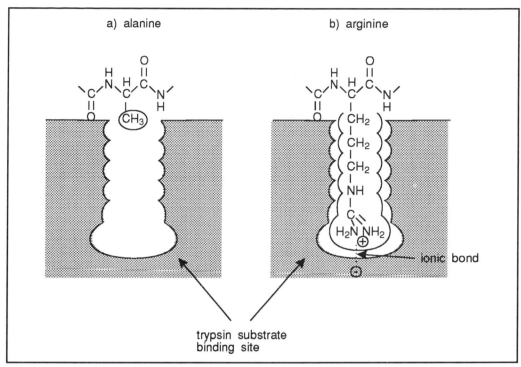

Figure 8.13 Complementarity between substrate and enzyme. a) The active site of trypsin would make very limited contact with the sidechain of alanine and there would be minimal tendency to bind. b) The sidechain of arginine (or lysine) makes numerous contacts (van der Waals radius); particularly important is an ionic bond at the bottom of the active site.

In contrast, that of arginine is much longer and matches the binding site much more closely. The crucial feature though, is the positive charge at the end of the arginyl side-chain. This is complemented by a negative charge at the bottom of the binding site and it is this which really enables trypsin to bind arginine (or lysine) tightly. Other amino acid side-chains would fail to bind through being too bulky, or having the wrong charge, or failing to make sufficient interactions.

∏ Try to design proteolytic enzymes (like trypsin) which would act on the peptide bond following: 1) glutamate, 2) phenylalanine or tyrosine. Hint: just think about these tasks in terms of the general shape, etc of the active site! You will need to consider the structure of the amino acids named in 1) and 2). Make a drawing of the active sites you have designed.

This may seem very daunting but it is not, if kept at a general level! We saw with our example of trypsin that the amino acid sidechain has to match the binding (recognition) site. If this is done the relevant peptide bond should be correctly positioned for hydrolysis. Thus your first task is to examine the sidechains of the proposed substrates. Glutamate has an acidic sidechain, with a carboxyl group which will be negatively charged at neutral pH ie:

Side chain of L-glu is -CH$_2$CH$_2$COO$^-$

Phenylalanine and tyrosine both have an aromatic ring in the sidechain, with tyrosine having a hydroxyl group on the ring (making it a phenolic group) ie:

sidechain of L-phe   sidechain of L-tyr

So what features do the binding sites need? For 1) an opposite charge would encourage strong binding and could enable the amino acid residue to be correctly positioned (see below). Thus we would expect to find a lysine or arginine sidechain at the bottom of a slightly less deep depression than that of trypsin (note that there are only 2 -CH_2- groups in the glutamate sidechain)

For 2) the sidechains are bulky and hydrophobic, because of the phenyl ring and the preponderance of hydrocarbon. The only polar character is in the tyrosine phenolic hydroxyl. The binding site would be expected to lack polar groups and be lined with non-polar hydrophobic groups which will 'fit' the shape of the phenyl group. Tyrosine would bind particularly well if there was something which its hydroxyl could hydrogen bond to. The sorts of features which might be expected in the binding sites are summarised in the diagram below.

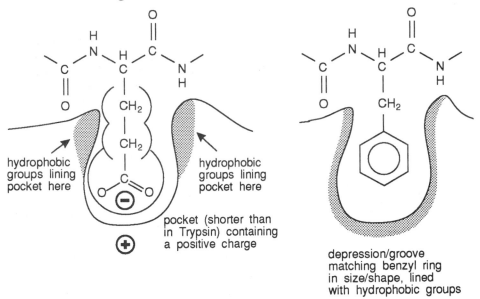

Before we move on to the next section let us summarise what we have learnt about enzymes as the 'workers' of cells. We have learnt that they can greatly speed up reactions and that each will only speed up a single or limited range of reactions. We have learnt that many can distinguish between different stereoisomers and carry out reactions in a stereospecific manner. We have also learnt that only a small part of each enzyme is directly involved in the catalytic process (ie the active site). The rest of the enzyme is, however, important in making certain that the active site is the correct shape and also, in some cases, to make certain that the enzyme is bound to the right structure (eg a membrane) in the cell. In our worker analogy, we can perhaps regard the active site as the hands of the worker. Hands alone, however, are of little value; we need the rest of the body to ensure that the hands are able to carry out their function. Enzymes are indeed rather like the workers in a large complex industrial process. Each is designed to carry out a specific task in a specific area of the factory.

Let us now turn our attention to the ways in which we can measure the activities of enzymes.

8.3 The assay of enzymes

8.3.1 General aspects

enzyme levels may reflect stage in development and disease states

— *Disease*

When we talk of assaying an enzyme, we are meaning the determination of how much of it is present. There are numerous reasons why this knowledge may be needed: some examples follow. Any detailed analysis of metabolism requires knowledge of the amounts of each component enzyme. Developmental changes in an organism are reflected in its enzyme complement and their relative levels. Many disease states are characterised by abnormal levels of particular enzymes. Thus appropriate assays may be relevant to both understanding of the condition and its diagnosis. Finally, if we wish to purify an enzyme, whether as part of a scientific investigation to improve our understanding of the biochemistry of a system or, in a technological context, to use the enzyme in analysis or industry, we must be able to determine how much is present in any given sample.

An enzyme assay must be specific to the particular enzyme: thus the assay should not be subject to interference from other enzymes and proteins present in a cell extract. For this reason assays of the total amount of protein present in an extract tell us nothing about how much hexokinase, for example, is present. The simplest specific assays are based on the catalytic activity of the enzyme. If we take the generalised case of substrate being converted to product thus:

$$S \xrightarrow[\text{Enzyme}]{} P$$

substrate loss product formed

we must either measure loss of substrate or appearance of product. Either can be used, depending on circumstances, although assays based on measuring the rate of product formation are often more satisfactory. Before we analyse how we might measure product formation, several general points should be made.

8.3.2 Ensure that rate of product formation is proportional to the enzyme concentration

Enzyme assays should enable us to quantitatively compare how much enzyme is present in various solutions. To be able to do this, the above condition must be satisfied, such that the following type of result is obtained experimentally (Figure 8.14).

ie an important criterion in measuring an enzyme is to ensure that the rate of reaction is proportional to the enzyme concentration.

We might express rate as, for example, μmol product formed/min.

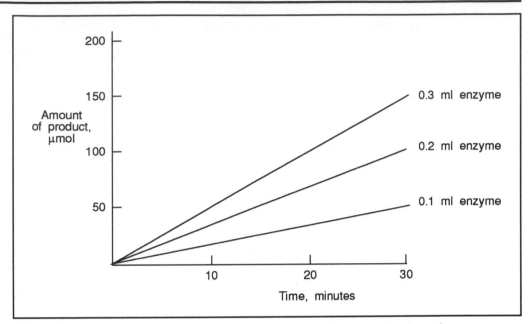

Figure 8.14 Effect of enzyme concentration on product formation. For an accurate assay of enzyme activity, doubling the volume of enzyme added must result in double the amount of product being formed in a given time.

If double the amount of enzyme is present, then the rate of product formation should be doubled. This requirement needs to be stressed because as we shall learn later several other factors influence enzyme activity. To assay an enzyme, we should ensure, as far as possible, that these other factors are not limiting the rate and that it is the enzyme concentration which is determining the rate. In practice, this means checking that rate is proportional to volume of enzyme solution used.

∏ Examine the following data, obtained in an attempt to assay an enzyme. Does it matter what volume of enzyme was used? If so, which volume do you recommend to best assay the enzyme? Why was incubation (A) included, in which no enzyme was present?

Incubation	Volume of Enzyme added (ml)	Amount of Product formed (μmol), after 10 min incubation
(A)	0	0
(B)	0.1	0.65
(C)	0.25	1.63
(D)	0.5	2.8
(E)	1.0	3.9

The purpose of an enzyme assay is to allow us to measure how much enzyme is present. To do this, the rate of the reaction must be proportional to the amount of enzyme present. This is most easily seen if the amount of product formed is normalised, for example to 1.0ml of enzyme. This gives the following:

Incubation	Vol. Enz. used (ml)	μmol Product formed	μmol Product formed per ml enzyme, assuming proportionality
(A)	0	0	0
(B)	0.1	0.65	6.5
(C)	0.25	1.63	6.5
(D)	0.5	2.8	5.6
(E)	1.0	3.9	3.9

For incubations (B) and (C) there is proportionality between rate and volume of enzyme: (C) has formed 2.5 x as much as (B). (D) should have formed 5 x as much as (B) but has not done so. 5 x that formed in (B) is 3.25. This under-production is shown up in the normalised figures in the table above. This assay condition would lead to a slight underestimate. (E) is more seriously flawed, giving only just over half of the true value.

Assuming the other assay conditions remain the same, a suitable volume is in the range of 0.1 to 0.25 ml since the amount of product formed is proportional to the amount of enzyme used within this range.

control   What was the purpose of incubation (A)? This is a control, to allow any product formed in the absence of enzyme to be allowed for. This will be discussed further shortly.

8.3.3 Ensure that the initial rate is measured

initial rate   If we examine the amount of product formed with increasing time, we often observe that the rate declines with time (Figure 8.15). The dashed line indicates a 'theoretical' result, in which product formation is maintained at a steady rate.

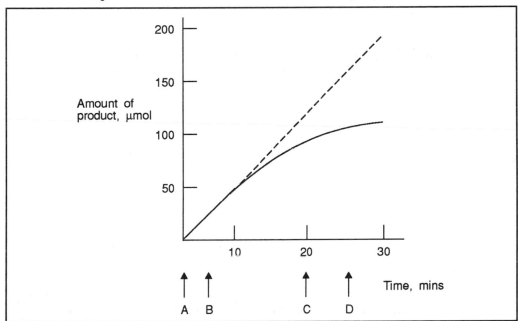

Figure 8.15 Progress of an enzyme-catalysed reaction with time. The amount of product formed is plotted on the y axis. The continuous line is a typical result; the dashed line is what should happen.

In practice, as shown by the continuous line, the reaction usually slows down. If we estimate the rate at different stages in the case above, we get very different results:

rate estimated over period A to B = 5 μmol/min

rate estimated over period C to D = 2 μmol/min

Of these, the higher value is the better estimate of enzyme activity. The second is seriously misleading and is a consequence of changes in the reaction mixture. There are various possible causes for this: any one of the following would be sufficient to produce the result shown:

- substrate is no longer saturating. We have already commented that only the enzyme concentration should be limiting in an enzyme assay. Consequently, we try to ensure that a high and saturating substrate concentration is used. With time, the substrate concentration may drop so that a lower reaction rate results (see Figure 8.19 in the next section, where the relationship between substrate concentration and velocity is described);

- product formed in the reaction may be inhibitory. We can visualise the reaction as follows:

$$E + S \rightleftarrows ES \rightleftarrows EP \rightleftarrows E + P$$

If product is allowed to build up, it may be temporarily binding to enzyme molecules, which will then not be available (however transiently) for reaction with substrate molecules. Hence the observed rate of reaction will slow down;

- the reverse reaction is occurring and equilibrium is being approached. Looking at the reaction scheme above, if we start with only enzyme and substrate, reaction can only be towards the right hand side. If product concentration builds up, the reverse reaction may begin to occur. Eventually an equilibrium would result, in which there was no net formation of product (reaction to the right proceeds at the same rate as reaction towards the left): the enzyme may be catalysing many millions of reactions but there would be no more product detected!;

- some change in reaction conditions may have occurred, leading to lessening of enzyme activity: this could include change of pH, such that the pH was no longer optimal, or the enzyme may have been denatured, so that fewer active enzyme molecules are present. We will learn later of the effects of pH and denaturation on enzymes.

initial velocity

This may seem a complicated problem for the experimenter, given the range of causes of slowing in rate. The solution is to avoid these problems by making measurements before the rate slows down. This is known as measuring the initial velocity (abbreviated as v_0 ie velocity at zero time), when conditions are still those predetermined by the experimenter (buffered at a known pH, with a known substrate concentration, no product present, controlled temperature). If this is done, all of the above problems will be avoided. In practice, one should check that the rate is linear over the chosen period of assay.

∏ More data to analyse! Attempts at assaying an enzyme yielded the data shown below. Recommend an assay duration for routine use (several hundred samples may have to be assayed per day!!), commenting on any circumstances when the experimenter may need to vary the strategy.

Incubation	Amount of product formed (μmol) after time (min)					
	0	1	2	5	10	20
(A) No enzyme	0	0	0	0	0	0
(B) 0.1ml enzyme	0	0.2	0.4	1.0	2.0	4.0
(C) 0.2ml enzyme	0	0.4	0.8	2.0	3.9	6.3
(D) 0.5ml enzyme	0	0.8	1.9	4.2	6.1	7.0

Hint : If in doubt how to analyse this, plot a graph!

The most important thing is to ensure that the initial velocity is being measured and that the reaction rate has not changed during the assay. Plot a graph of amount of product formed (y axis) against time (x axis). This shows us whether the rate is maintained for the full duration of the assay or not. This is shown here:

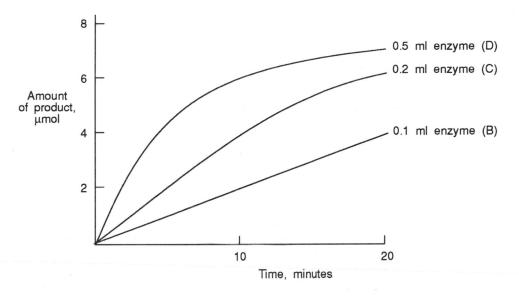

For incubation (B) there is no problem: the amount of product increases linearly with time. The duration of the incubation is a matter of convenience perhaps 5 or 10 minutes, because with shorter times, errors in timing will become more serious. In (C), the reaction is beginning to slow down, as seen by the curve between 10 and 20 minutes. For (D), this slowdown has occurred much earlier and net formation of product has slowed dramatically by 10 minutes.

Where does this lead us?

• Assays of 5 or 10 minutes, with 0.1 or $0.2cm^3$ enzyme solution, look suitable, providing the amount of product formed does not exceed 2-3μmol. If it does, we should either shorten the assay period or use a smaller volume of enzyme.

- We have assumed that the smaller amounts of product formed in incubation (B) are readily determined. If sensitivity is a problem (ie if it is difficult to measure small amounts), then larger volumes of enzyme or greater assay times may have to be used.

- Note that although we ran a no-enzyme control (incubation A), since no product was found in this, no correction to any other reading was needed. This will usually not be the case and controls must always be run.

8.3.4 Conduct appropriate controls

When we assay an enzyme we must ensure that we distinguish between the enzyme-catalysed reaction and any non-enzymatic product formation. For example, invertase hydrolyses sucrose to glucose and fructose; at low pH, acid hydrolysis will also occur. If we simply measure the amount of glucose formed, we will overestimate the enzyme activity if we do not correct for any non-enzymatic reaction (Figure 8.16).

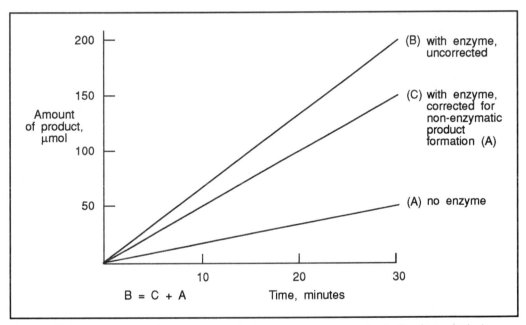

Figure 8.16 In an enzyme incubation, some product may arise non-enzymatically (line A); to obtain the true enzyme activity this must be deducted from that formed in the presence of enzyme (line B), giving the net product formation which is attributable to enzyme (line C).

boiled enzyme control

Various controls are conducted to allow non-enzymatic product formation to be deducted. The commonest is to use denatured enzyme (by boiling) to test for non-enzymatic reaction. Whatever product is found in this incubation is deducted from that found with the active enzyme. This approach ensures that any product present at the start of the assay (as a contaminant of the substrate or present in the enzyme extract), together with any formed non-enzymatically during the assay, is allowed for.

Π The data below illustrate the importance of controls. Calculate the enzyme activity, as μmol product formed per incubation per ml enzyme using the data from incubation (A). Then calculate the enzyme activity as μmol product formed per incubation per ml of enzyme, taking into account the additional information which incubation (B) provides.

		μmol Product at end of incubation
(A)	0.5 ml enzyme	15
(B)	0.5 ml boiled enzyme	8

If just the enzyme-containing incubation (A) had been conducted, we should have concluded that 1ml of enzyme would make 30 μmol product in the incubation. This gives an activity of 30 μmol product per ml enzyme.

If the boiled enzyme control (incubation B) is conducted, we must assume that this non-enzymatic product formation also occurred in incubation (A). We thus deduce that, of the 15 μmol formed in the presence of the enzyme in incubation (A), 8 μmol were formed non-enzymatically. Hence the 0.5 ml enzyme present only produced 15 - 8 = 7 μmol product, and its true activity was 14 μmol product per ml enzyme.

8.3.5 Use defined conditions

Ideally, enzyme assays are conducted at optimal conditions for the enzyme. This includes providing a substrate concentration which is saturating (ie not rate limiting), and any other compounds necessary for activity (eg Mg^{++} ions). The pH should be optimal and controlled by a suitable buffer. The temperature should also be held constant, preferably at a temperature which is unlikely to denature the enzyme such as 30°C or 37°C. Even if optimal conditions cannot be obtained, the conditions should be defined: only then can experiments be accurately described and repeated.

8.3.6 Enzyme units

International Unit

Enzyme activities are described in terms of units, where one International Unit (I.U.) is the amount of enzyme which will catalyse the transformation of 1 μmole of substrate per minute. Thus if 1 ml of extract caused the formation of 80 μmoles of product in 1 minute, we describe the enzyme activity as 80 U per ml of extract, or 80 IU ml^{-1}. A term which is widely used, especially during enzyme purification and when comparing enzyme activities, is specific activity. This expresses enzyme activities relative to the amount of protein present, specifically as enzyme units per mg protein (ie IU.mg protein^{-1}).

specific activity

8.3.7 Types of Assay

continuous and discontinuous assays

Two alternative approaches to assays are used. In one, samples of the incubation mixture are taken at predetermined times after adding the enzyme to a buffer-substrate mixture. The amount of product formed (or substrate lost) is determined by chemical analysis. This approach is often cumbersome (it is slow and separate chemical analyses are required) and there may be difficulty in measuring the initial rate. The second approach, the continuous assay, is preferred when it can be used. The first strategy described (sampling or discontinuous assay) is usually only used when a continuous assay is impossible. The continuous assay involves monitoring the product formation (or substrate removal) continuously, as it occurs. It is thus much easier to ensure that the initial velocity is measured and assays are generally faster and more convenient. Continuous assays rely on some measurable difference between substrate and product. We cite some examples here.

Assays involving changes in light absorbance are particularly convenient and widely used

Dehydrogenases are readily assayed by making use of differences in absorbance between oxidised and reduced forms of pyridine nucleotides. For example, lactate dehydrogenase catalyses the interconversion of lactate and pyruvate:

$$
\underset{\text{pyruvate}}{\begin{array}{c} COO^- \\ | \\ C=O \\ | \\ CH_3 \end{array}} \quad +NADH+H^+ \quad \rightleftharpoons \quad \underset{\text{L-lactate}}{\begin{array}{c} COO^- \\ | \\ HO-C-H \\ | \\ CH_3 \end{array}} \quad +NAD^+
$$

There are no suitable spectral differences between lactate and pyruvate but there are between NADH and NAD$^+$ (Figure 8.17). Whilst both absorb strongly at 260nm, only NADH displays absorbance at 340nm.

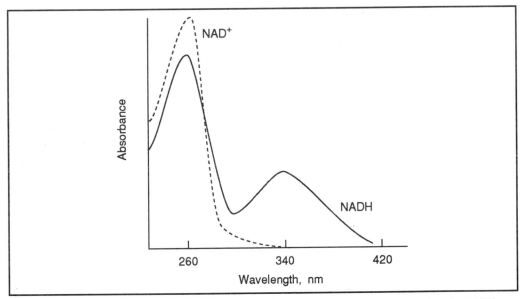

Figure 8.17 Absorbance spectrum for NAD$^+$ and NADH. The second absorbance peak given by NADH (with maximal absorbance at 340nm) allows [NADH] to be measured, even in the presence of NAD$^+$.

Thus, absorbance at 340nm is directly proportional to NADH concentration. In an assay, the rate of loss of NADH may be followed in a spectrophotometer (a device which allows light absorbance to be measured in both visible and ultraviolet wavelengths), by measuring the change in absorbance at 340nm. Thus the rate of change of absorbance ($\Delta A/\text{min}$) is proportional to enzyme concentration.

Artificial chromogenic substrate

artificial chromogenic substrate

When neither substrate nor product possesses convenient absorbance properties, assays based on absorbance change can sometimes still be used. We may be able to substitute an artificial chromogenic substrate for the natural substrate. For example, alkaline phosphatase normally removes phosphate groups from phosphorylated compounds in the cell, with no convenient absorbance change. The enzyme can also act on synthetic substrates such as 4-nitrophenyl phosphate, as follows:

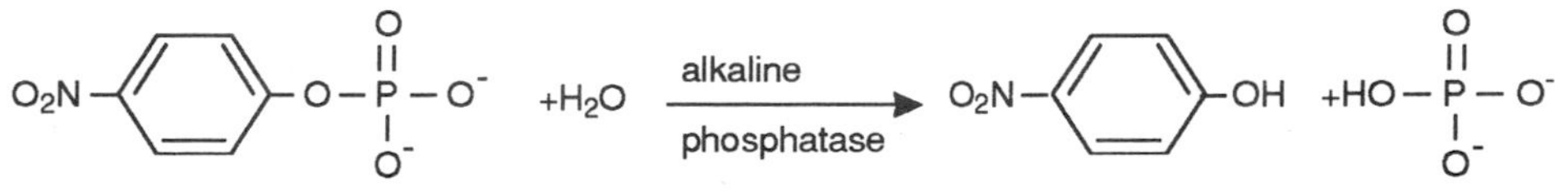

4-nitrophenylphosphate   4-nitrophenol   phosphate

4-nitrophenol absorbs strongly at 410nm, giving a yellow colour, whereas the substrate for this assay is colourless. Thus absorbance at 410nm increases in proportion to the amount of product formed. The 'chromogenic' of the title refers to the 'colour-generating' property on which the assay is based.

Coupled assays

An alternative approach is to use a coupled enzyme assay in which the enzyme reaction we wish to assay is coupled to a second, absorbance generating enzyme reaction. In outline terms, this may be depicted as:

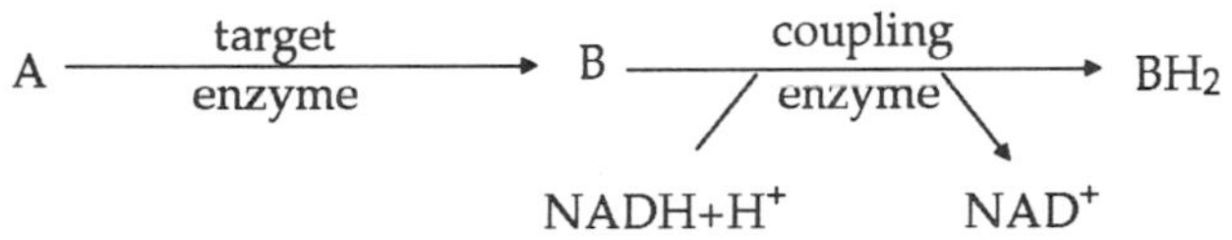

Providing the coupling enzyme and its cosubstrate (NADH) are present in excess, then as soon as 'B' is formed it will be converted to BH_2, with corresponding loss of NADH. Under these conditions, the rate of loss of NADH (and hence loss of absorbance at 340nm) will be determined by the rate of formation of B ie: ΔA_{340}/min is proportional to the activity of the target enzyme.

(NB ΔA_{340}/min is the rate of change in absorbance at 340nm).

Other methods

Numerous assays exist which do not depend on absorbance change. Thus change in proton concentration, $[H^+]$, gas uptake/release, or evolution of heat could be monitored. Conversion of radionuclide-labelled substrate to product and fluorescence assays are widely used, particularly when highly sensitive assays are required (eg when enzyme activity is very low). These assays are beyond the scope of this introduction.

SAQ 8.3

The following data was obtained during assay of an esterase; the reaction catalysed is as follows:

$$O_2N-\langle\!\!\bigcirc\!\!\rangle-O-\overset{\overset{O}{\|}}{C}-CH_2-CH_3 +H_2O \longrightarrow O_2N-\langle\!\!\bigcirc\!\!\rangle-OH +CH_3CH_2COOH$$

The reaction is monitored by the increase in absorbance at 410nm associated with the 4-nitrophenol produced (the substrate does not absorb at this wavelength). The incubation mixture contained the following ingredients:

	Volume (ml)
10mmol l^{-1} substrate, dissolved in water and adjusted to pH 7.0	2.0
Water	0.9

The sample assayed was an extract of a strain of the fungus *Aspergillus niger*. 5g wet mass of *A. niger* mycelium was homogenised with 25ml water. The homogenate was clarified by centrifugation at 20,000xg for 30 min at 4°C. The clear supernatant was used for the assay. The assay was started by addition of 0.1ml of the extract to the above incubation mixture. After 30 min at 25°C the entire 3ml of reaction mixture was analysed and found to contain 14 μmol of product.

1) Calculate the enzyme activity, in International Units ml^{-1} of extract.

2) If an independent protein assay showed the extract to contain 6mg protein ml^{-1} of extract, what is the specific activity of this enzyme solution?

3) Comment on any shortcomings in the conduct of the assay. How could the assay be improved?

8.4 Effect of substrate concentration on enzyme activity

In discussing enzyme assays, we mentioned that substrate concentration influences the rate of enzyme-catalysed reactions. In this section the precise relationship between these parameters will be explored.

first order reaction

If the rate of a chemical reaction (non-enzymatic) is analysed, it is often found to be proportional to substrate concentration (Figure 8.18a). This is referred to as a first order reaction, meaning that the rate is proportional to the concentration of a single compound (the substrate). ie rate α [S].

In contrast, early work with enzyme-catalysed reactions revealed a different relationship (Figure 8.18b).

∏ Examine Figure 8.18b and see if you can describe the differences.

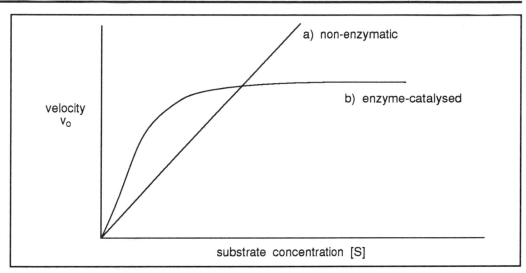

Figure 8.18 Relationship between rate and reactant concentration for a) non-enzymatic reaction; b) enzyme-catalysed reaction.

With enzymes, the rate increases in proportion to substrate concentration only at low substrate concentrations. At higher concentrations, the increase in rate lessens for unit increase in substrate concentration, eventually levelling off to a plateau. Under these conditions, further increase in substrate concentration has no effect on velocity. The rate is now said to be zero order, as rate is no longer determined by the precise substrate concentration. Between these two regions is a mixture of first and zero order kinetics.

zero order

Michaelis Menten

The mathematical relationship between initial velocity and substrate concentration was analysed in the early part of the twentieth century by Michaelis and Menten. They analysed the enzyme distribution on the basis of the following reaction scheme:

$$E + S \underset{k_2}{\overset{k_1}{\rightleftharpoons}} ES \underset{k_4}{\overset{k_3}{\rightleftharpoons}} E + P$$

where k_1, k_2 etc are rate constants. Catalysis was assumed to require the combination of enzyme (E) with substrate (S) to form an enzyme-substrate complex (ES). This reaction was reversible, but ES could also break down to form free enzyme (E) and product (P). In deriving an equation to describe the observed relationship between initial velocity and substrate concentration, they made several assumptions:

- that the concentration of enzyme, [E], was low, relative to that of substrate, [S]. This meant that formation of ES did not significantly alter the concentration of free substrate, [S], and hence simplified the analysis. Typically, this assumption is acceptable;

- that product was absent; as a consequence, the possible recombination of E and P to form ES could be ignored (thus any term involving rate constant k_4 was eliminated);

- that the concentration of ES was constant and was in equilibrium with E and S. This is referred to as the equilibrium assumption. Subsequent analysis showed this to be a restricted case. A wider generalisation none-the-less remained that [ES] was constant: a steady state exists between its formation (E + S $\rightleftharpoons$ ES) and its breakdown (ES $\rightleftharpoons$ E + S and ES $\rightleftharpoons$ E + P) and is established very soon after enzyme and substrate are mixed. The rate of the overall reaction (S → P) is given by k_3 [ES] and thus velocity is proportional to [ES]. The existence of ES was initially deduced by Michaelis and Menten, on the basis of the hyperbolic relationship between rate and substrate concentration; it has since been observed by electron microscopy and confirmed by spectroscopic techniques.

Michaelis Menten equation

These assumptions are used in the derivation of the Michaelis-Menten equation. At this stage you do not need to worry about deriving the relationship (which we therefore omit) but you should know and understand the Michaelis-Menten equation. This is usually given as:

$$v = \frac{V_{max}\,[\,S\,]}{K_M + [\,S\,]}$$

where v   =   initial velocity of the reaction ie v_o

[S]   =   substrate concentration in mol. l^{-1}

Vmax   =   maximum velocity, seen at (infinitely) high substrate concentration

K_M =   Michaelis constant, in mol. l^{-1}

We can now analyse the observed relationship between initial velocity and substrate concentration (Figure 8.19) in terms of the Michaelis-Menten equation.

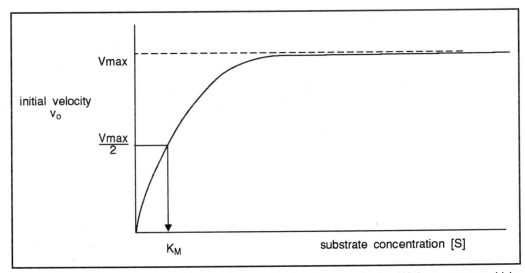

Figure 8.19 Relationship between initial velocity, v_o, and substrate concentration, [S], for an enzyme which obeys the Michaelis-Menten equation. The Michaelis constant, K_M, is the [S] at which half maximum velocity (1/2 Vmax) is displayed.

At low substrate concentration ([S] << K_M) the denominator approximates to K_M (because the contribution of [S] to K_M + [S] is negligible). Under these conditions, the equation becomes

$$v = \frac{V_{max}}{K_M} \times [S]$$

ie v α [S], which is consistent with our earlier comments at low substrate concentrations, where first order kinetics are observed.

At high substrate concentration, the contribution of K_M to the denominator can be ignored.

∏ Jot down in the margin the reason why the contribution of K_M can be ignored (we will give you an opportunity to check your reasoning later).

If we ignore the contribution of K_M at high substrate concentrations the equation becomes:

$$v = \frac{V_{max}\,[S]}{[S]}$$

ie velocity is maximal and zero order kinetics (with respect to substrate) will be displayed. Thus both features which we described qualitatively are consistent with the mathematical analysis.

What is the meaning and significance of the Michaelis constant K_M? We can answer this by analysing a third situation, when [S] is made equal to the K_M. The Michaelis-Menten equation then becomes:

$$v = \frac{V_{max}}{2} \qquad (\text{via} \quad v = \frac{V_{max}\,[S]}{[S] + [S]}\ )$$

Michaelis constant K_M The Michaelis constant, K_M, can be seen to be the substrate concentration which gives half-maximal velocity. This will be when half the enzyme active sites have substrate bound and are in the ES form. This is a useful operational definition for K_M. Other definitions are obtained from the derivation of the Michaelis-Menten equation; the interested reader may find these (together with the derivation of the equation) in standard biochemistry textbooks.

Now you can check your reasoning why K_M can be ignored at high [S] values. At high [S], the contribution of K_M to the denominator can be ignored because [S] >> K_M; thus

K_M + [S] approximates to [S].

The significance of the Michaelis constant is that knowledge of it enables us to predict over what substrate concentration range the activity of an enzyme will vary. The next in-text activity will amplify this point. Knowledge of the K_M of an enzyme also facilitates optimisation of assay conditions.

∏ What is the initial velocity of an enzyme-catalysed reaction at the following substrate concentrations, if K_M = 2mmol. l^{-1} and V_{max} = 100 μmol.min^{-1}? Assume that the enzyme obeys Michaelis-Menten kinetics.

	[S], mmol.l^{-1}
(A)	0.1
(B)	0.2
(C)	0.5
(D)	1.0
(E)	2.0
(F)	5.0
(G)	20.0
(H)	200.0

Try to do these calculations before reading on. It would be useful to plot a graph of the velocities (v) you have calculated against [S].

The initial velocities at the various concentrations are obtained by substitution into the Michaelis-Menten equation. For case (A), this gives:

$$v = \frac{V_{max} [S]}{K_M + [S]} = \frac{100 \times 0.1}{2.0 + 0.1} \text{ μmol.min}^{-1}$$

Therefore $v = \dfrac{10}{2.1} = 4.8$ μmol.min^{-1}

Corresponding calculations give:

	[S], mmol.l^{-1}	v, μmol.min^{-1}
(A)	0.1	4.8
(B)	0.2	9.1
(C)	0.5	20
(D)	1.0	33.3
(E)	2.0	50
(F)	5.0	71.4
(G)	20.0	90.9
(H)	200.0	99

If plotted as a graph, this will give the characteristic hyperbolic plot, with near proportionality between rate and [S] at low [S], and diminishing proportionality at higher [S] values, particularly above the K_M. Notice that the rate scarcely responds to a 10-fold increase in [S] when it is substantially (10x) above the K_M value. Note also that at very low [S], whilst the rate may change nearly proportionally with any change in [S] (look at (A) and (B)), it is none-the-less a small percentage of the maximum activity of the enzyme.

A low K_M means that an enzyme will bind substrate at low substrate concentrations. It is said to have a high affinity for its substrate. For example, if the K_M is 10^5 mol.l^{-1}, it will be 50% saturated when the substrate concentration is 10^5 mol.l^{-1}. It will be 91% saturated by a substrate concentration of 10^4 mol.l^{-1}.

A high K_M, in contrast, indicates that a relatively high substrate concentration will be required to achieve saturation. This is described as low affinity. As an example, a K_M of 10^2 mol.l^{-1} would require a substrate concentration of 10^1 mol.l^{-1} to display 91% of its maximum rate. At a substrate concentration of 10^4 mol.l^{-1}, it will only be working at 1% of its maximum rate! It may be impossible to achieve a high enough [S] to obtain Vmax.

Knowledge of K_M values is thus useful in understanding and predicting the behaviour of metabolic systems. As an example, consider the fate of a glucose molecule in the bloodstream. All cells obtaining glucose from the blood possess the enzyme hexokinase, which phosphorylates glucose according to the following scheme:

$$GLUCOSE + ATP \rightarrow GLUCOSE\text{-}6\text{-}PHOSPHATE + ADP$$

The phosphorylated glucose is now trapped within the cell (because of the negatively charged phosphate group it is no longer able to pass through the plasma membrane); this reaction is also the first step in glycolysis, leading to generation of ATP. Some cells, notably those of liver, possess a second enzyme which carries out this reaction, which is known as glucokinase. Glucose phosphorylated by this enzyme is predominantly stored (as glycogen) in the liver. Thus there are two alternative enzymes for the same reaction, one leading directly to energy generation, the other to a long-term energy store. What is the reason for this?

If we look at the relationship between glucose concentration and rate of reaction (Figure 8.20), we find marked differences between the two enzymes.

This is reflected in the K_M values for glucose (typically $0.7\text{-}5\times10^5$ mol.l^{-1} for hexokinase; 1.5×10^2 mol.l^{-1} for glucokinase). Hexokinases used for energy needed continuously are routinely saturated by the normal blood glucose concentration. If the blood glucose concentration rises (for example, after a meal), their activity is not affected: the substrate concentration had already made the reaction zero order with respect to glucose ([S] >> K_M). Activity of glucokinase, in contrast, does change as blood glucose concentration rises, leading to the excess glucose being taken up by liver and stored (until needed) as glycogen.

K_M values vary widely, typically from 10^6 to 10^2 mol.l^{-1}. Note also that they depend markedly on a variety of factors such as pH and temperature, as well as the particular substrate.

Vmax is the maximum rate that a given amount of enzyme can operate at and is achieved when substrate is saturating. A useful concept here is the turnover number of an enzyme. This is the number of substrate molecules transformed per second by a single enzyme molecule, when substrate is saturating; it is equal to the rate constant k_3. Turnover numbers vary widely, for example 2 molecules transformed per second for tryptophan synthetase to a remarkable 600,000 molecules per second for carbonic anhydrase.

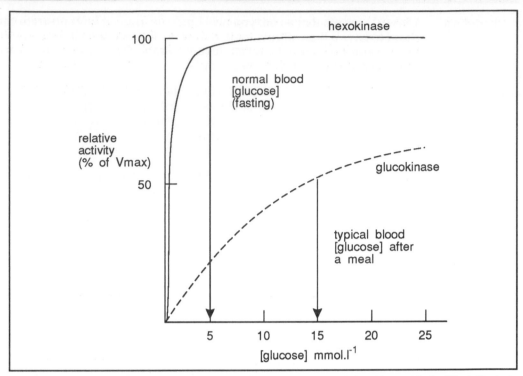

Figure 8.20 Effect of [S] on initial velocity for hexokinase (solid line) and glucokinase (dashed line). Typical blood glucose level during fasting (ie not immediately after a meal) is shown, as well as that after a meal

8.4.1 Determination of K_M and Vmax

When enzymes are being characterised and described, K_M and Vmax are routinely determined. How may this best be done? Experimentally, one measures initial velocity at a variety of substrate concentrations (having fixed other parameters such as pH and temperature). There may be a temptation to then plot the data as in Figure 8.19. From this direct plot, Vmax can, in principle, be obtained; K_M is then the substrate concentration which gives half maximal velocity. Resist this temptation! Whilst plotting the data in this way is useful as a check that Michaelis-Menten kinetics are occurring, satisfactory values for K_M and Vmax are unlikely to be obtained. If you look back to the in-text activity in which we calculated v at a variety of [S] values you will see that if the substrate concentration is 10x higher than the K_M, the rate is only 10/11th of Vmax. Even when the substrate concentration is 100x that of K_M, the rate is still only 99% of Vmax! It may thus be extremely difficult to experimentally obtain Vmax (it may be impossible, because the solubility of the substrate may not be adequate; it could also be very expensive!). If Vmax is not experimentally measured, then extrapolation of the hyperbola would be necessary, which could introduce additional error.

direct plot

difficult to obtain Vmax directly

These problems are avoided by various mathematical transformations of the data which result in linear plots (providing Michaelis-Menten kinetics are obeyed). These are obtained by taking reciprocals of the Michaelis-Menten equation, giving:

$$\frac{1}{v} = \frac{K_M + [S]}{Vmax\,[S]} = \frac{K_M}{Vmax} \times \frac{1}{[S]} + \frac{1}{Vmax}$$

Lineweaver-Burk
plot

This is the equation of a straight line (y = mx + c). Although various ways of plotting data have been devised, the most widely used plot is the Lineweaver-Burk plot (Figure 8.21), in which $1/v$ is plotted against $1/[S]$. For an enzyme which conforms to Michaelis-Menten kinetics, this will give a straight line, with slope of $K_M/Vmax$.

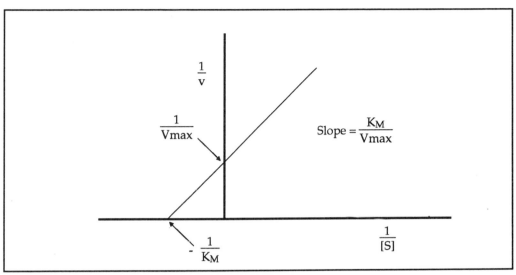

Figure 8.21 Lineweaver-Burk double reciprocal plot. For an enzyme which conforms to Michaelis-Menten kinetics, a straight line will be obtained. The intercept on the $1/v$ axis is $1/Vmax$, whilst that on the negative side of the $1/[S]$ axis equals $-1/K_M$.

The intercept on the $1/v$ axis equals $1/Vmax$; that on the $1/[S]$ axis equals $-1/K_M$. Since a linear relationship has been established, it is much easier and more accurate to extrapolate and the line of best fit can be obtained by regression analysis. This avoids the need to experimentally obtain Vmax. Indeed, data for a Lineweaver-Burk plot is generally most satisfactorily gained over the substrate concentration range $0.2\text{-}5.0\mathrm{x}\ K_M$.

SAQ 8.4

1) Calculate the K_M and Vmax for an enzyme, for which the following data is provided:

	[S], mmol.l^{-1}	Initial velocity, μmol.min^{-1}
(A)	0.2	0.57
(B)	0.33	0.85
(C)	0.5	1.18
(D)	1.0	1.82
(E)	2.0	2.5
(F)	5.0	3.23

You will need to plot a graph.

2) Assume that you have to accurately measure the K_M of an enzyme, for which preliminary estimates indicate that the K_M is about 2×10^{-3} mol.l^{-1}. Recommend what concentrations of substrate should be used for the careful analysis.

SAQ 8.5

Consider the following statements and then decide whether they are true or false, giving brief reasons.

1) If two enzymes have K_M values of 10^{-3} and 10^{-2} mol.l^{-1} respectively, that with K_M of 10^{-2} mol.l^{-1} will be more saturated when $[S] = 10^{-2}$ mol.l^{-1} than the other enzyme.

2) The K_M is the $[S]$ at which half maximal velocity is displayed.

3) K_M is most satisfactorily obtained from a plot of initial velocity against $[S]$.

4) In the Lineweaver-Burk double reciprocal plot, a straight line is indicative that Michaelis-Menten kinetics are being obeyed.

5) As a generalisation, a low K_M is characteristic of an enzyme with a high affinity for its substrate.

8.5 Effect of pH and temperature on enzyme activity

Both pH and temperature have profound influences on enzyme activity. The effect of pH is primarily related to the state of ionisation of particular groups in the enzyme and/or the substrate. We have already indicated that substrate specificity and substrate binding involve 'matching' or complementarity of charged groups. If the pH is such that the complementarity of charge no longer occurs, then some effect on enzyme activity should be expected. Whilst every enzyme has its own relationship between pH and velocity, a typical profile is the bell-shaped curve (Figure 8.22), in which there is a pH optimum at which the ionic states of enzyme and/or substrate are most appropriate.

pH optimum

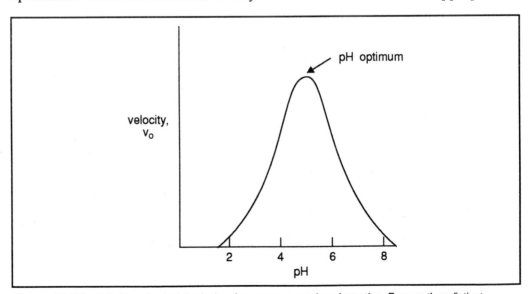

Figure 8.22 Effect of pH on the initial velocity of an enzyme-catalysed reaction. Frequently a distinct optimum pH is seen, although different enzymes have different pH optima.

effects of pH
on ionic state

As the pH is changed, a change in the ionic state of a key group occurs, such that it is no longer in the required form. For example, the side-chain of histidine will change from being 90% protonated (and hence positively charged) at pH 5.0 to 50% protonated at pH 6.0 (since pKa = 6.0 see Chapter 3) and 90% deprotonated (and thus uncharged) at pH 7.0 . Thus a decline in activity from the maximum at pH 5 to virtually zero at pH 7-7.5 indicates the involvement of a group with pKa around 6 and could mean that a protonated histidine is involved.

The bell-shaped activity curve we have shown is thus, in its simplest form, the consequence of changes in the ionic status of two crucial groups. If we examine the activity of lysozyme (which hydrolyses the peptidoglycan of bacterial cell walls), we observe a bell-shaped curve (Figure 8.23). X-ray diffraction analysis of the 3-dimensional structure, together with a range of other analyses, indicates that this results from changes in the side-chains of a glutamate residue (glu-35) and an aspartate residue (asp-52).

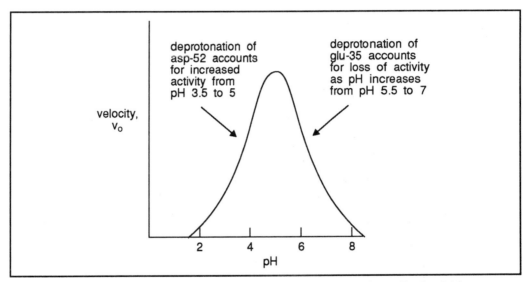

Figure 8.23 Effect of pH on the activity of lysozyme. The increase in activity from pH 3.5 to 5.0 is attributable to deprotonation of the sidechain carboxyl group pf Asp-52; The loss of activity from pH 5.5 to 7.0 is explained by deprotonation of the sidechain carboxyl group of Glu-35.

The increase in activity from pH 3.5 to 5.0 is consistent with deprotonation of asp-52 (which is required in its deprotonated -COO⁻ form), whilst the decrease in activity from pH 5.5 to 7.0 can be explained by deprotonation of glu-35 (which is required to be protonated, ie -COOH, for activity).

enzyme activity
over a broad
pH range

Not all enzymes display bell-shaped pH activity curves with distinct pH optima. Some, display constant activity over a wide range of pH: this simply indicates that no crucial changes in ionic state occur over this pH range (ie changes may occur but they do not affect residues involved in substrate binding and/or catalysis).

effects of pH
on
conformation The effects of pH referred to so far concern change in ionic state having little effect on the overall conformation of the enzyme. At extreme pHs, protein denaturation will occur, with loss of enzyme activity. This is frequently irreversible and represents a major change in the 3-dimensional shape (conformation) of the enzyme. Substantial changes

denaturation in charge occur, leading to unfolding, and ultimately exposure of previously buried (internal) parts of the protein, aggregation and precipitation. For enzymes which act at roughly neutral pHs, very low or very high pHs may be expected to cause denaturation and should be avoided. Because of the dependence of enzyme activity on pH and the

need for
buffers need to prevent denaturation, enzyme-containing solutions are routinely buffered. Care must be taken to ensure that the buffer does not interfere with the enzyme under study. It is equally clear why it is important for cells to strictly control the pH of their cytoplasm.

effect of
temperature Temperature affects the rate of most chemical reactions, so it is not surprising that the rate of enzyme-catalysed reactions increases with temperature. However, whilst increase in temperature accelerates the reaction (more substrate molecules possess the necessary activation energy), it can also cause denaturation of the enzyme, such that, with time, fewer active enzyme molecules are present. One must be careful not to confuse these two effects. Examine the data shown in Figure 8.24: this shows how the amount of product formed increases with time at various temperatures.

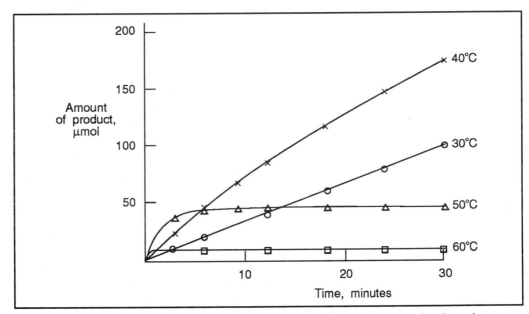

Figure 8.24 Effect of temperature on the amount of product formed in an enzyme-catalysed reaction.

At 30°C the rate is low; At 40°C it is about double that seen at 30°C (this is sometimes expressed as the temperature quotient, Q_{10}, which is the increase in rate over a 10°C rise in temperature; in this case $Q_{10} = 2$). However, denaturation is beginning to occur.

∏ What is our reason for saying that denaturation is beginning to occur at 40°C? Look carefully at Figure 8.24. Jot down the reason in the margin.

The rate of reaction is no longer linear with time at 40°C. The increase in amount of product, per minute, is less towards the end of the experiment than at the beginning, at

40°C. At higher temperatures, denaturation is even faster. There is apparently no denaturation at 30°C over the period of assay, as shown by the linear increase in product with time.

At 50°C reaction is faster still, but complete denaturation has occurred within a few minutes; at 60°C reaction was extremely rapid - and extremely short-lived!

The purpose of this lengthy description is to emphasise the importance of measuring initial velocity when assaying enzymes and to establish the different effects of temperature on enzyme activity. At elevated temperatures, unless great care is taken, an enzyme assay may not measure the true initial rate but may be affected (dominated?) by enzyme denaturation.

∏   Look now at Figure 8.25, which shows the calculated rate plotted against temperature for various assay periods.

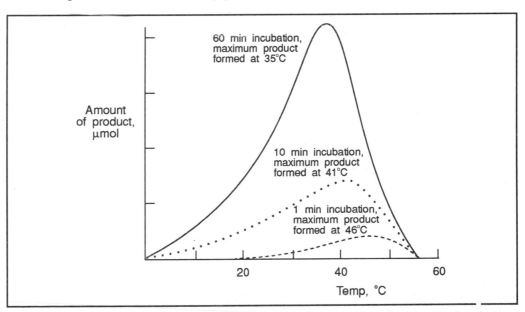

Figure 8.25 Apparent enzyme activity at different temperatures, calculated from the amounts of product formed over various time periods at the given temperatures. Note how the apparent 'optimum' temperature depends markedly on how long the assay was conducted for.

optimum
temperature

If a longer period is chosen, lower temperatures appear to be best (because the enzyme keeps going for longer) whereas if short assay periods are used (and the contribution of denaturation is thus avoided) higher temperatures appear best. These arguments show that the term 'optimum temperature', which is sometimes used, should be treated with great care, since it is likely to depend markedly on the way the assays were conducted. You have been warned!

We have seen in this section that extremes of pH and high temperatures can cause denaturation of enzymes. Other agents which can cause denaturation include organic solvents (ethanol, acetone), trichloroacetic acid, detergents and high concentrations of urea. Whilst major changes in conformation will eventually occur, loss of enzyme activity is frequently one of the first indications that denaturation is occurring. In working with enzymes, one tries to avoid conditions which will cause denaturation.

SAQ 8.6

Carefully consider the 5 statements which follow; decide whether they are true or false and jot down any comments or qualifying remarks which you think are relevant.

1) All enzymes have a precise temperature optimum.

2) Change in enzyme activity with pH is caused by change in the charge of active site groups.

3) The only effect that increasing temperature has on enzymes is to increase their activity.

4) In assaying an enzyme, you do not need to worry about denaturation.

5) Raising the temperature means that more substrate molecules possess the necessary activation energy and can therefore react.

8.6 Effect of inhibitors on enzyme activity

reversible and irreversible inhibitors

Enzyme inhibitors are compounds which diminish or eliminate the activity of an enzyme. For convenience, they are categorised as either reversible or irreversible inhibitors. As the name implies, reversible inhibition may be reversed, such that full enzyme activity is restored when the inhibitor is removed. In contrast, with irreversible inhibition, active enzyme is not regenerated, because removal of the inhibitor is impossible.

8.6.1 Reversible inhibition

Any compound which can bind to an enzyme in a reversible manner, and reduce its activity when bound, is classified as a reversible inhibitor. Such compounds may interact with different parts of the enzyme. They are designated, and distinguished, in the following ways. Reversibility is confirmed by recovery of enzyme activity on removal of the inhibitor, by dilution or by dialysis. Dialysis involves placing the enzyme solution in a sac made of a semi-permeable membrane (like cellophane) which allows low molecular weight compounds to cross, whilst retaining high molecular weight compounds like enzymes (the membrane acts as a molecular sieve). If this sac is placed in a large volume of buffer (Figure 8.26), the inhibitor will dissociate from the enzyme and become dispersed throughout the buffer in the vessel.

The concentration of inhibitor within the sac (and thus in the vicinity of the enzyme) will be much reduced.

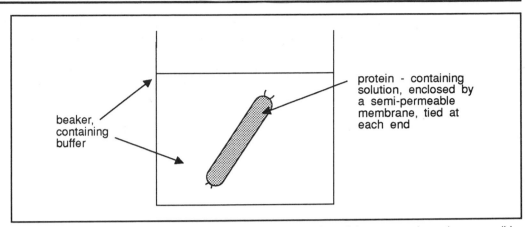

Figure 8.26 Dialysis of an enzyme solution to remove low molecular weight compounds, such as reversible inhibitors. Low molecular weight compounds can pass through the 'pores' of the semi-permeable membrane, whereas high molecular weight proteins cannot.

Competitive inhibition

If substrate and inhibitor compete for the active site of the enzyme, competitive inhibition results. This competition may be visualised as depicted in Figure 8.27, whereby if one compound is bound, then the other cannot bind.

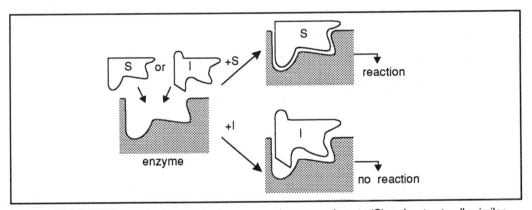

Figure 8.27 Competition for the active site of an enzyme between substrate (S) and a structurally similar compound (I).

It is helpful to think of the distribution of enzyme molecules in this way: in the absence of inhibitor, only the following occurs:

$$E + S \rightleftarrows ES \rightarrow E + P$$

With a competitive inhibitor present, an enzyme-inhibitor complex (which does not form product) is formed:

$$E + S \rightleftharpoons ES \longrightarrow E + P$$

$+ I \big\updownarrow$

EI

In presence of I, some E is present as EI, thus [ES] is lowered, therefore the rate of the reaction S → P is decreased.

Thus some of the enzyme is 'siphoned off' into a non- productive form which cannot form the ES complex whilst inhibitor is bound to it. Less enzyme is available for substrate binding, and [ES] is therefore lowered. Since the rate is proportional to [ES], the reaction is inhibited.

factors influencing the extent of inhibition

What factors determine the extent of inhibition? Looking at the enzyme distribution, increasing [I] will result in increased [EI], whereas increasing [S] will, conversely, result in increased [ES]. Thus both [S] and [I] will influence the enzyme distribution and hence the rate. In addition, the affinity of the enzyme for the inhibitor and substrate is important. The K_M serves, in general terms, as an indication of the enzyme-substrate affinity (ie how tightly do E and S bind to each other?). That for the inhibitor is equivalent to the dissociation constant of the enzyme- inhibitor complex. This is given by the expression

$$K_I = \frac{[E]\,[I]}{[EI]}$$

effects of inhibitors on K_M and Vmax

High substrate concentration will 'pull' the enzyme distribution over towards the ES form. Hence, even though inhibitor is present in the solution, it may still be possible for all enzyme molecules to bind substrate (ie form ES). Thus Vmax is not affected by a competitive inhibitor, although higher [S] will be needed to obtain it than in the absence of the inhibitor. The need for higher [S] will be reflected in an increase in the K_M value; this is shown in Figure 8.28a.

Experimental determination of the type of inhibition requires a similar approach to that used to obtain the Michaelis constant, K_M. The reaction rate is measured at various substrate concentrations (perhaps 6 different values ranged around the K_M value) in the presence and absence of a constant inhibitor concentration (or with 2 inhibitor concentrations). Direct analysis of velocity against [S] (Figure 8.28a) should not be used.

Instead, a Lineweaver-Burk double reciprocal plot is made (Figure 8.28b). The intercepts formed are characteristic of the type of inhibition. With competitive inhibition, the intercept on the 1/v axis is not altered (Vmax remains the same), whereas that on the 1/[S] is changed (K_M is increased).

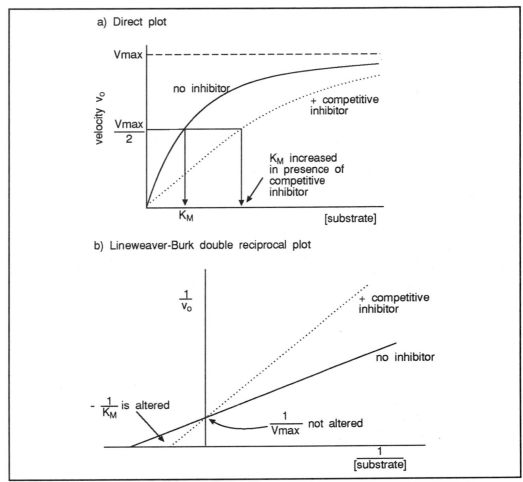

Figure 8.28 Effect of a competitive inhibitor on the initial velocity of an enzyme-catalysed reaction. a) Direct plot of v_o against [S]. b) Lineweaver-Burk double reciprocal plot.

Competitive inhibitors are usually compounds which have similar structures to the substrate. They can thus bind to the active site but do not undergo reaction. Krebs made use of such an inhibitor in his classical studies on the citric acid cycle : the step involved is that catalysed by succinate dehydrogenase:

$$\begin{array}{l} CH_2 - COO^- \\ | \\ CH_2 - COO^- \end{array} \xrightarrow[\text{dehydrogenase}]{\text{succinate}} \begin{array}{l} HC - COO^- \\ || \\ {}^-OOC - CH \end{array} \ +2H$$

malonate $\begin{array}{c} COO^- \\ \diagup \\ CH_2 \\ \diagdown \\ COO^- \end{array}$ can compete with succinate for the enzyme's active site

Malonate (also shown in the scheme above) has only one methylene (-CH$_2$) group but none-the-less binds to the active site of succinate dehydrogenase. There is no -CH$_2$-CH$_2$-bond from which a pair of hydrogen atoms can be removed (ie oxidation), therefore no reaction can occur. It is, however, a potent inhibitor because it prevents succinate binding and thus no ES complex is formed.

Non-Competitive inhibition

An alternative type of reversible inhibition occurs when a compound binds to the enzyme, at a site away from the active site, but still prevents reaction.

A non-competitive inhibitor does not prevent substrate binding: indeed, they can both be bound at the same time. As a consequence, substrate cannot displace the inhibitor and high [S] does not alleviate the inhibition. Inhibition is presumably caused by interfering with the catalytic steps, rather than substrate binding. The enzyme distribution can be summarised as follows for non-competitive inhibition:

$$E + S \rightleftharpoons ES \longrightarrow E + P$$
$$+ I \qquad\qquad + I$$
$$EI + S \rightleftharpoons EIS$$

Since substrate cannot displace the inhibitor, [S] no longer has any influence on the extent of inhibition. The amount of inhibition is solely determined by the concentration of the inhibitor, [I], and its affinity for the enzyme (described by the dissociation constant, K_I).

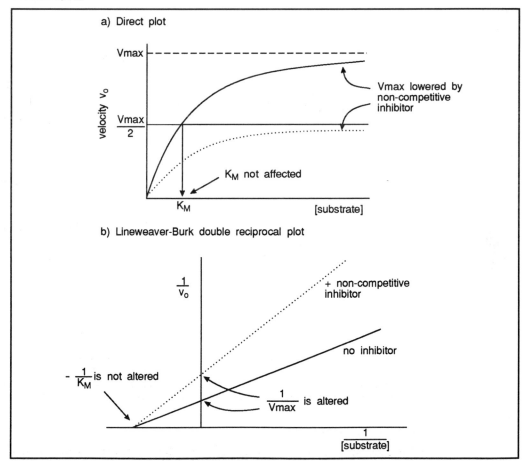

Figure 8.29 Effect of a non-competitive inhibitor on the initial velocity of an enzyme-catalysed reaction. a) Direct plot of v_0 against [S]. b) Lineweaver-Burk double reciprocal plot.

effects on K_M
and Vmax A plot of velocity against [S] for an enzyme-catalysed reaction in the presence of a non-competitive inhibitor is shown in Figure 8.29a. When inhibitor is present, some enzyme molecules are non-productive. It does not matter whether they are in the form EI or EIS. Thus less active enzyme is present and Vmax is lowered. Substrate binding by the enzyme associated with I is not affected (K_M is not altered).

As with competitive inhibition, identification of non-competitive inhibition is based on the pattern given by a Lineweaver-Burk double reciprocal plot (Figure 8.29b). In this case the $1/v$ intercept is altered (Vmax is lowered), reflecting the fact that some enzyme molecules are out of action or are working less efficiently. The intercept on the $1/[S]$ axis, from which K_M is obtained, is unchanged: this is consistent with the inhibitor having no effect on substrate binding.

Non-competitive inhibitors are less easily visualised than competitive ones. One way is to think of the inhibitor causing a conformation change (Figure 8.30a) such that the enzyme is less or in-active. Heavy metal ions (especially Hg^{++} and Pb^{++}) are often inhibitory because they bind tightly to sulphydryl groups (ie cysteine sidechains). This can occur away from the active site yet still disrupt catalysis, giving non-competitive inhibition.

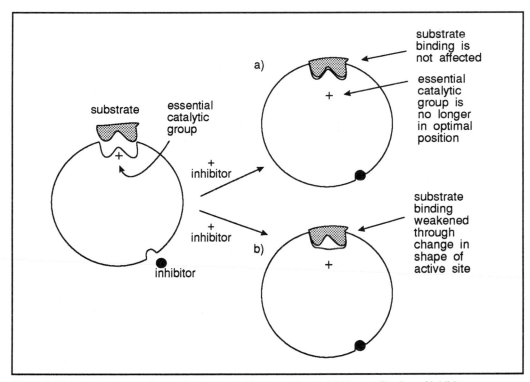

Figure 8.30 Possible mechanisms of non-competitive and mixed inhibition. a) Binding of inhibitor causes a conformation change which does not affect substrate binding, but adversely affects catalysis. b) Binding of inhibitor influences both substrate binding and catalysis, although substrate binding is not prevented.

Mixed inhibition

mixed inhibition Whilst the two types of reversible inhibitor described are widespread, mixed inhibition, involving a combination of the two types, can occur. For example, imagine an enzyme which has an important sulphydryl group (Figure 8.30b); if an Hg^{++} ion binds to it, its activity is diminished. If the bound Hg^{++} also interferes with substrate binding, then the apparent K_M will also be raised. Thus both Vmax and K_M will be affected, and the observed Lineweaver-Burk pattern will not be consistent with either 'pure' competitive or 'pure' non-competitive inhibition.

SAQ 8.7

Which of the following statements are correct; write down reasons for rejecting the other statements as incorrect.

A competitive inhibitor:

1) lowers Vmax and raises K_M;

2) lowers both K_M and Vmax;

3) raises K_M and leaves Vmax unaltered;

4) lowers Vmax, whilst K_M is unchanged;

5) results in complete abolition of enzyme activity.

8.6.2 Irreversible inhibition

effects on K_M and Vmax

Irreversible inhibition involves interaction between enzyme and inhibitor such that dissociation of the inhibitor does not occur. This is accomplished by agents which react covalently with an important functional group of the enzyme. Since reaction is covalent, neither dilution nor dialysis will remove the inhibitor, nor will addition of excess substrate affect the situation. If active site residues are involved, causing complete inactivation upon reaction, then the proportion of active enzyme molecules is reduced. Non-reacted enzyme molecules show normal kinetics. K_M is therefore unaffected whilst Vmax is lowered. Irreversible inhibition is distinguished from reversible non-competitive inhibition by failure to restore full activity upon dialysis. In addition, reversible inhibition is usually instantaneous; irreversible, covalent inhibition is progressive and may take minutes or hours to reach its maximum effect.

iodoacetate as an alkylating agent

A classical example is iodoacetate, which alkylates sulphydryl groups (Figure 8.31a). If a cysteine sidechain which is essential for activity is modified, enzyme activity will be irreversibly lost. It has been known for many years that the glycolytic enzyme glyceraldehyde-3-phosphate dehydrogenase is subject to inhibition by alkylation, due to an essential, active site, sulphydryl group. This was used experimentally to inhibit glycolysis and thereby facilitated studies on intermediary metabolism.

active site directed inhibitors

Certain irreversible inhibitors are said to be active site directed (or active site specific). In some cases, this is because of the very high reactivity of particular amino acid residues. Thus diisopropylfluorophosphate reacts with only one serine residue in chymotrypsin (Figure 8.31b), despite the presence of a further 27 serine residues in the protein. The reactive serine is located in the active site and is particularly reactive because of its unique location.

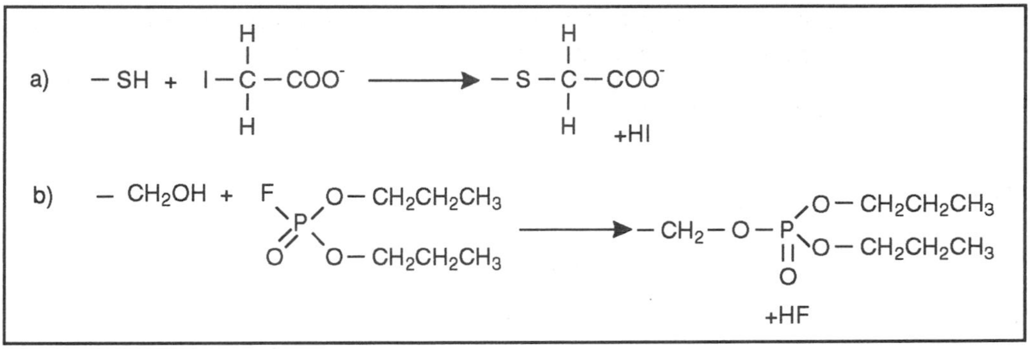

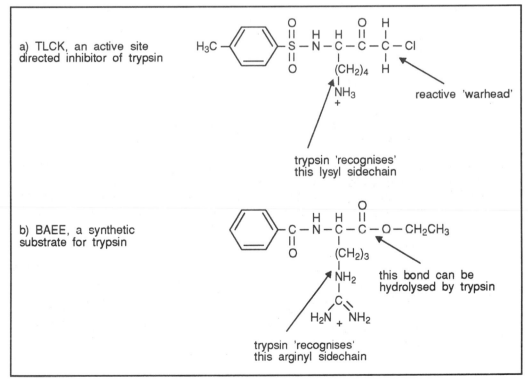

Figure 8.31 a) Alkylation of the sulphydryl group of cysteine by iodoacetate. If the -SH group is essential for activity, the enzyme will be irreversibly inactivated. b) Covalent modification of the serine sidechain in the active site of chymotrypsin by diisopropylfluorophosphate.

Other active site directed inhibitors have been developed by analogy with artificial substrates. They contain a 'recognition' site, which leads to binding to the enzyme. However, once bound, the enzyme cannot act. Instead, a 'warhead' is now appropriately positioned to react covalently with the enzyme, leading to irreversible inhibition. An example is TLCK (Figure 8.32a), an inhibitor of trypsin; this is strikingly similar to synthetic substrates of trypsin, such as BAEE (Figure 8.32b). TLCK is tosyl-L-lysine chloromethyl ketone and BAEE is benzoyl-L-arginine ethyl ester.

Figure 8.32 Active site directed inhibitors. These are designed to resemble substrates, such as BAEE (structure b); the inhibitor does not have a susceptible bond, but instead has a reactive 'warhead', as in TLCK (structure a).

Both substrate (BAEE) and inhibitor TLCK are 'recognised' by the enzyme; BAEE contains a susceptible bond which is hydrolysed by trypsin; instead, TLCK has a highly reactive chloromethyl ketone group. This reacts with a crucial histidine in the active site. Whilst diisopropylfluorophosphate reacts with several enzymes, all of which contain active site serine residues (acetylcholinesterase, trypsin, chymotrypsin, elastase), TLCK only reacts with trypsin. Such reagents have been widely used to study enzyme active sites and have enormous potential as drugs.

SAQ 8.8

The following data was obtained in an attempt to determine the type of inhibition produced by a particular compound. Enzyme assays were conducted in the presence and absence of a fixed concentration of the inhibitor at various substrate concentrations. In all other respects, the assays were identical. The data shows the [S] used and the initial velocity in the absence (column A) and presence (column B) of inhibitor.

Substrate Concentration, [S], mmol.l^{-1}	Initial Velocity, μmol.min^{-1},	
	A (- Inhib)	B (+ Inhib)
1.0	2.33	1.20
1.43	2.94	1.52
2.0	3.60	1.85
3.0	4.44	2.27
6.0	5.71	2.94

1) Using whatever procedure you consider to be most appropriate, determine what type of inhibition is occurring.

2) Are there any additional experiments which you recommend should be conducted to confirm your conclusions. Briefly outline these.

8.6.3 Significance of enzyme inhibitors

product inhibition, feedback inhibition, end product, allosteric regulation

The activity of numerous enzymes is inhibited by the presence of the product of the reaction. This is competitive, reversible and is known as product inhibition. It means that if the product is not being utilised (and as a consequence its concentration builds up), then further synthesis is inhibited. A special type of inhibition, called feedback inhibition, involves inhibition of the first enzyme of a pathway by the end-product of the pathway. This is an example of allosteric regulation, which is considered further towards the end of this chapter.

Many antibiotics, drugs and toxins, whether natural or synthetic, have their effects through inhibition of enzymes. A selection of these are grouped in Table 8.1, with comments on their mode of action and significance. Pharmaceutical companies spend considerable resources on discovering, designing and developing enzyme inhibitors.

Compound	Enzyme inhibited	Comments
Glyphosate	5-enolpyruvylshikimate-5-phosphate synthase (in pathway of aromatic amino acid synthesis)	Active ingredient of Tumbleweed* and Roundup* weedkillers; non-toxic to humans
Penicillin	Glycopeptide transpeptidase (synthesis of peptidoglycan cell wall of bacteria)	Penicillins and derivatives form basis of many bactericidal treatments
Iproniazid	Monoamine oxidase (MAO)	Anti-depressant, through inhibition of monoamine oxidase
6-mercaptopurine	Various enzymes of purine biosynthesis	Anti-cancer agent
Malathion, and other organo-phosphorus compounds	Anticholinesterase	Potent insecticides and nerve poisons, through covalent modification of acetylcholinesterase

Table 8.1 A selection of enzyme inhibitors.
*NB Tumbleweed and Roundup are trademarks used by the manufacturers of these weedkillers.

roles of protein inhibitors of enzymes

Enzyme inhibitors which are themselves proteins also occur widely in nature. Examples are shown in Table 8.2. Whilst their precise function is often unknown, they are presumed to either be part of a metabolic regulation system (eg invertase inhibitors preventing breakdown of sucrose by invertase) or have a defence role against pests. For example protease (trypsin) inhibitors in bean seeds may stop pests larvae from digesting the food they have eaten and therefore they will die of starvation. In humans, elastase activity in the lungs is normally inhibited by a protein inhibitor called (confusingly!) α_1-antitrypsin. If the level of α_1-antitrypsin is low (either hereditary or by inactivation of the inhibitor by smoking), elastase is no longer controlled and causes destruction of the elastic tissue of the alveoli of the lungs. This is the basis of emphysema.

basis of emphysema

Enzymes inhibited	Source
Trypsin, chymotrypsin and other proteases	Legumes and cereal seeds
α-Amylase	Cereal seeds
Invertase	Sugar beet
Elastase	α_1-antitrypsin, in serum
Thrombin	anti-thrombin III

Table 8.2 Naturally occurring enzyme inhibitors which are themselves proteins.

Π Glyphosate is one of the most commercially successful weedkillers. It is the active ingredient of the weedkillers sold under the tradenames of Tumbleweed and Roundup. We noted in Table 8.1 that a side-attraction of glyphosate was that it was of very low toxicity to man and other mammals. Write down in the margin why this might be so. This is a very open-ended question, which should allow your developing biochemical knowledge to show itself! If you don't know where to begin, just speculate - try to come up with at least two possible reasons.

- the glyphosate is metabolised (ie detoxified) before it can do any harm;
- it is not absorbed by humans;
- it is bound by a different protein, without adverse effects;
- the corresponding enzyme in humans is insensitive to the inhibitor;
- the corresponding enzyme in humans is absent.

In fact, the last of the possibilities above is the correct explanation. We lack this enzyme (and the rest of the pathway). As a consequence we require tryptophan and phenylalanine in our diet (they are essential amino acids), but at least we are safe from glyphosate!

This exercise is not entirely frivolous. Two points are significant:

- comparative biochemistry and physiology can help us design 'safe' drugs/toxic agents;

- our list above contained, as a possibility, a form of the enzyme which was insensitive to glyphosate. Possession of such an enzyme by a plant would confer resistance against glyphosphate. Precisely such an approach has been used by the American companies Calgene and Monsanto to produce (by genetic engineering) glyphosate-resistant plants.

8.7 Effect of cofactors

cofactors

Cofactors are non-proteinaceous compounds which are essential for the activity of some enzymes. Not all enzymes require cofactors, many are complete as one (or more) polypeptide chains. However, many enzymes do need an additional compound to be catalytically active. Cofactor is the collective term for these. They may be loosely categorised in one of 3 sub-groups.

8.7.1 Prosthetic groups

tightly bound

FAD
biotin

These are organic cofactors which are tightly bound (sometimes covalently) to the enzyme, such that they are normally present after isolation of the enzyme and are not removed by dialysis. Some, however, may be released on denaturation. Examples include the flavin adenine dinucleotide (FAD) of succinate dehydrogenase and biotin of carboxylases such as pyruvate carboxylase.

8.7.2 Coenzymes

NAD^+
$NADP^+$

These are organic cofactors which are more easily removed from the enzyme, for example by dialysis. Many coenzymes, such as NAD^+ and $NADP^+$ used by many dehydrogenases, can be viewed as though they are a second substrate of the enzyme reaction. The reason they are called coenzymes is partly historical, in that they were originally detected as thermally stable compounds which restored enzyme activity when added back to dialysed enzyme solutions (hence 'co'-enzyme). Coenzyme is a useful term because such compounds (eg NAD^+) act in the same way in many reactions. In the case of NAD^+, it accepts reducing equivalents from a variety of compounds, depending on the particular enzyme (eg malate with malate dehydrogenase, glutamate with glutamate dehydrogenase, lactate with lactate dehydrogenase). Unlike the individual products of the various enzymes, which are further metabolised, the

individual products of the various enzymes, which are further metabolised, the coenzymes are converted back to their starting form. There is a small pool of each coenzyme in the cell.

8.7.3 Metal Ions

haem

Some enzymes require a metal ion to be catalytically active. They may be loosely or tightly bound. In some cases they are associated with a prosthetic group, for example Fe^{++} or Fe^{+++} in the haem group of the cytochromes. Otherwise, the free metal ion is required for activity. Examples include Zn^{++} in carboxypeptidase A and Cu^{++} in cytochrome oxidase. The presence of a catalytically-essential metal ion can make enzymes susceptible to non-competitive inhibition by chelating agents, such as EDTA.

8.7.4 Vitamins and cofactors

vitamins may act as cofactors

Vitamins are organic compounds which many animals require in small amounts in their diets in order to remain healthy. They are needed because the animal fails to synthesise adequate quantities itself. In many cases, the requirement for the vitamin is because it (or a derivative) is a cofactor. Table 8.3 contains several of the vitamins which have important roles as cofactors, together with examples of their roles.

Vitamin	Example of cofactor (Prosthetic group/coenzyme)	Example of role
Nicotinic acid	NAD^+, $NADP^+$	Oxidation/reduction reactions involving transfer of hydride (H^-) ion
Riboflavin (vitamin B_2)	Flavin adenine dinucleolide (FAD)	Oxidation/reduction reactions
Folic acid	Tetrahydrofolate	1-carbon transfers
Biotin (vitamin H)	Biotin	Carboxylation reactions
Pyridoxine (vitamin B_6)	Pyridoxal phosphate	Amino acid metabolism: transaminations and decarboxylations
Thiamin (vitamin B_1)	Thiamin pyrophosphate	'Active' acetaldehyde transfer (in transketolase)

Table 8.3 Vitamins which act as cofactors.

Vitamin deficiency usually leads to characteristic symptoms, which can ultimately be traced to deficiency of particular enzymes.

holoenzymes

apoenzyme

Enzymes which require cofactors for activity are sometimes called 'holoenzymes' when in their complete (active) form. Removal of the cofactor causes loss of activity. The remaining (catalytically inactive) protein moiety is then known as an 'apoenzyme' (Figure 8.33).

Activity of an apoenzyme may often be restored by addition of cofactor: this has for many years been a way of assaying cofactors.

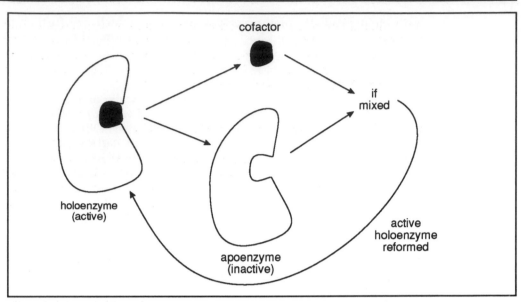

Figure 8.33 In cases where cofactors are needed for enzyme activity, removal of the cofactor converts active holoenzyme to an inactive form; the inactive protein is known as an apoenzyme.

Π Match the following experimental outcomes (1-3) with the most likely explanation from those given (a-c). Each explanation is only appropriate once!

1)   Dialysis of an enzyme solution led to loss of activity. Addition of dialysate (ie the fluid surrounding the protein-containing dialysis sac) regenerated activity, providing it had not been boiled.

2)   Treatment with EDTA caused inhibition.

3)   Dialysis had no influence on enzyme activity. Treatment with 6 mol.l^{-1} urea prior to dialysis caused loss of activity.

Possible explanations

a)   An organic prosthetic group is required for activity

b)   An organic coenzyme, which is not stable to boiling, is needed.

c)   For activity, an inorganic ion is required.

1)   The reversible loss of activity upon dialysis indicates that a low molecular mass compound is dissociating from the enzyme. Since the ability to restore enzyme activity is lost on boiling, this compound is presumed to be organic. Thus b) is the explanation (coenzymes readily dissociate; prosthetic groups do not).

2)   EDTA is a chelating agent, which binds metal ions such as Mg^{++}. Inhibition by EDTA strongly suggests that some metal ion is required for activity. c) is the answer.

3) $6mol.l^{-1}$ urea is a powerful protein denaturant, which will unfold most enzymes (they will adopt random coil conformations). Since dialysis only caused inactivation after unfolding by urea, we must assume that there is a tightly (but not covalently) bound factor which is essential for activity. This is known as a prosthetic group and a) is the correct response.

8.8 Mechanisms of enzymatic catalysis

In Section 8.2, we identified the basis of enzymatic catalysis as being the provision of an alternative 'reaction-route', for which less activation energy was required. This meant that a greater proportion of the population of substrate molecules possessed the required level of energy and could therefore react. In this section, we shall explore how this overall effect is accomplished, in molecular terms. This is a somewhat simplified treatment.

relationship
between k and
activation
energy

Before studying the ways by which enzymes lower the activation energy, it is worth establishing what effect reduction in activation energy actually has on the rate. In Figure 8.4 (Section 8.2), the enzyme-catalysed route is shown with an activation energy which is about a third of the size of the uncatalysed activation energy. Does this relatively modest change in activation energy really lead to the massive increases in reaction rates which enzymes achieve? We remind you that the overall rate of reaction = k[ES] where k is the rate constant for the breakdown of ES. It can be shown that the relationship between this rate constant k and activation energy is exponential. Thus a small reduction in activation energy produces a large increase in rate. This is shown in Table 8.4. The rate of reaction increases by a factor of about 2.4×10^4 for every drop of 25 kJ mol^{-1} in activation energy.

$\Delta G°$, J.mol^{-1}	k
100,000	1.84×10^{-5}
75,000	4.4×10^{-1}
50,000	1.06×10^4
25,000	2.5×10^8

Table 8.4 Effect of change in activation energy on enzymatic rate enhancement.

An enzyme able to reduce the activation energy from 100 to 50 kJ mol^{-1} would increase reaction rate by a factor of about 570 million!

identification of
amino acids of
the active site

In studying enzyme mechanisms, we need to establish, in chemical terms, precisely what the enzyme does. We also need to establish which amino acid residues are important in the active site and their three dimensional relationship. Only then can we make plausible suggestions as to how rate enhancement is achieved. Active site residues can be identified by chemical modification of particular residues. Active-site directed inhibitors are particularly useful. We have already noted that diisopropylfluorophosphate reacts only with serine-195 in chymotrypsin. This was established by covalent modification, followed by isolation and amino acid sequence analysis of the single peptide containing the modified serine. This also enables the sequence of amino acid residues around one of the key catalytic groups to be identified. With chymotrypsin, other methods have implicated histidine- 57 as another crucial residue.

X-ray diffraction

The precise 3-dimensional geometry of the active site can only be determined by X-ray diffraction analysis. With this information, model-building studies enable the positioning of substrate within the active site to be predicted. In some cases, X-ray diffraction analysis can be carried out with a substrate analogue bound to the active site. These approaches allow a detailed chemical mechanism to be proposed, describing the precise role of the catalytic groups of the enzyme.

Detailed descriptions of individual enzyme mechanisms are beyond the scope of this book. We conclude this section by summarising the main ways by which enzymes are believed to achieve catalysis. Note that different enzymes use different ways to achieve rate increases.

8.8.1 Proximity

close contact

Most reactions involve more than one substrate, although in hydrolytic reactions one of the reactants is the enzyme's solvent, water. For reactions to occur, the two substrates must come close enough (or collide) to react. One way by which enzymes promote reaction is by facilitating the necessary close contact. Since both substrates bind to the enzyme they are brought into close proximity with each other (Figure 8.34). Another way to think of this is that their localised concentration is much higher than it was in free solution. As we have already noted (Section 8.4), chemical rates of reaction increase when substrate concentrations rise.

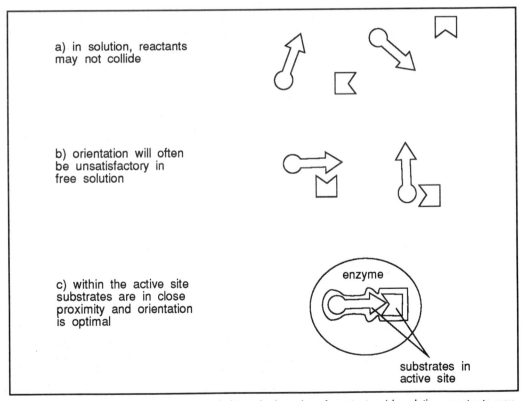

Figure 8.34 Effect of an enzyme on the proximity and orientation of reactants. a) In solution, reactants may not become close enough to react. b) Collisions between substrates in free solution are often unproductive because the wrong parts of the molecules meet; for reaction to occur they must be correctly oriented. c) The precise geometry of the active site means that when the two substrates are bound, they are not only close together but also meet in the optimal orientation.

8.8.2 Orientation

orientation

transition state

For reaction to occur between two molecules, not only must they collide, they must do so such that the reactive parts of each molecule are correctly positioned or 'oriented'. Only then will the so-called 'transition state' (an intermediate form between substrate and product) be entered. This requirement means that, when reaction occurs in free solution, most collisions will be unproductive, because the 'wrong' parts of the molecules meet (Figure 8.34b).

orbital steering

Even a small departure from the optimal alignment may mean that the energy required to enter the transition state is greatly increased. In contrast, the active site of an enzyme binds substrates in a precise position. It is likely that , through evolution, enzyme active sites bind substrates so that they are optimally oriented for reaction (Figure 8.34c). Another way in which this may be thought of is that correct alignment of the electronic orbitals of the reacting molecules is vital. The enzyme achieves this, during binding of substrates, by 'orbital steering' ie steering the reactants into the correct orbital positions.

8.8.3 Strain

bond distortion

One way by which the activation energy requirement is lowered is by the enzyme only binding a distorted ('strained') form of the substrate. This is accomplished by the 'complementarity' between active site and substrate being correct except in the region of the bond to be reacted upon. Here, for binding to occur, the substrate's shape must be altered (perhaps because, unless this distortion occurs, two atoms would be forced too close together). This distortion has the effect of raising the energy level of the reactant, such that the difference between this level and that of the transition state is lessened. This is certainly one of the mechanisms used by the enzyme lysozyme, where it probably contributes a rate enhancement of approximately 3000.

8.8.4 Covalent catalysis

reaction intermediate

Amino acids in the active site can provide a number of reactive groups, including R-COO⁻ and R-OH. These groups can react with electrophilic (ie electron deficient) parts of the substrate forming a covalent bond between the enzyme and the substrate. This results in a reaction intermediate being formed, which is less stable and will react more rapidly than would the reactant alone. Thus the enzyme is providing an alternative reaction route (as compared to the uncatalysed reaction) which has a lower activation energy requirement. Rate acceleration will occur.

Chymotrypsin is a particularly well-studied enzyme which utilises a covalently-bound intermediate to achieve catalysis.

8.8.5 Acid-base catalysis

proton donation

This concept helps to demonstrate the importance of the correct positioning of catalytic functional groups in the active site. These groups may be able to donate or accept protons (or electrons) to/from the substrate. This may contribute to reaction. Thus acid or base catalysis is an important component of rate enhancement. For example, in lysozyme, glutamate-35 is in an environment which means that its side-chain will still be protonated at the pH at which lysozyme is active. It is also perfectly positioned to 'donate' this proton to the substrate, thereby promoting bond cleavage. This is an example of acid catalysis.

There is also here an element of the effect on concentration discussed earlier. Proton concentrations are not high in cells (at pH 7.0, the [H⁺] is only 10^{-7} mol.l⁻¹). The ability of an active site group to provide a proton in this way may mean that non-availability of protons is eliminated.

8.8.6 Solvent effects and reactivity

changed pKa

The local environment of an active site can lead to enhanced reactivity of groups. In the case we have just mentioned, the side-chain of glu-35 in lysozyme is still protonated at pH 5-5.5 (the pH optimum of the enzyme). This should seem surprising since the pKa for the side-chain carboxyl group is normally 4.3. The pKa of 4.3 is for an aqueous solution. In a hydrophobic environment, the pKa will be raised (this is because the interaction between the -COO⁻ and the H⁺ is strengthened in a region of low dielectric constant). In lysozyme, glu-35 is surrounded by hydrophobic amino acid residues, such that the pKa becomes about 6.5. The side-chain of glu-35 is thus fully protonated at pH 5 and able to donate a proton. This type of effect is often known as a 'solvent effect'.

Enhanced reactivity is also seen with trypsin and chymotrypsin where (as we have already noted) an active site serine is particularly reactive. Although there are 34 serine residues in trypsin, only the active site one, because of its local environment, reacts with diisopropylfluorophosphate.

8.8.7 Stabilisation of the transition state

Anything which promotes the formation of the transition state will enhance the rate of reaction. One way in which this can be achieved is by having active site groups perfectly positioned to stabilise charges which arise on the transition state. One example where this occurs is lysozyme, where a positive charge is believed to occur in the transition state (there being no charge on the substrate). This positively charged intermediate is stabilised by the presence nearby of a negatively charged aspartate side-chain (asp-52). Thus the active site is even more complementary (sterically and electrostatically) to the transition state than to the substrate. As a consequence, enzymes bind the transition state even more tightly than the 'un- activated' substrate. We can summarise this by saying that the enzyme stabilises the transition state and this will have the effect of promoting its formation. This approach has important implications for the design of drugs, since we have seen that the transition state is bound more tightly than the substrate. Analogues of the presumed transition state should bind more tightly (and hence be more effective inhibitors) than analogues of the substrate of an enzyme.

∏ This has been quite a long section in which we have discussed seven different methods by which enzymes may speed up reaction. Will you be able to remember all seven? It might be helpful to you to list them on a piece of paper and pin this up where you will see it.

8.9 Allosteric enzymes and regulation of enzyme activity

So far we have explained how enzymes work and have described them earlier as the 'workers' of cells. But can they be controlled? In this section we examine one aspect of regulating enzyme activity.

multiple
subunits

Whilst many enzymes display a hyperbolic relationship when initial velocity is plotted against substrate concentration, and thus obey Michaelis-Menten kinetics, others do not. Instead, a plot of initial velocity against [S] gives a sigmoidal curve (Figure 8.35). Such enzymes usually consist of several subunits, although possession of multiple subunits is not a prerequisite for the sigmoidal relationship. What must occur, however, is communication between binding sites. This is termed a cooperative effect or cooperativity, and it is this which leads to the sigmoid curve. What cooperativity means is that the binding of one substrate molecule facilitates or enhances the binding of subsequent substrate molecules. Enzymes which display sigmoid velocity against substrate concentration curves are called allosteric enzymes.

cooperative
effect or
cooperativity

allosteric

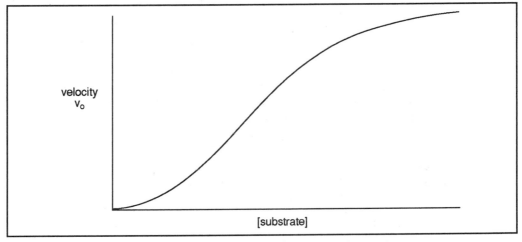

Figure 8.35 Initial velocity against substrate concentration for an allosteric enzyme.

R and T
conformation

How is cooperativity achieved? Let us assume several subunits (often 4) occur, each of which has a substrate binding site for substrate. Each subunit is believed to be able to adopt two alternative conformations, originally called R (for 'relaxed') and T (for 'tight'; Figure 8.36a). These alternative conformations differ in their kinetic properties, particularly in terms of their affinity for substrate(s). The T form has low affinity for substrate(s), whereas the R form has high substrate affinity. The sigmoid curve results if the enzyme normally exists in the T form, but can switch to the R form.

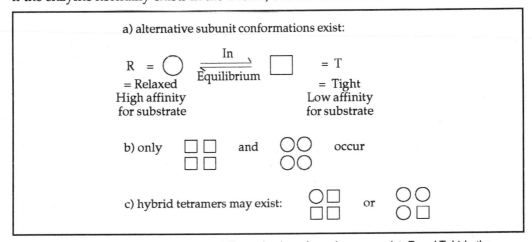

Figure 8.36 Models for allosteric enzymes a) Two subunit conformations can exist, R and T. b) In the MWC model, no hybrid states are permitted. c) The KNF model permits hybrid states.

Two models have been proposed to explain allosteric behaviour. The 'concerted symmetry' model proposed by Monod, Wyman and Changeux (MWC model) assumes that hybrid states (ie a mixture of R and T forms in one tetramer) do not occur (Figure 8.36b). Although the low- affinity T form predominates in the absence of substrate, binding of a substrate molecule to one subunit causes this subunit to switch to the high-affinity R form. Alternatively, although the T form predominates in the absence of substrate, it is in equilibrium with the high-affinity R form, which therefore occurs, albeit transiently. When the substrate binds to it, it is locked into the R form and the equilibrium will thus shift towards the R form. The presence of one subunit locked into the R form by the bound substrate forces adjacent subunits to also adopt the high-affinity R form. Since they have higher affinity for substrate than the T form, it is now easier for other substrate molecules to bind. This results in a sigmoid velocity against substrate concentration plot. The degree of sigmoidicity depends on the relative affinities of the two forms for substrate and the equilibrium between T and R in the absence of substrate.

The second model, proposed by Koshland, Nemethy and Filmer (the KNF model), allows both conformations (R and T) to exist within a tetramer (Figure 8.36c). This makes for a more complicated, but also more general, model, which can accommodate observations which the MWC model cannot. For example, in the MWC model, substrate binding leads to the high affinity form being adopted. It cannot explain cases (which exist) where initial substrate binding impedes additional binding.

What is the significance of sigmoidal binding? Let us examine the degree of saturation of an enzyme at various substrate concentrations (Figure 8.37). With Michaelis-Menten kinetics, to increase the rate from 20% to 80% of Vmax requires a concentration increase of 16 times - you can verify this for yourself, using the Michaelis-Menten equation!

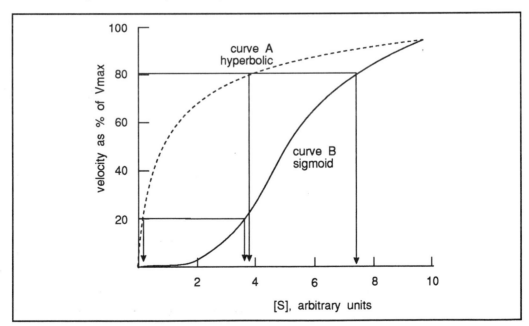

Figure 8.37 Influence of sigmoid and hyperbolic v against [S] relationships on the responsiveness of the enzyme to change in [S]. To increase from 20% to 80% Vmax requires a 16-fold increase in [S] with Michaelis-Menten kinetics (curve A); with the sigmoid plot depicted (curve B) for an allosteric enzyme, only a 2-fold increase in [S] is needed.

Inspection of Figure 8.37, which includes a 'typical' sigmoid curve, shows that the same relative increase only requires a 2-fold increase in substrate concentration. This means that modest increases in [S] cause more marked rate increases with allosteric enzymes. Allosteric enzymes are thus more responsive to change in [S].

Let us illustrate this with a quite different example, examine Figure 8.37 as you read this next section.

O₂ binding by haemoglobin

Oxygen binding by haemoglobin behaves like the sigmoid curve in Figure 8.37 (with the rate axis being % saturation with oxygen and the [S] axis being oxygen partial pressure). The oxygen partial pressure in the alveoli would correspond to the right hand side of Figure 8.37, so haemoglobin will be essentially fully oxygenated. After transport to muscle, the oxygen tension will be lower (around 4-5 on the [S] axis of Figure 8.37) and oxygen will therefore be released by haemoglobin - thereby being made available for muscle. Myoglobin, which stores oxygen in muscle and does not have 4 subunits, displays a hyperbolic binding curve (much like that shown in Figure 8.37 for the Michaelis-Menten-type enzyme), and is thus able to bind the oxygen released by haemoglobin. Note that if single subunits of haemoglobin were analysed, their oxygen binding would be hyperbolic and much less oxygen would be released.

allosteric regulation

activation inhibition

A second aspect of allosteric enzymes is also important, especially for regulation. The term allostery literally means 'other site' and refers to the occurrence of binding sites other than substrate binding sites. With many enzymes, these bind compounds which alter or 'regulate' the activity of the enzyme. Allosteric regulation in this way can include activation and/or inhibition. Compounds which have such effects are then referred to as allosteric activators or allosteric inhibitors. Their effect is to shift the velocity against [S] plot as shown in Figure 8.38.

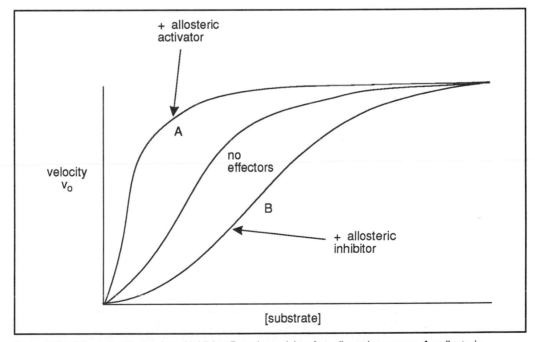

Figure 8.38 Effect of activator A and inhibitor B on the activity of an allosteric enzyme. An allosteric activator gives a higher rate at a given [S]; an allosteric inhibitor lowers the activity at a given [S].

Since the v against [S] plot can be changed by inhibitors and/or activators, and the curve is not hyperbolic, Michaelis-Menten kinetics do not apply and K_M is not very helpful. The sigmoid relationship means that a non-linear Lineweaver-Burk double reciprocal plot will be obtained. Rather than attempting to obtain a meaningless K_M value, an indication of the relationship between velocity and [S] under particular conditions is given by $[S_{0.5}]$, which is the [S] giving half-maximal velocity.

benefits of allosteric regulation

The benefit of allosteric regulation is that compounds other than substrate or product can regulate the activity of a crucial enzyme. Typical allosteric inhibitors are the products of a metabolic pathway acting to inhibit the first step (often called the committing step) of the pathway. When adequate levels of the end-product of the pathway (which is why the pathway exists!) are present, it switches off or lessens further synthesis (Figure 8.39). This is obviously beneficial to an organism, preventing an unnecessary build-up of unneeded compounds.

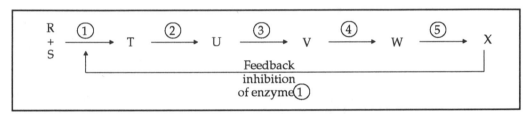

Figure 8.39 Feedback inhibition of the first enzyme of a pathway by the product of the pathway.

feedback inhibition

This type of regulation is widely used and is referred to as feedback inhibition. The end product binds to a site other than the substrate binding site (it would thus not compete with substrate for the enzyme). This is perhaps not surprising since, if the pathway is long, the end product (X in Figure 8.39) may bear little resemblance to the substrate(s) of the first enzyme of the pathway (R and S in Figure 8.39). In terms of the models to explain allosteric behaviour (Figure 8.36), allosteric inhibitors are seen as 'locking' the enzyme into the low-affinity T form, thereby making substrate binding less likely.

∏ Examine this metabolic pathway.

$$A \xrightarrow{\text{enzyme 1}} B \xrightarrow{\text{enzyme 2}} C \xrightarrow{\text{enzyme 3}} D \xrightarrow{\text{enzyme 4}} \underset{\text{(End Product)}}{E}$$

Circle the enzyme which is most likely to be subject to allosteric inhibition and the compound which will inhibit its activity?

Usually (but not always) the first enzyme of a pathway is inhibited by the end product. We would expect you to have circled enzyme 1 and E because we would anticipate that E would act as an inhibitor. Structurally E is likely to be quite different from A so it is probable that E will not bind to the same site as A therefore enzyme 1 is likely to be an allosteric enzyme.

There may be cases where the end product of one pathway is needed to react with the product of a second pathway (Figure 8.40). What if there is an imbalance in the levels of these two compounds? One resolution of this is if the more abundant compound

allosteric activation

activates the synthesis of the compound which is deficient. This can be accomplished by an allosteric activator, acting on the first (and rate-limiting) enzyme of the less active pathway (Figure 8.40). The effect of this is to raise the rate at a given [S], as shown in

Figure 8.38. This is achieved by binding to the enzyme and 'locking' it into the high-affinity R form.

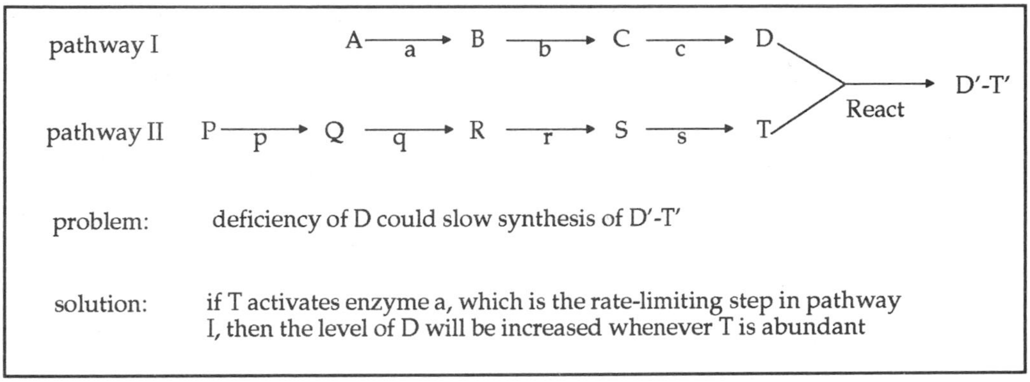

Figure 8.40 The activity of two pathways may need to be coordinated. Deficiency of the end-product of one pathway can be rectified by activation of the first enzyme of this pathway by the product of the second pathway.

These activation and inhibitory effects are exemplified by the enzyme aspartate transcarbamylase (ATCase). This enzyme catalyses the condensation of carbamyl phosphate and aspartate to form carbamyl aspartate, which is the first step in the biosynthesis of pyrimidines (Figure 8.41). Subsequent reactions result in the formation of cytidine triphosphate (CTP). CTP is a highly effective allosteric inhibitor of ATCase, thereby shutting off entry of material into the pathway when CTP is abundant.

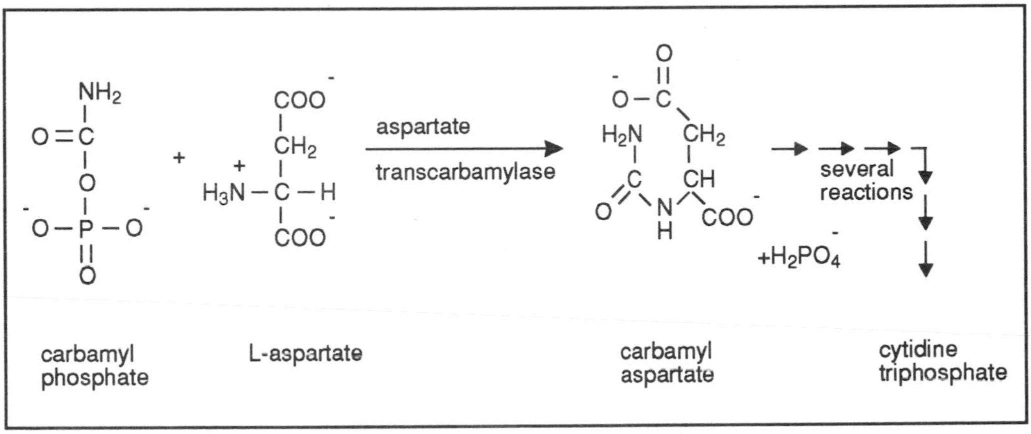

Figure 8.41 Reaction catalysed by aspartate transcarbamylase (ATCase). L-aspartate and carbamyl phosphate condense to form carbamyl aspartate. Further reactions result in synthesis of cytidine triphosphate.

Binding of CTP to ATCase causes a decrease of affinity for the substrates, without affecting Vmax. Inhibition (resulting from the decrease in affinity) can be as high as 90%. This is much more efficient than waiting for product inhibition of each step to feed back down the pathway (high [CTP] inhibits the enzyme which made it; the substrate of this enzyme builds up and then inhibits the preceding enzyme, etc).

ATP as activator   ATP also acts as an activator of ATCase. It increases the affinity of the enzyme for its substrates, whilst having no effect on Vmax. ATP can be thought of as shifting the equilibrium of the enzyme in favour of the high-affinity R form. Both ATP and CTP bind at the same site. They therefore compete with each other, and high levels of ATP prevent CTP inhibiting the enzyme. We have already seen that CTP inhibition will occur when [CTP] is high. ATP activation (which will offset or override the effect of CTP) occurs when [ATP] is high and therefore energy is available for growth. Growth will require the synthesis of nucleic acids. These can only be synthesised if pyrimidine nucleotides are available. Hence the desirability of positive control by ATP, to ensure that purine and pyrimidine nucleotides are simultaneously available. Re-examine Figure 8.40 carefully for it provides the model of the ATCase system.

loss of regulating features

desensitisation

C and R subunits effectors

The most remarkable aspect of ATCase emerged from structural studies on the enzyme. Treatment with mercury-containing reagents, (such as p-hydroxymercuribenzoate) which react with sulphydryl groups, resulted in loss of regulation by ATP or CTP. In addition, the relationship between rate and [S] was now hyperbolic (ie the cooperativity had been lost), although the modified enzyme was still fully active. This loss of regulation by effectors is called desensitisation. Analysis by ultracentrifugation revealed that treatment with the mercurial reagent had caused dissociation of the enzyme into smaller particles. These were found to be either trimers of a subunit of 34000 daltons (ie 3 x 34000) or dimers of 17000 dalton subunits (ie 2 x 17000; Figure 8.42). The 34000 dalton subunits are catalytically active but do not bind ATP or CTP. They are called C subunits. The smaller subunits bind ATP and CTP (competitively) but do not interact with the substrates. These are known as R subunits. Thus in this case there are 2 distinct types of subunits. The sigmoid binding relationship must be caused by substrate binding to one C-type subunit causing conformation changes in other C subunits. Superimposed on this, binding of effectors (ATP and CTP) to the regulatory (R) subunits also influences the conformation of the catalytic subunits. A complex system indeed! If the mercurial is removed (by treatment with 2-mercaptoethanol) then reassociation of C and R subunits occurs to form native enzyme:

$$2\ C_3 + 3\ R_2 \rightarrow C_6R_6$$

The reformed complex displays similar allosteric properties to those of untreated enzyme.

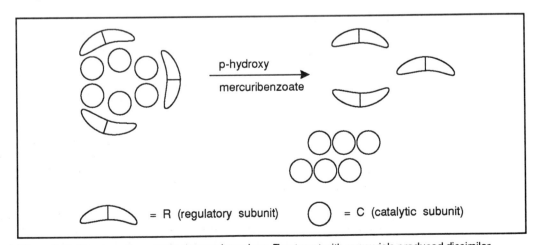

Figure 8.42 Structure of aspartate transcarbamylase. Treatment with mercurials produced dissimilar subunits. C-type subunits are catalytically active but do not bind ATP or CTP. R-type subunits bind ATP and CTP (competitively) but do not bind substrates.

∏ Calculate the molecular weight of the native ATCase enzyme.

You should have come to the conclusion that the molecular weight of the native enzyme is 306,000 daltons (6 x 34,000 + 6 x 17,000).

Feedback inhibition, involving allosteric regulation at a site other than the substrate binding site, is a widely used strategy in metabolism. It commonly, as we have seen with aspartate transcarbamylase, acts at the first enzyme of a pathway. This allows a pathway to be regulated without having excessive concentrations of intermediates. This, of course, only works if the regulated enzyme is the rate-limiting step. One simple way in which this can be visualised is via a pipe analogy. If you think of water flowing through a pipe, which has regions of varying diameters, it is the region of narrowest diameter which determines the flow rate (Figure 8.43).

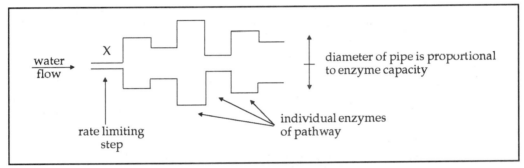

Figure 8.43 The pipe analogy. Pipe diameter represents enzyme capacity: the step with the smallest capacity will be rate-limiting; affecting this step (either activation or inhibition) will have most effect on the overall throughput of the pathway.

enzyme capacity

rate-limiting step

If the water pressure remains the same, then varying the diameter anywhere other than point X will have minimal effect. If we look at a metabolic pathway, thinking of the enzyme capacity at each stage as the diameter of a pipe, then step X in Figure 8.43 is clearly the 'rate-limiting step'. Activation of this enzyme (widening of the pipe) will allow faster throughput, whereas activating any other step will not affect the overall flux through the pathway. We can thus summarise this area by stating that:

• in a metabolic pathway, the enzyme present with least activity will be the rate-limiting step. The activity of an enzyme will result from the level of gene expression and the efficacy of the enzyme as a catalyst;

• if one wishes to regulate the flux of material through the pathway, it is this step which would be expected to be regulated. We should, however, note that this model is a simplification. Actual pathways are often found to have more complicated and subtle control mechanisms than this;

• allosteric regulation (whether inhibition or activation) is a widely used method for achieving 'second-by-second' adjustments in enzyme activity: this is known as 'fine control' or 'fine tuning' and is distinct from altering the rate of synthesis of the enzyme, which is called 'coarse control'.

SAQ 8.9

Why do allosteric enzymes display a sigmoid rate against [S] plot, whereas other enzymes display a hyperbolic relationship?

Choose appropriate answers from the following, jotting down why you think each statement does, or does not, provide the explanation.

1) Because they are more responsive to changes in substrate concentration.

2) Because they have sites for inhibitors to bind to, distinct from the substrate binding site.

3) Because they have more than one substrate binding site per active enzyme molecule.

4) Because they are tetramers of dissimilar subunits.

5) Because they have several subunits, which can exist in two forms of differing affinity; substrate binding influences the equilibrium between these two forms.

8.10 Other methods of regulating enzyme activity

Methods other than allosteric regulation exist.

8.10.1 Phosphorylation of enzymes

phosphorylation

One approach is by covalent modification of the enzyme, frequently by phosphorylation. This is used in a reversible manner to convert an enzyme from a high-activity form to one of lower activity. One example of this is the enzyme phosphorylase, which catalyses the release of glucose-1-phosphate from glycogen. This enzyme exists in two forms (Figure 8.44), one of low activity (known as phosphorylase b) and a second of higher activity (phosphorylase a).

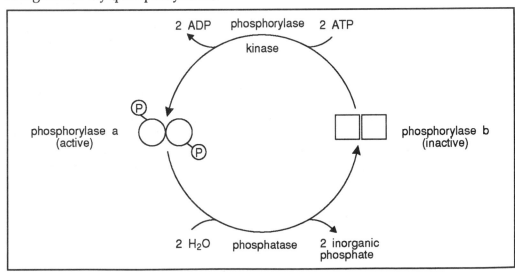

Figure 8.44 Enzyme activation by covalent modification. Phosphorylase is converted from a low activity form (phosphorylase b) to a high activity form (phosphorylase a) by phosphorylation.

A specific enzyme (protein kinase) converts phosphorylase b to phosphorylase a by phosphorylating the hydroxyl side-chains of particular serine residues. The reverse process (conversion of phosphorylase a to b) is catalysed by a phosphatase which removes the phosphate groups. This is part of an elaborate cascade control mechanism by which adrenaline and insulin regulate glycogen metabolism.

8.10.2 Isoenzymes

In the context of regulation of enzymes and metabolic pathways, the existence of isoenzymes (or isozymes for short) should be mentioned. These are different molecular forms of a particular enzymatic activity, which are present within a cell or tissue. We have already discussed one case of isoenzymes when we reviewed the roles of hexokinase and glucokinase. You may recall that both enzymes catalyse the same reaction but with different kinetic properties. This allows each enzyme to be suited for different circumstances in the metabolism of a cell or organ. Isoenzymes are widespread and commonly arise through the occurrence of alternative subunits in an enzyme possessing a quaternary structure. Thus lactate dehydrogenase in mammals consists of a tetramer, in which the subunits may be of two types (Figure 8.45). In this case the subunits are designated H (for heart type) or M (for muscle type).

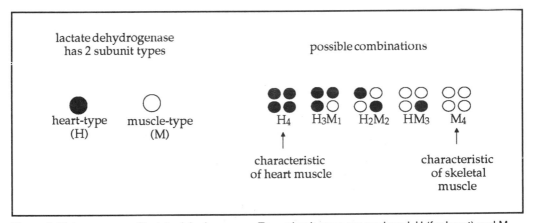

Figure 8.45 Isoenzymes of lactate dehydrogenase. Two subunit types are produced, H (for heart) and M (for muscle). These differ in their amino acid composition (and hence charge) and in their kinetic properties. Since the enzyme exists as a tetramer, 5 possible combinations exist. They may be separated by electrophoresis.

isoenzymes
and tissue
difference

The H_4 tetramer is the form typically found in heart whilst M_4 is characteristic of skeletal muscle. Intermediate forms occur. 5 potential forms exist (ie H_4; H_3M_1; H_2M_2; H_1M_3, M_4) and different tissues display characteristic ratios of each of the 5 isoenzymes. The different subunits have different kinetic characteristics, which are reflected in the properties of the 5 isoenzymes. The particular isoenzyme complement of each tissue or cell is presumed to reflect the different metabolic circumstances or needs of different cells.

8.11 Enzyme classification

-ase

We have hitherto identified enzymes by using 'trivial' names which roughly describe their activity. This involves adding '-ase' to the name of the substrate for example lactate dehydrogenase oxidises (ie 'dehydrogenates') lactate; maltase acts on maltose. In some cases, the name does indicate what reaction the enzyme catalyses (as in our first example above). In other cases this is not clear. To avoid ambiguity over the labelling and activity of enzymes, there is a classification system by which enzymes can be unequivocally identified. This consists of a Systematic Name, which describes the function of an enzyme more clearly than the trivial names in common use. As an example, the systematic name for lactate dehydrogenase is L-lactate:NAD^+ oxidoreductase. This tells us that the reaction involves both oxidation and reduction (if one compound is oxidised, then something else must be reduced!) and indicates that L-lactate and NAD^+ are the reactants. Systematic names, whilst informative, are long and awkward and the trivial names tend to be used on a day-to-day basis.

systematic
name

International
Enzyme
Classification
number

Associated with the systematic name is a International Enzyme Classification number. This groups enzymes into one of 6 categories, according to the type of reaction catalysed. These are:

1)   Oxidoreductases: oxidation-reduction.

2)   Transferases: transfer of groups.

3)   Hydrolases: hydrolysis reactions.

4)   Lyases: additions to double bonds.

5)   Isomerases: isomerisation reactions.

6)   Ligases (synthetases): joining of 2 molecules, using energy supplied by ATP.

These numbers provide the first of a 4-digit number which should unambiguously identify an enzyme. Thus all transferases begin with a 2.

The second number tells us in more detail what the substrate is. The third figure indicates what the acceptor is (for example, of reducing equivalents from the substrate), or provides more details of the group acted on. The fourth figure is a number given to each enzyme according to the order in which they were discovered (within each class). Table 8.5 should make this more clear; it contains part of the classification system for groups 1 and 2 (oxidoreductases and transferases, respectively).

Π   Give the systematic name and deduce the first three numbers of the enzyme which catalyses the following reaction:

$$C_2H_5OH + NAD^+ \rightleftarrows CH_3CHO + NADH + H^+$$

Note: in cases where oxidation and reduction of $NADH/NAD^+$ are involved, by convention the reaction is thought of as proceeding from left to right as written above.

---

1. Oxidoreductases

1.1 Acting on CH-OH group of donor

    1.1.1 with NAD^+ or $NADP^+$ as acceptor

    1.1.2 with a cytochrome as acceptor

    1.1.3 with oxygen as acceptor

1.2 Acting on aldehyde or keto group

    1.2.1 with NAD^+ or $NADP^+$ as acceptor

    1.2.2 with a cytochrome as acceptor

1.3 Acting on the CH-CH group of donor

    1.3.1 with NAD^+ or $NADP^+$ as acceptor

1.4 Acting on the CH-NH2 group of donor

2. Transferases

2.1 One-carbon groups

    2.1.1 methyl transferases

    2.1.2 hydroxymethyl and formyl transferases

    2.1.3 carboxyl transferases

2.2 Aldehyde or keto groups

2.3 Acyl groups

2.4 Glycosyl groups

2.5 Alkyl or aryl groups, other than methyl

2.6 Nitrogen-containing groups

2.7 Phosphorous-containing groups

2.8 Sulphur-containing groups

Table 8.5 Partial classification for enzymes of groups 1 (oxidoreductases) and groups 2 (transferases). (Note, we have not included all of the subdivisions of these two groups).

This looks forbidding but isn't! The reaction is the oxidation of (or removal of hydrogens from) ethanol and is known colloquially as alcohol dehydrogenase. Its systematic name is thus 'Alcohol:NAD^+ oxidoreductase' (based on the 1972 Enzyme Nomenclature). As for its enzyme classification number, it can be deduced as follows:

1st number must be 1, since it is an oxidoreductase

2nd number must be 1, since a CH-OH group is the donor

3rd number must also be 1, since NAD^+ is the acceptor, giving so far 1.1.1.

In fact, the 4th number is also 1, as it was the first enzyme discovered in its class!

So its classification number is 1.1.1.1.

This section was included so that you are aware that there is a formal nomenclature and classification system. You should note that these formal identification systems exist, but the trivial, descriptive names will continue to be used by practising biologists!

8.12 Enzyme technology

The study of enzymes has been of immense importance in increasing our understanding of how living organisms operate at the molecular level. The knowledge obtained has also been of value in discovering the molecular basis of many illnesses and has been used in both diagnosis and treatment of illness and in the design of drugs. Along with understanding of the structure, role and properties of enzymes has come the realisation that these remarkable catalysts could have a role in technology: that their catalytic capabilities could be used to achieve desired objectives. The purpose of this section is to briefly review some of the uses that have been made of enzymes; these are collectively described as enzyme technology.

We have already identified the factors which make enzymes attractive for technological purposes. These are:

- specificity - both in terms of substrate and the reaction performed;

- high catalytic rates;

- frequently, use of mild reaction conditions;

- ability to be controlled - reaction can be stopped by changing the pH or temperature;

- wide range of reactions catalysed.

The use of enzymes as industrial/commercial catalysts is discussed in detail elsewhere in this series (Technological Applications of Biocatalysts). Operational areas are summarised in Table 8.6.

Area	Use of Enzymes/Product
Food industry	Amino acids/high fructose syrup, baking, brewing, wine industries.
Pharmaceutical industry	Production of semi-synthetic penicillins. Transformation of steroids.
Detergents	Use of proteases and amylases
Analytical	Glucose estimation. Detection of toxins.
Medical	Diagnostic use; biosensors.
Environmental	Detoxication
Textiles	Desizing

Table 8.6 Applications of enzymes in industry.

Note that before any of these applications could be commercialised, it was necessary to isolate, purify, characterise and develop the relevant enzyme. In many cases enzymes from many sources were evaluated in order to optimise enzyme characteristics to the process requirements. We shall highlight several of these areas.

8.12.1 Enzymes in production and biotransformation

HFS

High fructose syrup. Starch is plentiful, and hence cheap, whereas there is a large demand for sweetening agents, which are (relative to starch) expensive. Conversion of starch to the higher-value low molecular-mass sugars (which taste sweet) thus makes economic sense. Although treatment with acid will hydrolyse starch, side reactions

α-amylase

occur. Enzymatic hydrolysis using α-amylase is preferred; a mixture of various glucosidases is then used, depending on the particular food industry. Glucose is not as sweet as sucrose (table sugar) but fructose is rather sweeter. Conversion of glucose,

Rglucose isomerase

produced by hydrolysis of starch, to a 'high-fructose syrup' is accomplished by glucose isomerase immobilised in large reactors. Routinely a glucose/fructose mixture containing 42% fructose is obtained.

Enzymes are also extensively used for the production of L-amino acids used as food supplements. The particular attraction here has been their ability to produce only the desired L-amino acids, whereas chemical methods would yield mixtures of both D- and

L-aminoacylase

L-forms. The enzymes used are also fairly stable: L-aminoacylase, used in production of L-amino acids, only loses about 30% of its activity after continuous operation at 65°C for 30 days. (Case study approaches to the production of high fructose syrup and L-amino acids are presented in the BIOTOL text 'Biotechnological innovations in food processing').

semi-synthetic penicillins

Semi-synthetic penicillins. Only a few penicillins are produced in high yields directly from fermentation. Benzyl- penicillin (Penicillin G) is the usual product of a commercial penicillin fermentation. Whilst useful as an antibiotic, an individual penicillin (such as penicillin G) may have disadvantages (for example, acid lability, making it unsuitable for oral use, or penicillin G- resistant bacteria). These problems can be overcome if the natural penicillin is modified. Every penicillin consists of 6-aminopenicillanic acid (6-APA; Figure 8.46a) to which a sidechain is attached. With penicillin G this is a benzyl group (Figure 8.46b). If this is removed, the 6-APA produced can be used to synthesise other, semi-synthetic penicillins, with altered properties (for example, acid resistance or activity against penicillin G-resistant bacteria). One such is ampicillin, with a D-phenylglycine sidechain (Figure 8.46c). Penicillin acylase, which removes the benzyl

penicillin acylase

sidechain from penicillin G, has been of crucial importance in the development of the semi-synthetic penicillins.

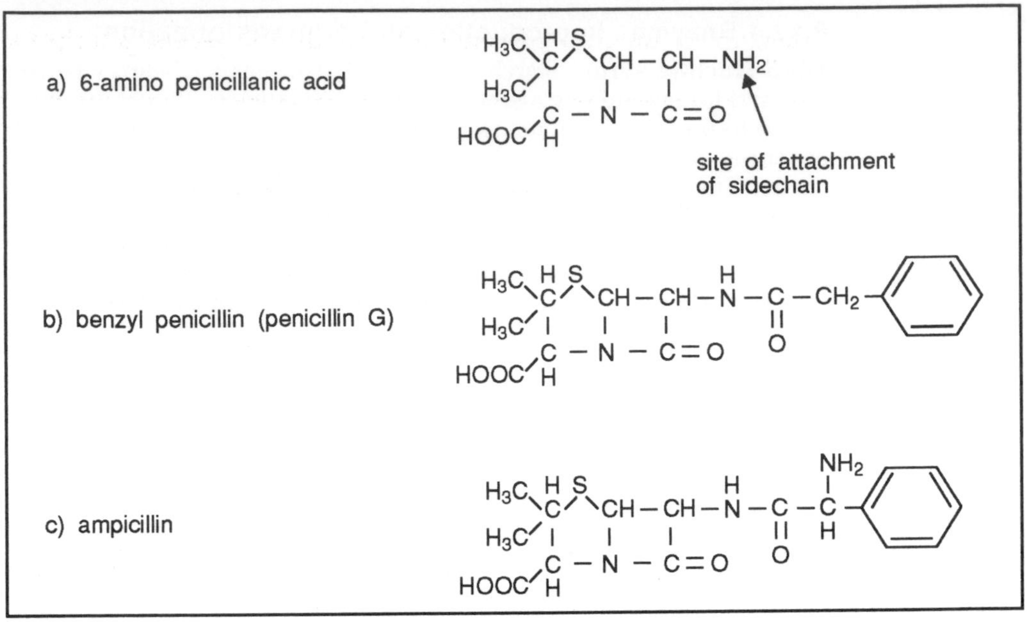

Figure 8.46 Semi-synthetic Penicillins. a) 6-amino penicillanic acid (6-APA), present in every penicillin; a sidechain is attached at the position arrowed. b) In benzyl-penicillin (Penicillin G), the sidechain is a benzyl group. c) With a D-phenylglycine group as the sidechain, ampicillin is produced.

8.12.2 Enzymes in analysis and enzyme electrodes

Enzymes are widely used in analysis of metabolites because their specificity enables a precise assay in the presence of similar compounds. Under these circumstances, chemical analysis would give falsely high values due to reaction with structurally similar compounds. Thus chemical determination of pyruvate by reaction with phenylhydrazine, whereby a yellow phenylhydrazone is formed which may be estimated by absorbance at 410nm, would suffer from reaction with other keto acids such as oxaloacetate or α-ketoglutarate (also called α-oxoglutarate). Enzymatic determination, using lactate dehydrogenase, would not suffer from this lack of specificity. One very common assay is that of glucose, using the enzyme glucose oxidase and a second coupling enzyme as follows:

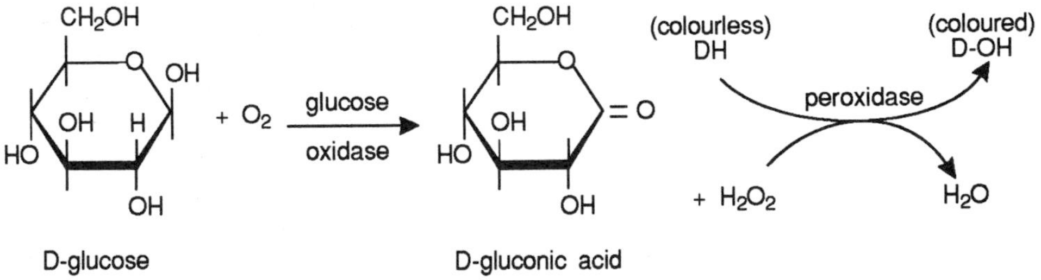

D-glucose   D-gluconic acid

This can be carried out so as to produce an absorbance, or with the enzymes immobilised onto a pad, which is then used as a 'dipstick'.

enzyme electrodes

biosensors

An area of immense potential is that of 'enzyme electrodes', also known as 'biosensors'. These combine the benefits of electrodes as measuring devices with the specificity of enzymes. They consist of an immobilised (ie fixed) enzyme in close contact with an electrochemical sensor (ie an electrode which responds to the concentration of a chemical by producing an electrical signal). The general principle is shown in Figure 8.47.

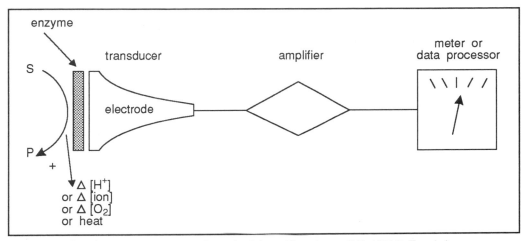

Figure 8.47 Components of an enzyme electrode. Adapted from Lowe, C.R. (1984). Trends in Biotechnology 2, 59-64, Elsevier Science Publishers.

The electrode acts as a transducer to convert the product of enzymatic reaction (heat; change in [ion] eg [H^+]; change in [O_2]) to an electrical signal which can be easily processed and analysed. To date, although many enzyme electrodes have been described (based on appropriate enzymes coupled to pH and other ion- selective electrodes, oxygen electrodes, thermistors and light-sensing devices), only that for glucose has been widely commercialised. None-the-less, biosensors are likely to be widely used in the future for analysis, diagnosis and process control.

8.12.3 Enzyme engineering

alteration of amino acid sequence

Following the development of genetic engineering, it has become possible to deliberately alter single amino acid residues at particular sites within enzymes. For example, it is possible to change one particular amino acid residue within the active site of an enzyme. This technique of Enzyme Engineering (also known as protein engineering), achieved through a procedure called site-directed mutagenesis, raises the possibility of altering a protein in a deliberate manner to alter one property (eg, turnover number, K_M, thermostability, substrate specificity). Thus with the enzyme tyrosyl-tRNA synthetase from *Bacillus stearothermophilus*, the K_M for ATP was lowered by a factor of 100 by converting the threonine residue at position 51 in the amino acid sequence to proline. As a second example of the power and potential of this technique, we have already referred to the role of α_1-antitrypsin in controlling elastase activity in the lungs. Loss of this control leads to destruction of the elastic tissue of the alveoli and the crippling condition of emphysema results. Inactivation of α_1-antitrypsin by smoking is believed to arise through oxidation of a crucial methionine residue (Met-358). Use of site-directed mutagenesis to convert Met-358 to Val results in a protein which is still active as an elastase inhibitor and is also resistant to oxidation. This could form the basis of a therapy for emphysema. The potential benefits from enzyme engineering are enormous.

SAQ 8.10

Write down 5 features of enzymes which are important in their use in the food and healthcare industries, citing, for each feature, one case where this feature is clearly important.

Summary and objectives

This has been a long chapter. But now you have completed it, you will have learnt much about enzymes. In particular you should be able to:

- describe the catalytic ability of enzymes in terms of reducing the activation energy required for reactions to take place;

- explain why enzymes show specificity both in terms of the substrates they will interact with and the nature of the reactions they will catalyse;

- use the Michaelis-Menten equation to calculate K_M and to explain the meaning of K_M;

- explain the principles underpinning enzyme assays, with particular reference to measuring the initial reaction rate and the importance of suitable controls;

- explain the effects of pH and temperature on enzyme activity;

- explain how competitive and non-competitive inhibition of enzymes may be distinguished experimentally;

- identify, in general terms, how enzymes achieve catalysis;

- describe how enzyme activity may be regulated especially by feedback inhibition and explain regulation by allostery, phosphorylation and isoenzyme mechanisms;

- describe why and for what purposes enzymes are useful in biotechnology.

Responses to SAQs

Responses to Chapter 1 SAQs

1.1   1) protoplasm.

2) the bacterium *Salmonella typhi*. Prokaryotic organisation is restricted to the bacteria. The others listed are all eukaryotes.

1.2

nucleolus	site of ribosome construction
Golgi apparatus	processing of macromolecule for export
rough endoplasmic reticulum	processing of the products of protein synthesis
chloroplast	harvesting the energy of light
mitochondria	site of terminal oxidation of organic nutrients
lysosomes	breakdown of macromolecules
nucleus	houses the genetic material
centriole	helps in organising the process of cell division
ribosome	carries out protein synthesis

1.3   Plant. The tell-tale signs are: the presence of chloroplasts, the presence of vacuoles, the thick cell wall and the plasmodesmata.

1.4   1) True. Mitochondria are typical of eukaryotic cells.

2) False. Ribosomes are used by both prokaryotes and eukaryotes to carry out protein synthesis. They are, however, quite different in the two types of cells.

3) False. Mature red blood cells have lost their nuclei, mitochondria and rough endoplasmic reticulum and are no longer capable of growth and cell division.

4) False. Histones are only associated with the genetic material in eukaryotes. They are not found in prokaryotes.

5) False. The eubacteria are chemically more similar to the eukaryotes than are the archaebacteria.

6) True. Typically prokaryotic cells are 2-5μm long and have a diameter of about 0.5-2μm. Eukaryotic cells often have dimensions of the order of 10-100μm. In terms of volume therefore eukaryotic cells are often 200-10,000 times bigger than prokaryotic cells.

Responses to Chapter 2 SAQs

2.1   Glycine (Figure 2.5) is the one amino acid found in proteins which does not exist in D- and L- forms. For such stereoisomers to occur, there must be an asymmetric centre in the molecule. With amino acids this requires there to be 4 different substituent groups attached to the central (ie α-) carbon atom. As the sidechain of glycine is -H, there are only 3 different substituent groups (ie -COO⁻, $-NH_3^+$ and -H). The structure of glycine is:

$$COO^-$$
$$|$$
$$H-C-H$$
$$|$$
$$^+NH_3$$

and the 'sidechain' is of course indistinguishable from the other H atom attached to the α-carbon atom.

2.2   1) Amino acids a) and b) are non-polar, hydrophobic amino acids; they are phenylalanine and alanine respectively. Their sidechain atoms, being solely hydrogen and carbon, show no tendency for charge separation, which would lead to a polar character.

2) Amino acids c) (lysine) and e) (serine) have polar sidechains. In the case of lysine, the sidechain will be positively charged under physiological conditions. That of serine does not ionise but partial charge separation occurs, creating a dipole. This makes the group polar, with a tendency to form hydrogen bonds.

3) Structure d) is not an α-amino acid. In an α-amino acid, a carboxyl and an amino group must be attached to the same carbon atom. In this case they are attached to different carbon atoms. This molecule is known as β-alanine. Notice that it has the same atomic composition as alanine; it is therefore a structural isomer of alanine. β-alanine occurs naturally, although it is not incorporated into proteins.

2.3   Structure d) will predominate at pH 6.0. This can be deduced from the pKa values in the following way:

• for the α-carboxyl group (pKa 2.1) complete deprotonation will have taken place, thus this group will be present as -COO⁻;

• although the pKa of the sidechain carboxyl group is higher (4.1), none-the-less at pH 6.0 this will also be fully deprotonated ie -COO⁻;

• the pKa of the amino group (9.5) is well above this pH, hence this group will still be protonated ie $-NH_3^+$

Structures b) (amino group deprotonated whilst the carboxyl groups are protonated) and c) (α-carboxyl protonated when the γ- carboxyl is not) are most unlikely to occur under any conditions. Structure a) will occur at high pH (10.5); structure e) occurs at low pH (1.0).

2.4   Expected results are as follows:

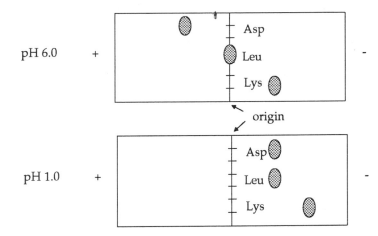

At pH 6.0, the ionic state of the relevant groups will be as follows:

Aspartic acid: both carboxyl groups deprotonated; amino group protonated, hence net charge = -1, hence migration towards anode.

Leucine: α-amino group still protonated and α-carboxyl group deprotonated. Net charge = 0, and leucine remains at the origin.

Lysine: α-carboxyl group present as -COO⁻; both α- and ε-amino groups will still be protonated. Therefore net charge of +1 and migration to the cathode (by a similar distance to that migrated by aspartate to the anode).

pH 1.0: This is just low enough for all ionising groups to be fully protonated. Thus the various groups will be in the following forms:

	α- carboxyl	α- amino	sidechain	charge
Asp	-COOH	$-NH_3^+$	-COOH	+1
Leu	-COOH	$-NH_3^+$		+1
Lys	-COOH	$-NH_3^+$	$-NH_3^+$	+2

Thus aspartic acid and leucine would migrate similar distances towards the cathode (leucine slightly less as it is more bulky). Lysine would be expected to migrate further (approximately double the distance) towards the cathode.

2.5   The figure which follows shows the titration curves which glycine, histidine and arginine would give. Detailed comments are as follows:

1) Glycine will behave in a manner similar to alanine. You will have had to guess its' pKa values - hopefully you chose 2-2.5 for the α-carboxyl group and 9.5-10 for the α-amino group (precise pKa values are 2.35 and 9.8 respectively). There will be two plateau regions, centred on the pKa values, at which buffering is occurring. One equivalent of base will be required to reach pH 6.0 and 2 equivalents to reach pH 12.

2) The sidechain of histidine means that there is a third region of buffering, centred around the sidechain pKa value (6.0). This pH will be reached after 1.5 equivalents of base have been added (1 equivalent to titrate the α-carboxyl group; 0.5 equivalent to half titrate the sidechain imidazole group and thus reach the pKa value). Three equivalents of base will be required to reach pH 11-12.

3) The sidechain of arginine only deprotonates at very high pH (pKa = 12.5). Up to pH 10-11, this titration will be similar to that of glycine. A third region of buffering would be found from pH 11.5-13.5.

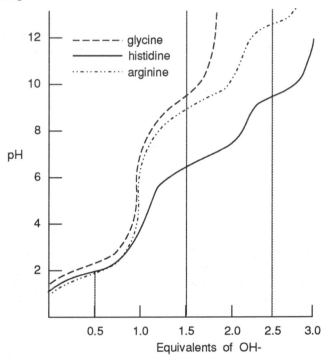

2.6   The answer is 2. Since the histidine is in the middle of a polypeptide chain only its' sidechain can be involved. For an ionic bond to form, this imidazole group must be protonated - and the pH must therefore be lower than the pKa, which is 6.0. Choice 3 is thus unsuitable, since the histidine sidechain will be deprotonated and have no charge.

At pH 1.0, both α-amino and α-carboxyl groups of L-alanine will be protonated, since the pH is sufficiently below their pKa values. L-alanine will have a net charge of +1 and would show no tendency to bind to the positively charged histidine sidechain. Thus 1 is unlikely.

At pH 4, the α-carboxyl group of L-alanine will have deprotonated and will be negatively charged. Providing that it approaches the histidine with the correct orientation (and this always applies to substrate binding by an enzyme), then an ionic bond can form between the α-carboxyl group of L-alanine and the imidazole sidechain of histidine.

Responses to Chapter 3 SAQs

3.1   1) lysine (lys), the last amino acid residue of the polypeptide.

2) glutamic acid (glu), the first amino acid residue.

3) val-11.

The key to answering this question is to remember that primary sequences are normally written beginning with the N terminal amino acid (ie the one with the free α-amino group) and finish with the C terminal amino acid (ie the one with the free α-carboxylic acid group. The amino acids are listed from the N terminal.

3.2   1) a), c) and d) are polar.

2) c) and d) are hydrogen bonds (note hydrogen bonds also involved polarity).

3) b) and f) in both cases hydrocarbon (ie non-polar) sidechains are involved.

4) e) the sulphydryl sidechains of two cysteine residues have been oxidised to form a disulphide bridge.

3.3   The recovery of only 7% of the initial activity indicates that the active native conformation is being reformed by only this proportion of the insulin chains. The explanation is provided by thinking about the structure of insulin when it is originally folded. This is as a longer inactive form, proinsulin, in which there is an extra 30 amino acid residue C-chain, which is subsequently removed. Thus the 'active' shape is dictated by the intact proinsulin primary structure. Disulphide bridges then 'lock' the shape, which is largely retained following removal of the C-chain. If the experiment were to be repeated with proinsulin, much higher recovery would be expected.

If you missed this, re-read part e) of Section 3.3.2.

Responses to Chapter 4 SAQs

4.1

1) Structures b) and e) are pyrimidines; they are thymine and uracil, respectively.

2) Only d), which is adenine, is a purine.

3) Structure e), uracil, does not occur in DNA.

4) Neither a), which is phenol, nor c), which is ribose, are nitrogenous bases.

4.2

1) Adenosine-5'-monophosphate: Structure d).

2) Guanosine: Structure e).

3) Thymine: Structure f).

4) Deoxythymidine-5-monophosphate: Structure c).

5) Uridine-5-monophosphate: Structure a).

Hence structure b) is left over, which is thymidine (ie the nucleoside consisting of thymine and ribose). If you had any problem with this, check that you can distinguish purines from pyrimidines, ribose from deoxyribose and that you know the difference between bases, nucleosides and phosphorylated nucleosides (ie nucleotides).

4.3

You should have produced the following structures:

1)

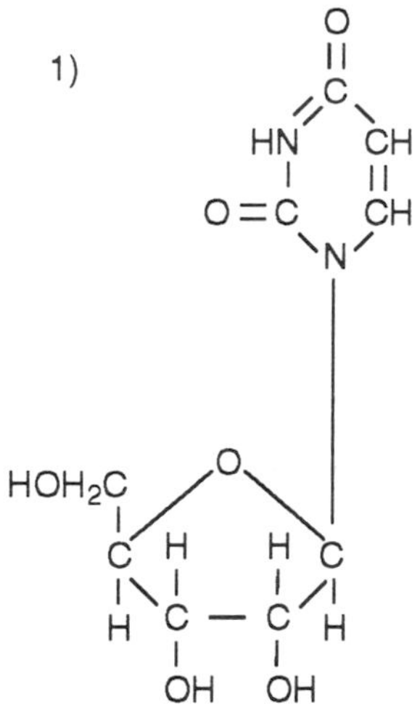

2)

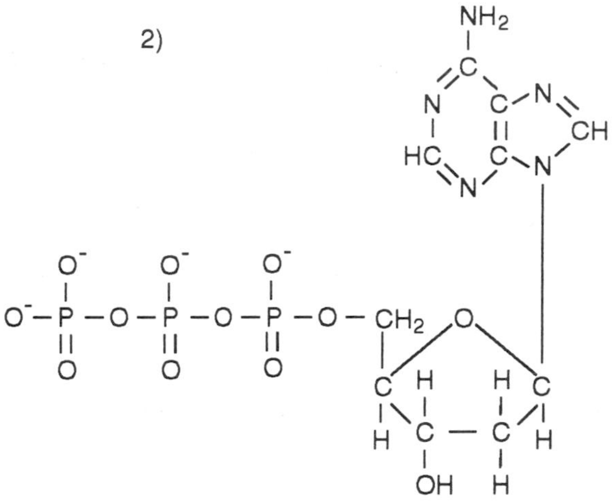

3)

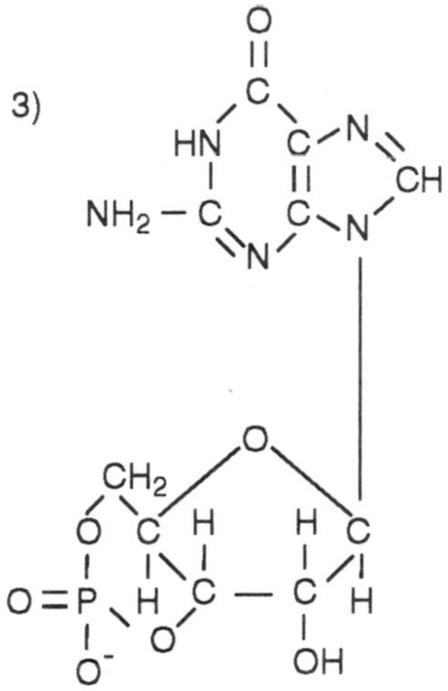

For 2) we have drawn ATP; GTP was equally acceptable. If in doubt re-read Sections 4.2 to 4.4.

262

4.4   1) The structure should look that given below.

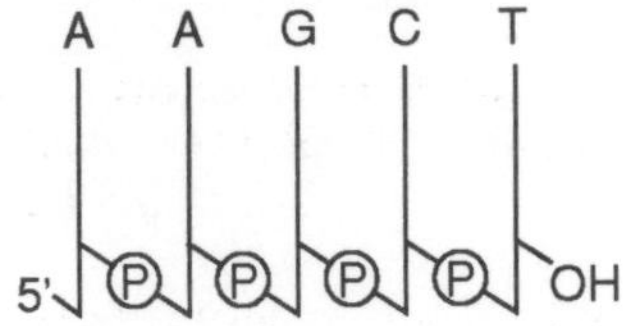

A   A   G   C   T

or

adenine

OH
|
CH₂   O
|
C H   H C
| \ | | / |
H C — C H
     | |
     O   H
|
O⁻—P=O
|
O
|
CH₂   O
|
C H   H C
| \ | | / |
H C — C H
     | |
     O   H

adenine

|
O⁻—P=O
|
O
|
CH₂   O
|
C H   H C
| \ | | / |
H C — C H
     | |
     O   H

guanine

|
O⁻—P=O
|
O
|
CH₂   O
|
C H   H C
| \ | | / |
H C — C H
     | |
     O   H

cytosine

|
O⁻—P=O
|
O
|
CH₂   O
|
C H   H C
| \ | | / |
H C — C H
     | |
     OH   H

thymine

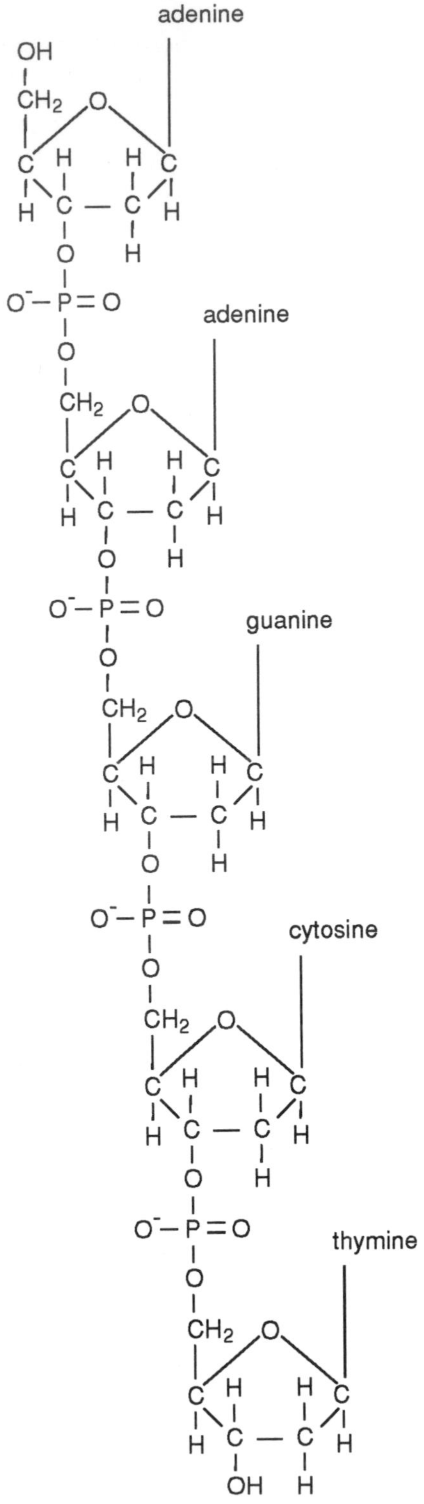

The base at the 5' end is adenine; that at the 3' end is thymine. Note the lack of -OH at the C-2' positions.

2) b), 5'AGCTT3' is the complement of that given as structure 1) above. This is because the strands in the DNA double helix are anti-parallel ie: 5'AAGCT3'
3'TTCGA5'.

3) b), that containing GC, will have the higher melting temperature, Tm. This is because GC basepairs have 3 hydrogen bonds, whereas AT basepairs only have 2 hydrogen bonds. Thus more energy, ie a higher temperature, is required to rupture GC basepairs.

4) The GC content is about 62%; hence AT content is 38%. The GC is read off Figure 4.19, which serves as a calibration curve.

4.5   The completed central dogma should be as follows:

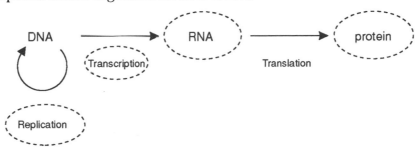

None of the other words are relevant in this context. If you had any difficulty with this, re-read Section 4.1, and perhaps the rest of the chapter!

4.6   Errors will mean that the wrong base then occurs in the DNA sequence. If this is the coding region of a protein, an incorrect amino acid may be inserted, giving rise to a protein with a different primary structure. This may not function properly, or at all; however the alteration may not affect the functioning of the protein. Note that the 'correct' amino acid may still be inserted; this is the benefit of the degeneracy of the genetic code.

Such an error in nucleotide incorporation by DNA polymerase represents a mutation. If the frequency were high, then the genetic information would not be stable and preserved during DNA replication. The survival of the organism would be severely threatened - a higher proportion of its proteins would be non-functional.

4.7   Below are listed the sorts of differences between DNA and RNA which might be used to help to distinguish them.

	DNA	RNA
1) Base composition	ATGC	AUGC + unusual bases
2) Sugar	deoxyribose	ribose
3) Hydrolysis by NaOH?	No	Yes
4) Chromogenic test	Diphenylamine	Orcinol
5) DNA-ase sensitive?	Yes	No
6) RNA-ase sensitive?	No	Yes
7) Denaturation	Sharp Tm, 40% increase in absorbance	Poorly defined Tm
8) Size	Typically $> 10^8$	Usually $< 2 \times 10^6$

Notes: 1) and 2) would be revealed by chromatographic analysis following hydrolysis. 3) provides a rapid method of distinguishing RNA from DNA. 4) Specific chromogenic assays are based on the sugar present. 5) and 6) could be helpful, providing the enzymes were known to be specific and not contaminated. 7) The lack of extensive double helix means that RNA shows a modest hyperchromic shift. 8) Whilst clear differences in size do occur, both polymers are easily damaged (broken) during isolation. This would be of limited discriminative use unless great care was taken.

4.8

1) The DNA base composition is A = 30%; T = 30%; C = 20%; G = 20%. The data regarding the Tm strongly suggests that the DNA is double stranded. This being the case, then A must equal T (whenever A occurs in one strand, T must be present in the opposite position), hence A = T = 30% each. G and C will be present in equal proportions (because of obligatory basepairing), therefore G = C = 20% each.

2) You cannot deduce anything further regarding the base composition of this RNA. Remember that it was made on a template which was only one strand of the DNA. The A = T and G = C rules only hold for double helical DNA: they do not hold within one strand of a DNA double helix (or single stranded DNA - remember φx174). Thus it will not hold in the RNA transcript of a single DNA strand.

4.9

The correct answer is 2). The RNA transcript is synthesised in an anti-parallel manner, giving 5' -AAU-UCC-GUC-AGG-UCU- 3' as the relevant part of the mRNA, split up into codons. These are read from the 5' end, the first being AAU, giving asn-ser-val-arg-ser as the peptide product. The asparagine will be the N-terminal end of this peptide. If you thought that mRNA was produced in a parallel manner, you would get mRNA of this sequence: 5' -UCU-GGA-CUG-CCU-UAA 3'. If this is translated in 5' → 3' order you get Ser-Gly-Leu-Pro; only 4 amino acids residues, because UAA = stop; this was sequence 1). If you produced the correct mRNA (ie anti-parallel), but then read it in the wrong order ie 3' → 5', you should also have got the same as 1).

3) Wrong! Possible reasons are mis-reading of the genetic code; reading the mRNA overall in the correct direction (5' → 3') but each codon in the wrong order (eg from the mRNA AAU-UCC etc, taking AAU as the first codon, but reading it as UAA).

Responses to Chapter 5 SAQs

5.1   The family of 5 carbon aldoses should look something like this:

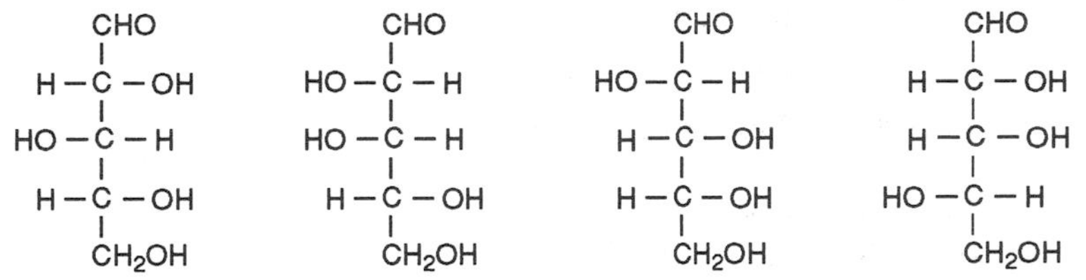

With 5 carbon aldose sugars, there will be 3 asymmetric centres, at each of which two configurations can exist. There will thus be 8 (ie 2x2x2) stereoisomers. Half will be D-isomers, where the hydroxyl furthest from the aldehyde group is to the right of the main carbon chain. These are the first four structures shown above. The remaining will be L-sugars. Satisfy yourself that these 8 sugars are all different molecules, although they are of course very closely related.

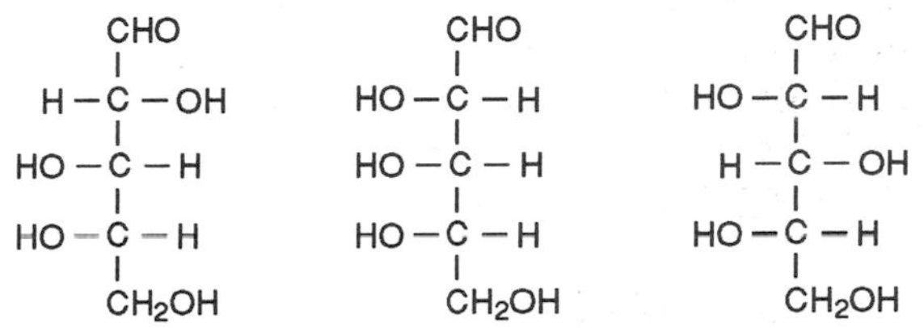

5.2

1) a), c), and e) are all stereoisomers of D-glucose, since they are all 6-carbon aldose sugars. They differ only in the distribution of the hydroxyl groups around carbon atoms 2-5. c) is an L-sugar; a) and e) are both D-sugars and are D-galactose and D-mannose, respectively; both of these are of widespread occurrence.

2) a) and b) are in the D configuration, c) is in the L configuration. The key is to look at the asymmetric centre furthest from the aldehyde or keto group. To be an asymmetric centre, a carbon atom must have four different substituents. Thus compound a) and c) have 4 asymmetric centres (carbon atoms 2-5); b) only has 3 (carbon atoms 3-5). The absolute configuration is D if the hydroxyl is to the right of this critical asymmetric centre, L if it is to the left. Thus a) and b) are D-sugars; c) has absolute configuration L.

3) Aldoses possess an aldehyde (-CHO) group. Neither b) (which has a keto group) nor d) and f) (which have carboxylic acid groups) conform to this. Thus b), d) and f) are not aldoses.

4) Compound c) is a 6-carbon aldose sugar. Structural isomers will have the same number of particular atoms and the same relative molecular mass, but will differ in which 'functional' groups are present. In this case, b) (a keto-containing compound) and d) (a carboxylic acid-containing compound) are structural isomers of c) (an aldose sugar). If in doubt, check carefully how many of each atom are present. f) is not a structural isomer of c) or any of the other compounds. It has one more oxygen atom than any of the other compounds.

5.3 a) and d) are maltose. b) is cellobiose, c) is lactose.

You should have quickly decided that structure c) was not maltose, since it contains different monosaccharides and maltose consists of two glucose units. c) is in fact lactose, consisting of glucose and galactose joined by a β-1,4-bond. Similarly, although b) only contains glucose, the two monosaccharides are joined by a β-1,4-bond; in maltose the glycosidic bond is α-1,4; b) is cellobiose.

a) and d) may have caused more of a problem. They both contain two glucose units joined by an α-1,4-bond: the only difference lies in the position of the hydroxyl at the C-1 of the right hand glucose unit. They are thus both examples of maltose, but in the alternative α-(structure a) and β-forms (structure d). They would both be expected to occur in solution.

5.4 If you said yes, good! If you said no, look again at the reasoning applied to maltose and lactose. The basis of the reducing property is a free/potentially free aldehyde group (ie one which is not involved in a glycosidic bond). Although in each of these disaccharides described so far, the C-1 of the left hand sugar unit is involved in a glycosidic bond, the C-1 of the right hand unit (as written) is not. It can therefore open out to the open chain form, it will display mutarotation and it will have reducing properties.

5.5
1) Glucose units a) and c) are non-reducing ends of the polymer; in each case, the potentially reducing aldehyde group at C-1 is involved in a glycosidic bond.

2) The only reducing terminus is glucose f), where the C-1 is not involved in a glycosidic bond. This glucose can open out into the straight chain form and will then display reducing properties.

3) The glycosidic bond between units e) and f) is between the C-1 of e) and the C-4 of f), with the C-1 in the β-configuration. This is thus a β-1,4-glycosidic bond.

4) Unit b) is linked via its C-1 (in an α-configuration) to the C-6 of glucose e). Thus the bond between b) and e) is an α-1,6-glycosidic bond.

5) Bonds between a) and b), c) and d), and between d) and e) are all α-1,4-glycosidic bonds.

5.6   There are quite a lot of different ways you could answer this question. We believe the
following points are important and should have been included in your answer.

1) The position and type of glycosidic bond has a dramatic effect on the shape that a
polymer adopts. Polymers of α-1,4-linked glucose, without branching, form a
hydrated gentle helical structure, which is loose and permits solvent penetration. If
β-1,4-linkages occur, the structure becomes highly linear, and close packing of dozens
of similar chains is permitted. Insoluble rigid fibrils result, giving an insoluble
structural polysaccharide.

2) Numerous branches means that the polymer becomes globular and compact, whilst
permitting solvent penetration. Numerous ends occur, at which additional
monosaccharide units may be added or from which they may be removed. This is
advantageous for an energy store, facilitating access by the cell to its energy reserves.

Responses to Chapter 6 SAQs

6.1   The answer to 1) is c) whilst the answer to 2) is b). Our reasoning is as follows.

Based on the behaviour of the fatty acids, a) would probably be solid at room temperature, because the mean of the melting points is about 32°C. We would expect a) to be liquid at 40°C.

Palmitic acid has a melting point of 63°C, b) would be expected to be solid at both room temperature and 40°C.

The combination of 2 polyunsaturated fatty acids and a shorter chain saturated fatty acid should give a neutral fat with a low melting point; c) should be liquid at room temperature.

6.2   1) Your triglyceride must avoid very short chain saturated fatty acids. Any triglyceride with fatty acids with at least 10 carbons should be solid at room temperature. For example:

$$
\begin{array}{l}
\quad\quad\quad\quad\; O \\
\quad\quad\quad\quad\; \| \\
CH_2 - O - C - (CH_2)_{12} - CH_3 \\
\;\; | \quad\quad\quad\; O \\
\quad\quad\quad\quad\; \| \\
HC \;\; - \; O - C - (CH_2)_{14} - CH_3 \\
\;\; | \quad\quad\quad\; O \\
\quad\quad\quad\quad\; \| \\
CH_2 - O - C - (CH_2)_{12} - CH_3
\end{array}
$$

2) Now, long chain fatty acids must also be avoided. A triglyceride with about C-10 to C-12 fatty acids should melt by 37°C. For example:

$$
\begin{array}{l}
\quad\quad\quad\quad\; O \\
\quad\quad\quad\quad\; \| \\
CH_2 - O - C - (CH_2)_8 \; - CH_3 \\
\;\; | \quad\quad\quad\; O \\
\quad\quad\quad\quad\; \| \\
HC \;\; - \; O - C - (CH_2)_{10} - CH_3 \quad \text{or} \\
\;\; | \quad\quad\quad\; O \\
\quad\quad\quad\quad\; \| \\
CH_2 - O - C - (CH_2)_8 \; - CH_3
\end{array}
\qquad
\begin{array}{l}
\quad\quad\quad\quad\; O \\
\quad\quad\quad\quad\; \| \\
CH_2 - O - C - (CH_2)_8 \; - CH_3 \\
\;\; | \quad\quad\quad\; O \\
\quad\quad\quad\quad\; \| \\
HC \;\; - \; O - C - (CH_2)_8 \; - CH_3 \\
\;\; | \quad\quad\quad\; O \\
\quad\quad\quad\quad\; \| \\
CH_2 - O - C - (CH_2)_8 \; - CH_3
\end{array}
$$

or a simple triglyceride only containing C-10 fatty acids.

3) This must contain at least one unsaturated fatty acid, probably two, and might be represented by the following:

$$CH_2 - O - \overset{\overset{\textstyle O}{\|}}{C} - (CH_2)_7 - CH = CH - CH_2 - CH = CH - (CH_2)_4 - CH_3$$

$$HC \;\; - \;\; O - \overset{\overset{\textstyle O}{\|}}{C} - (CH_2)_7 - CH = CH - (CH_2)_7 - CH_3$$

$$CH_2 - O - \overset{\overset{\textstyle O}{\|}}{C} - (CH_2)_7 - CH = CH - CH_2 - CH = CH - (CH_2)_4 - CH_3$$

Note that this is a rather simplistic analysis designed to illustrate, in general terms, the influence of fatty acid composition on properties of neutral fats.

6.3   1) Structure e) is a triglyceride. Three fatty acids are joined to glycerol by ester linkages, formed by condensation between an alcohol group and a carboxyl group, as shown:

$$-CH_2OH \;\; + \;\; HOOC-(CH_2)_x-CH_3 \;\; \longrightarrow \;\; -CH_2-O-\overset{\overset{\textstyle O}{\|}}{C}-(CH_2)_x-CH_3 \;\; + \;\; H_2O$$

Structure d) contains a phosphate group; structure a) has an ether linkage at one position; thus neither of these are triglycerides.

2) Structure d) is a phospholipid: it is based on glycerol, with 2 fatty acids attached by ester linkages. The third position of the glycerol has a phosphate group which is itself substituted with a polar, alcohol-containing group. The one shown is serine. Hence it is phosphatidyl serine.

3) The characteristic feature of a plasmalogen is the presence of an ether linkage at the C-1 position of the glycerol. The answer is thus structure a), which is the only structure with this bond.

4) Simple! Structure c) is the unsaturated fatty acid.

5) Steroids have a characteristic and highly distinctive structure consisting of 4 fused rings. The only remotely suitable candidate is structure b)!

6.4   The answer is quite complex; 2, 4, 6 and 9 are all made from isoprene units. The sidechains of vitamins E and K are also derived from isoprene units. The key to identifying if a compound is formed from isoprene units is to see if you can identify repeats of the basic isoprene structure in the molecules. This structure can be drawn as:

$$\overset{\overset{\textstyle CH_3}{|}}{H_2C = C} - CH = CH_2$$

It might be helpful for you to go back through the text and draw circles around the 'isoprene' units of the molecules that have been illustrated.

Sphingosine 1 and prostaglandin 3 are produced from fatty acids. Palmitic acid 5 is, itself, a fatty acid.

Responses to Chapter 7 SAQs

7.1

1) An amphipathic lipid is one containing 2 regions of contrasting character. Part of it is hydrophobic ('water hating') and nonpolar: this is provided by long hydrocarbon chains. The other part is hydrophilic ('water loving') and is polar. For this, groups which are ionised (eg phosphate or amino groups) and those with dipolar character (which form hydrogen bonds) are common.

2) a) and c) are amphipathic lipids, since each contains a polar region (the phosphate-containing group, shaded below, in a); the galactose moiety shaded in c). Although there are no charged groups with C, the numerous hydroxyl groups make the galactose hydrophilic.

b) is a neutral triglyceride which lacks any clear polar character. It is not amphipathic. c) is an example of a glycolipid, which are particularly common in chloroplast membranes.

3) Amphipathic lipids such as a) and c) will form characteristic structures when mixed with water. The polar 'head' region interacts strongly with water, whilst the nonpolar 'tails' aggregate together, 'protected' from water by the polar head groups. This leads to the formation of lipid bilayers and other structures described in Section 7.3. Non-amphipathic lipids, such as neutral fats, will simply form a fat globule. Here, the hydrophobic fatty acid groups coalesce so that water is excluded and an insoluble globule results.

7.2

1) General composition: B is devoid of cholesterol. Prokaryotes characteristically lack cholesterol thus this preparation might be of prokaryote origin. A, B and C all have substantial amounts of protein, whereas D has relatively little. D may thus be a relatively 'inert' membrane whose role is primarily insulating (as in myelin). From the data given we would find it difficult to identify these membranes any further.

2) Effect of salt treatment: Treatment with salt will tend to remove peripheral proteins which are only attached to the membrane by ionic interaction with the polar heads of phospholipids. It will not tend to remove those proteins which make significant contact with the hydrophobic interior of the membrane (and are hence integral). The protein content of A has been substantially reduced, suggesting that much of this protein was peripheral. With preparation C, in contrast, there has been little loss of protein, suggesting that most of the protein is integral; B is intermediate. With D, only a small amount remains after salt treatment, suggesting very little integral protein.

Response to Chapter 8 SAQs

8.1 1) False: the substantial energy barrier between A and B (the activation energy) means that few molecules of A will possess sufficient energy to react.

2) False: the first part of this is correct. The activation energy requirement is now less and more molecules will possess at least this level of energy. However, the enzyme does not alter the overall energy yield of the reaction A → B.

3) False: enzymatic catalysis is based on the general principle that rate increases result from a lowering of activation energy.

4) True: B has a lower energy content than A: conversion of A to B will result in release of energy (amounting to the difference in energy contents of A and B) and the reaction will tend to proceed spontaneously. The high activation energy requirement may, however, make the rate of reaction negligible.

5) True. Re-read the responses to questions (2) and (3) if in doubt.

6) False. As a generalisation, the lower the activation energy, the faster the reaction will be. This is because more molecules will possess the necessary energy (look again at Figure 8.5). Enzymes lower the activation energy and a greater proportion of the population of molecules will then have sufficient energy. Thus, on energy considerations alone, we would expect enzyme X to produce greater rate enhancement than enzyme Y.

8.2 1) This is untrue: catalysts alter the rate of a reaction without being consumed or changed in it, but do not alter the equilibrium position. This is precisely what enzymes do and thus they are catalysts.

2) True: enzymatic catalysis is achieved by lowering the activation energy of a reaction. This is accomplished by providing an alternative reaction route. Enzymes must bind to their substrates and form an enzyme-substrate (ES) complex for catalysis to occur.

3) False: look again at Figure 8.9 and the associated comments in the text. Binding of 2 groups of the substrate may not be sufficient to guarantee distinction between 2 stereoisomers. With only 2 recognition sites, both stereoisomers could bind, as shown below; even better, try it with atomic models.

272

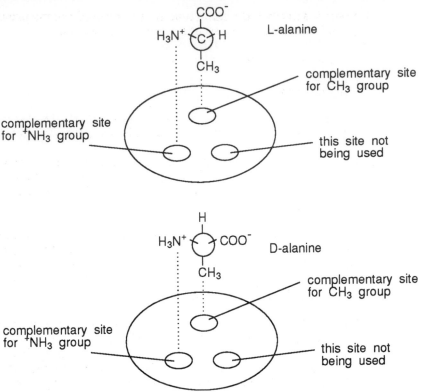

4) False: enzymes are flexible and may change shape as a consequence of binding substrate (known as induced fit). Many proteins can change their conformation; one notable case is haemoglobin, which changes shape upon binding or releasing oxygen.

5) True: a substrate only binds reasonably tightly and correctly positioned for catalysis if there is a good match between the shape and charge distribution of active site and substrate. This is how enzymes discriminate between different molecules. The next section expands on this discriminative ability.

8.3
1) An International Unit (I.U.) is the amount of enzyme which transforms 1 μmole of substrate per minute under defined conditions. Since 0.1ml of enzyme solution was present, and assuming that no non-enzymic product formation occurred, we conclude that the 0.1ml enzyme present was responsible for forming 14 μmol product in 30 minutes.

Hence 0.1ml enzyme → 14μmol in 30 min

$$= \frac{14}{30} = 0.47 \, \mu mol.min^{-1}$$

1ml would have transformed $0.47 \times \frac{1.0}{0.1} \, \mu mol.min^{-1}. \, ml^{-1}$

$$= 4.7 \, \mu mol.min^{-1}. \, ml^{-1}$$

Since 1 IU = 1 μmol transformed per min, the activity is 4.7 IU. ml^{-1}

2) Specific activity is the enzyme activity expressed per mg protein. Since there are 6mg protein per ml enzyme solution, and 4.7 IU per ml.

the specific activity is 4.7 IU/6mg protein

$= 0.78 \text{ IU.mg}^{-1}$

3) You should have noticed several omissions or problems with the assay as reported. Here are the more obvious, with suitable changes to improve the quality of the assay.

a) No control for non-enzymic hydrolysis of the substrate appears to have been done. Thus we are not sure that the enzyme was responsible for all of the product formed: an overestimate of enzyme activity would result. An incubation using boiled enzyme would help here.

b) The assay was run for 30 minutes: there is no indication whether reaction was linear for the entire incubation period. It may have slowed down, as shown in Figure 8.15; this would lead to an underestimation of activity. For reasons given below (points c and d) this is highly likely in this case. Steps to ensure that the initial rate is being measured should be taken (see point e) below).

c) Neither the extract nor the reaction mixture was buffered; the substrate was merely adjusted to pH 7.0. Thus the pH may have changed during the assay, which could have caused a change of rate. In fact, this is highly likely, because one of the products is propionic acid; the reaction as shown is correct stoichiometrically but at pH 7.0 (the starting pH) the propionic acid (also known as propanoic acid) formed would dissociate, as shown:

$$CH_3CH_2COOH \rightleftharpoons CH_3CH_2COO^- + H^+$$

Thus the protons produced would cause acidification and lowering of pH. The original substrate concentration [S] is 6.7mmol.l^{-1} (substrate solution makes up $\frac{2}{3}$ of the final reaction volume; hence initial [S] $= \frac{2}{3}$ of 10mmol.l^{-1}). As we show below, about $\frac{2}{3}$ of the substrate has been hydrolysed in the 30 min reaction. Thus 5mmol.l^{-1} [H$^+$] will have been produced, giving a pH of about pH 3.0!

The solution is to provide adequate buffering of both substrate and the extraction medium, and lessen H$^+$ production by using a shorter time.

d) Substrate depletion will have occurred, leading to a slowing of the reaction; product inhibition is also possible. A simple calculation will show why this is likely. We know that 14 µmol of product were formed in the 30 min reaction. How much substrate was present at the start? 2ml of 10mmol.l^{-1} substrate solution was used:

ie $\frac{2}{1000}$ x 10 mmol = 0.02 mmol = 20µmol substrate present

In the assay 14 µmol were consumed (the reaction stoichiometry is 1:1), thus only 1/3 of the initial [S] remains at the end of the assay. This may well only support a lower reaction rate (see section 8.4).

274

These problems are avoided if the true initial rate is measured, such that only a small proportion of the initial [S] is removed. Incidentally, we do not know if the 6.67mmol.l⁻¹ [S] (ie the [S] which the enzyme encounters) is sufficient to saturate the enzyme and give maximum reaction velocity.

e) The assay appears to be based on a single analysis. Duplicates (at least) should always be conducted to increase reliability. The mean value is likely to be closer to the true value!

There are clearly major shortcomings with the assay as described. We hope you noticed the above points, although we would not have expected you to do the quantitative analysis of points c and d. Let us conclude with recommendations which address all of the problems noted.

The assay described is a chromogenic assay, with a useful change in absorbance. There is no need to rely on a 30 min incubation, with all the problems that brought. Use a much shorter time, preferably following the reaction as it occurs, by running the assay in a cuvette in a spectrophotometer or colourimeter. With a continuous assay (as this would then be), it is easy to see if the rate is slowing down. The initial rate must be measured. Use buffered extraction medium and include a buffer in the assay. Check (experimentally) that [S] is adequate; run controls and conduct the assay several times.

8.4

We hope you resisted any temptation to plot v against [S], then attempting to extrapolate the line to get Vmax (along the lines of Figure 8.19). Instead, the data should be transformed such that a straight line will be formed (providing Michaelis-Menten kinetics apply; assume so at this stage). Whilst there are various ways to do this, we have described the Lineweaver-Burk plot. Reciprocals of v and [S] have to be calculated, giving $1/v$ and $1/[S]$, respectively:

	$[S]$ mmol.l⁻¹	$\frac{1}{[S]}$ l.mmol⁻¹	v µmol.min⁻¹	$\frac{1}{v}$ min.µmol⁻¹
A	0.2	5.0	0.57	1.75
B	0.33	3.0	0.85	1.17
C	0.5	2.0	1.18	0.85
D	1.0	1.0	1.82	0.55
E	2.0	0.5	2.5	0.4
F	5.0	0.2	3.23	0.31

These are then plotted as shown below:

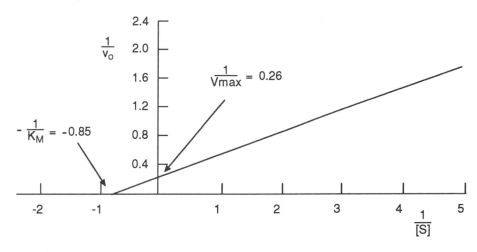

The line of best fit is determined (by eye; better by regression analysis) and the K_M and Vmax obtained as:

a) the intercept on the $1/v$ axis = $1/V$max. In this case it is 0.26, thus $1/V$max = 0.26 and Vmax = 3.85 μmol.min^{-1}

b) K_M is determined from the intercept on the x axis; this is -0.85 in this case, thus -$1/K_M$= -0.85 K_M = $1/0.85$ = 1.18mmol.l^{-1}

2) The crucial data is obtained from substrate concentrations around the K_M value, particularly in the range 0.2 - 5.0x K_M. Inspection of Figure 8.19 shows that significant change in rate occurs over this range. Assays at very low [S] may cause problems, because they tend to be more prone to error; additionally, when expressed as $1/[S]$, they give very high values on the x axis. Within the range given above, try to arrange substrate concentrations which will give a good spread of points when plotted as reciprocals. Note how this was done with the numerical data above: Whilst it was a highly non-linear distribution in mol.l^{-1} terms, they were more evenly spaced as reciprocals.

Suitable values would be:

[S], mmol.l^{-1}	giving $\frac{1}{[S]}$, l.mmol^{-1}
0.4	2.5
0.55	1.8
0.8	1.25
2.0	0.5
5.0	0.2
10.0	0.1

8.5

1) False. The lower the K_M, the greater the degree of saturation at a given [S]. This question can be analysed by calculating v, for the 2 K_M values, as a % of Vmax. Try it if you haven't already. You should find:

	$K_M = 10^{-3}$ mol.l^{-1}	$K_M = 10^{-2}$ mol.l^{-1}
% saturation when [S] = 10^{-2} mol.l^{-1}	91	50

2) True! If you make [S] = K_M, and then calculate v from the Michaelis-Menten equation, you obtain:

$$v = \frac{Vmax\,[S]}{K_M + [S]} = v = \frac{Vmax\,[S]}{2\,[S]} \quad v = \frac{Vmax}{2} \text{ when } [S] = K_M$$

You may prefer to substitute values for Vmax, [S] and K_M; for example set

Vmax = 100 µmol.min^{-1}
[S] = 5 mmol.l^{-1}
K_M = 5 mmol.l^{-1}

then $v = \frac{100 \times 5}{5+5}$ µmol min^{-1} = $\frac{500}{10}$ = 50 µmol min^{-1}

ie half Vmax.

3) False. The main problem with this approach is that unless you actually achieve high enough [S] to give Vmax, you have to extrapolate a hyperbola in order to estimate Vmax. Using this approach, Vmax must be obtained to accurately calculate K_M; there may be considerable error introduced by doing this. Instead, transform the equation to give a linear relationship: this is easier to extrapolate without introducing error, and is also easily analysed by regression analysis.

4) True. The Lineweaver-Burk plot requires a graph of 1/v against 1/[S]. This justification comes from taking reciprocals of the Michaelis-Menten equation

ie $v = \frac{Vmax\,[S]}{K_M + [S]}$ gives $\frac{1}{v} = \frac{K_M + [S]}{Vmax\,[S]}$

this may be written as $\frac{1}{v} = \frac{K_M}{Vmax\,[S]} + \frac{1}{Vmax}$

(ie the [S] cancels in the second term on the right hand side). This is the equation of a straight line (y = mx + c).

5) True. Since K_M is the [S] giving 1/2 Vmax, the lower the K_M the lower the [S] needed to saturate the enzyme. Low K_M thus means that the enzyme binds substrate even at low [S], which indicates high affinity.

8.6

1) No, they do not. Under particular conditions, and over a particular period of time, one can determine the temperature at which most product is formed. However, if

the duration of the incubation is changed, then the apparent optimum temperature will also change. This is a consequence of the two conflicting effects of temperature on enzymes: the effect on catalytic rate (higher rate at higher temperature) and on denaturation (higher temperatures are more likely to denature; denaturation will be more rapid).

2) True, although change in charge on the substrate may also be important and could be the crucial effect. The requirement for substrate binding and/or catalysis is that appropriate amino acid residues are in the 'correct' ionic form (protonated or deprotonated) to fulfil their role. If they are not, the rate will be less. Thus if a carboxyl group must be deprotonated (and hence negatively charged) in order to interact with a positively charged group on the substrate, lowering the pH in the region of the pKa of the carboxyl group will lead to its protonation and lessening (or complete loss) of activity.

3) False: we have already noted (in response 1 above) that denaturation occurs as the temperature is raised.

4) False. Of course you do! When assaying enzymes you should try to measure the initial rate, before the reaction rate changes. It will frequently slow down: one cause of this is denaturation of the enzyme, whereby, with time, some of the enzyme molecules are denatured; as a consequence the rate slows since there are fewer active enzyme molecules and rate α [E] (or should be!). Denaturation is most likely to be a problem at elevated temperature.

5) True. This is why reactions (both enzyme- catalysed and non-enzymatic) proceed faster at higher temperatures.

8.7   Only statement 3) is correct. Competitive inhibition means that less active enzyme is available when inhibitor is bound to any enzyme molecules. However, substrate can displace inhibitor and, if [S] is raised sufficiently, the same Vmax can still be achieved - although higher [S] will be needed. Thus K_M is raised, whilst Vmax is not affected.

1) would be expected for mixed inhibition, where adverse effects on both substrate binding and catalysis are seen.

2) is an interesting case, in which inhibitor makes it easier for substrate to bind, whilst none-the-less lowering enzyme activity. This occurs if inhibitor can bind to ES but not to E. This is a form of inhibition which is known as uncompetitive inhibition. It is rare with single substrate enzymes but can occur when more than one substrate is involved (for example, imagine a case where substrates must bind in a particular order (known as compulsory order):

$$E + A \rightleftarrows EA + B \rightleftarrows EAB \rightleftarrows EPQ \rightleftarrows EP + Q \rightleftarrows E + P$$

A competitive inhibitor for substrate B will behave as an uncompetitive inhibitor with respect to substrate A because it will convert EA to EAI; since EA is being removed, the equilibrium/steady state between E + A and EA will shift towards EA).

4) is the result expected for non-competitive inhibition.

5) is consistent with extreme damage to the enzyme, such as denaturation or perhaps irreversible inhibition.

8.8

1) Whilst plotting v against [S] may give some indication of the type of inhibition, it is unlikely to distinguish between the possibilities. The data needs to be linearised, as in the Lineweaver-Burk plot. Thus $1/v$ and $1/[S]$ for both inhibited and non-inhibited assays must be calculated, giving:

[S]	1/[S]	(A) - Inhib.		(B) + Inhib.	
mmol.l^{-1}	l.mmol^{-1}	v	1/v	v	1/v
1.0	1.0	2.33	0.43	1.20	0.84
1.43	0.7	2.94	0.34	1.52	0.66
2.0	0.5	3.6	0.28	1.85	0.54
3.0	0.33	4.44	0.225	2.27	0.44
6.0	0.167	5.71	0.175	2.94	0.34

These are then plotted as $1/v$ against $1/[S]$, and straight lines can be fitted, as shown:

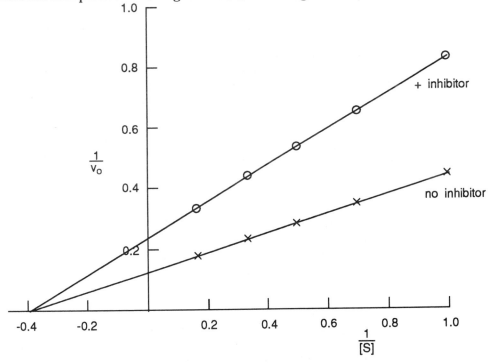

It is apparent that the intercept on the $1/v$ axis is raised in the presence of inhibitor: thus Vmax has been reduced (intercept on $1/v = 1/Vmax$). The intercept on the negative side of the $1/[S]$ axis is unaltered: K_M has not been affected (since this intercept $= -1/K_M$). This pattern is characteristic of non-competitive inhibition, since when inhibitor is bound, enzyme molecules are inactive (or at least less active) and Vmax is thus lowered; substrate binding is not affected.

2) All experiments should be repeated several times before conclusions are made. In this case, if reversible non-competitive inhibition is suspected, then regeneration of activity after careful removal of the inhibitor must be demonstrated. If we cannot regenerate activity, then irreversible inhibition must be suspected. Additional experiments which address this question should be conducted.

8.9   Statement (1) This is true - but it is a consequence of the sigmoidal relationship, not the cause. Thus it is not the answer you are looking for!

Statement (2) This is also true, at least in some cases, but is not the answer. The existence of allosteric inhibition means that activity at a given [S] can be lessened - but this does not cause the sigmoidal curve (it could, in principle, occur with a hyperbolic curve, with the line in the presence of inhibitor simply being lower ie non-competitive inhibition)!

Statement (3) Again, true, but this is only a partial explanation. Some enzymes, such as malate dehydrogenase, have more than one subunit but do not display a sigmoid rate curve. Thus presence of subunits does not, on its own, result in allosteric effects.

Statement (4) No; some enzymes occur as isoenzymes or multiple molecular forms. With lactate dehydrogenase they arise through the existence of two types of subunits (H-type and M-type), which combine to form a tetramer. All 5 possible combinations occur. Although dissimilar subunits are present, allosteric behaviour is not displayed.

Statement (5) This is the real reason for allosteric phenomena and a sigmoid rate against [S] plot. Several binding sites are required, which can switch between low- and high-affinity forms. Binding of substrate to one subunit 'helps' binding of substrate to other subunits in the group (if we think of the 4 subunits of a tetramer as a group). This communication or 'cooperativity' is vital.

8.10   This question is clearly open-ended and you may have thought of different reasons to those given below, or different ways of describing the same factors. Thus our list is not necessarily comprehensive but does cover the most important areas; there are, of course, many possible examples in each case.

1) Specificity, especially stereospecificity: this is vital in many areas (not all - enzymes in detergents are simply required to degrade proteins and/or starch; proteases which are specific to particular amino acid sequences would be of limited use!). Suitable examples are production of L-amino acids for food supplements (by aminoacylase) and the use of enzymes in analysis.

2) High catalytic rates: all cases!

3) Mild reaction conditions: in the production of high fructose syrup, the use of enzymes avoids the need for acid hydrolysis; acid hydrolysis causes browning and side-reactions. Thus product quality is improved by the use of enzymes.

4) Stability: good stability of enzymes means they can be used for extended periods, making for a cheaper process. Thus aminoacylase used for production of L- amino acids can be used continuously for 30 days before needing to be replenished.

5) Easily controlled: for example, by changing the pH or by heat treatment. Think what would happen to a loaf of bread during baking if the enzymes leading to CO_2 production in yeast were not denatured!

Appendix 1

Ionisation and weak acids

Many compounds exist in solution in an ionised form. As an example, sodium chloride dissolves in water to form Na^+ ions and Cl^- ions. Compounds which ionise with release of protons (H^+) are called acids. A convention introduced by Brönsted is that any group which donates a proton is termed an acid. This means that not only traditional acids such as hydrochloric acid (HCl) and acetic acid (CH_3COOH) but also ammonium ions (NH_4^+) can be described as acids. The unifying feature is that they can all dissociate, in a reversible way, to release a proton, for example:

$$HCl \rightleftarrows Cl^- + H^+$$

$$CH_3COOH \rightleftarrows CH_3COO^- + H^+$$

$$-NH_4^+ \rightleftarrows NH_3 + H^+$$

For convenience, we can summarise this behaviour as

$$AH \rightleftarrows A^- + H^+$$

This terminology is equally applicable whether the reaction is

$$AH \rightleftarrows A^- + H^+ \text{ or}$$

$$YH^+ \rightleftarrows Y + H^+$$

Dissociation of an acid thus results in release of a proton; the deprotonated form of the acid is known as its conjugate base (a base, according to the Bronsted definition, is simply a proton acceptor. The reactions given above are completely reversible; in going from right to left we see a base accepting a proton to form an acid). Hence the relationship between an acid and its conjugate base can be summarised as

$$ACID \rightleftarrows CONJUGATE BASE + H^+$$

strong acids

fully dissociated

What determines the extent to which this process occurs? How can we predict the ionic state of a given compound which can ionise? Let us examine this by comparing acetic acid and hydrochloric acid. Hydrochloric acid is known as a strong acid. This is because it is fully dissociated. In terms of the possible forms:

$$HCl \rightleftarrows Cl^- + H^+$$

weak acid

the equilibrium is far to the right and essentially only H^+ and Cl^- are present. Acetic acid, (also known as ethanoic acid), in contrast, is much less completely dissociated and as a consequence is termed a weak acid. An equilibrium exists between the non-dissociated acid and its conjugate base, ie:

$$CH_3\,COOH \rightleftarrows CH_3COO^- + H^+$$
$$\text{'acid'} \qquad \text{'conjugate base'}$$

We can measure the concentrations of the various molecular species and so determine their ratios. An equilibrium constant can be obtained, where:

$$K_{eq} = \frac{[CH_3COO^-]\,[H^+]}{[CH_3COOH]}$$

However, since this is an acid dissociation constant it is referred to as Ka. For acetic acid this is found to have the value, at 25°C, of 1.86×10^5. This indicates that only a tiny proportion of acetic acid molecules have dissociated, hence the term 'weak acid'.

Although the acid dissociation constant Ka provides information on the extent of dissociation of a weak acid, it is even more useful in a modified form. This involves taking the logarithm (to base 10) of the reciprocal of Ka: the product of this is known as the pKa.

ie $pKa = \log \dfrac{1}{K_a}$ or $pKa = -\log Ka$

This means that there is precisely the same relationship between Ka and pKa as there is between $[H^+]$ and pH, since

$$pH = \log \frac{1}{[H^+]} \text{ or } pH = -\log [H^+].$$

Modern calculators make calculation of log values simple. None-the-less, check that, on the basis of a Ka of 1.8×10^5, the pKa of acetic acid is 4.74!

∏ What are the pKa values of the following:

- pyruvic acid, $Ka = 3.98 \times 10^{-3}$
- NH_4^+ $Ka = 5.0 \times 10^{-10}$

You should have calculated that the pKa of pyruvic acid = 2.4 and that for NH_4^+ = 9.3

Using the equation $pKa = -\log Ka$ or $\log \dfrac{1}{K_a}$

hence pyruvic acid $pKa = -\log 3.98 \times 10^{-3}$

$= -(-2.4) = 2.4$

for NH_4^+, $pKa = -\log 5 \times 10^{-10}$

$= -(-9.3) = 9.3$

Ⅱ Before reading on, calculate 1) the pH of a solution whose [H⁺] is 5×10^{-6} mol.l⁻¹. 2) the [H⁺] of a solution of pH 8.2.

Your answers should have been for 1) pH = 5.3 and for 2) [H⁺] = 6.3×10^{-9} mol.l⁻¹

In more detail, for 1) pH = - log [H⁺] (or $\log \dfrac{1}{[H^+]}$)

therefore pH = - log 5×10^{-6} = - (-5.3) = 5.3

2) - log [H⁺] = 8.2

therefore [H⁺] = antilog - 8.2 = 6.3×10^{-9} mol.l⁻¹

Each group which can ionise has (at 25°C, in aqueous solution) a particular pKa. Tables of pKa values are often presented for organic and amino acids. Their usefulness is that knowledge of pKa values allows us to predict the extent of ionisation of a group at a particular pH. We can thus deduce the charge on a molecule, which will be a consequence of the ionic state of its ionising groups.

What does the pKa value indicate? It is derived from the acid dissociation constant, so it is clearly related to the extent of dissociation. Let us go back to the original equilibrium expression for the general case of a weak acid HA

$$HA \rightleftarrows A^- + H^+$$

from which Ka = $\dfrac{[A^-][H^+]}{[HA]}$

If we make [A⁻] equal [HA], then the expression (through cancellation of [A⁻] and [HA]) becomes

Ka = [H⁺]

Since the relationship between Ka and pKa, and between [H⁺] and pH is identical, this means that

When [HA] = [A⁻], then pKa = pH.

Thus, simply by inspection of the equilibrium expression, we can deduce that the pKa is the pH at which 50% dissociation of a group has occurred. This must be so, since A⁻ originates from dissociation of HA. We now have a precise meaning for pKa. Extensive use is made of pKa values, since they give us the pH at which a group is half ionised.

Further analysis is justified. Starting from the equilibrium expression, we can rearrange this as follows:

Ka = $\dfrac{[A^-][H^+]}{[HA]}$ ∴ [H⁺] = Ka $. \dfrac{[HA]}{[A^-]}$

If we now take reciprocals, we obtain

$$\frac{1}{[H^+]} = \frac{1}{Ka} \cdot \frac{[A^-]}{[HA]}$$

If we now take logarithms, we get

$$\log \frac{1}{[H^+]} = \log \frac{1}{K_a} + \log \frac{[A^-]}{[HA]}$$

We already know alternative expressions for $\log \frac{1}{[H^+]}$ and $\log \frac{1}{K_a}$, namely pH and pKa respectively. Substitution gives

$$\boxed{pH = pKa + \log \frac{[A^-]}{[HA]}}$$ This equation is worth remembering.

Henderson-Hasselbalch equation

It is known as the Henderson-Hasselbalch equation. This is an extremely useful equation as it allows us to:

- predict the ratio of dissociated (A⁻) to protonated (HA) form if we know the pH and pKa.

- calculate pH if pKa and the ratio of A⁻ and HA are known

- determine pKa values.

The first two of these areas are particularly important in:

- preparation of buffers

- predicting the behaviour of molecules during separation and purification

- the relationship between biological activity and pH.

Let us now use the Henderson-Hasselbalch equation to establish how ionisation changes with pH. First, a worked example:

What is the ratio of [A⁻] to [HA] when the pH is one pH unit higher than the pKa?

For ease, we will provide actual values - let pKa = 7.0 and pH = 8.0. From the Henderson-Hasselbalch equation,

$$pH = pKa + \log \frac{[A^-]}{[HA]} \therefore 8.0 = 7.0 + \log \frac{[A^-]}{[HA]}$$

$$\therefore \log \frac{[A^-]}{[HA]} = 8.0 - 7.0 = 1.0$$

thus the actual concentration ratio must be

$$\frac{[A^-]}{[HA]} = \text{Antilog } 1.0 = 10$$

ie if the pH is one unit above the pKa, the ratio of [A⁻]: [HA] is 10: 1 ie 10 out of every 11 molecules (91%) are in the form A⁻. Now explore for yourself how the ratio changes as the pH is changed.

$\prod$ Calculate the ratio of [A⁻]: [HA] if the pKa is 7.0 and the pH is 1) 6.0 2) 9.0.

1) From $pH = pKa + \log \frac{[A^-]}{[HA]}$ $\therefore \log \frac{[A^-]}{[HA]} = pH - pKa$

$\text{Log } \frac{[A^-]}{[HA]} \text{ now} = 6.0 - 7.0 = -1.0$

$\therefore \frac{[A^-]}{[HA]} = \text{Antilog } -1.0 = 10^{-1} \text{ or } 0.1 \text{ or } \frac{1}{10}$

the ratio [A⁻]: [HA] is now 1: 10 ie the protonated form predominates to the extent of 91%.

2) From $\log \frac{[A^-]}{[HA]} = pH - pKa = 9.0 - 7.0 = 2.0$

$\therefore \frac{[A^-]}{[HA]} = \text{Antilog } 2.0 = 100$

The ratio [A⁻] : [HA] is now 100: 1 ie approximately 99% is A⁻.

Notice from calculation 1) that the change in ionic state is symmetrical around the pKa value. When the pKa equals the pH, the ratio of [A⁻] to [HA] is 1:1.

Notice also that the ratio goes from 10:1 in favour of HA or 10:1 in favour of A⁻ in going from pH=pKa-1 to pH=pKa+1. Thus the vast majority of ionisation change (9% A⁻ to 91% A⁻) occurs over the pH range pKa ±1. This is of vital importance in understanding how the behaviour of biological molecules is affected by pH.

The calculation also shows that when the pH is 2 pH units away from the pKa, there is an insignificant amount of the less abundant form (eg when pH=pKa+2, then A⁻ = 99% and HA = 1%).

There is one other general rule which can be clearly seen. If pH is less than pKa, the protonated form (HA) predominates. If the pH is greater than the pKa, then the deprotonated form (A⁻) will predominate. This always holds true. It should not be a surprise! At low pH, [H⁺] is relatively high and, given the reversible reaction

$$HA \rightleftharpoons A^- + H^+$$

we would expect, as [H⁺] increases, that the equilibrium will be pushed towards the left (ie HA predominates). At high pH (low [H⁺]) dissociation towards A⁻ is to be expected.

We are now in a position to interpret titration curves of weak acids. These show the pH of a solution of a weak acid when base (sodium hydroxide) is progressively added. The figure below shows titration curves for acetic acid and for the ammonium ion. In the case of acetic acid, the pH is fairly low initially, rises rapidly at first, but then the rate of change of pH with addition of base slows.

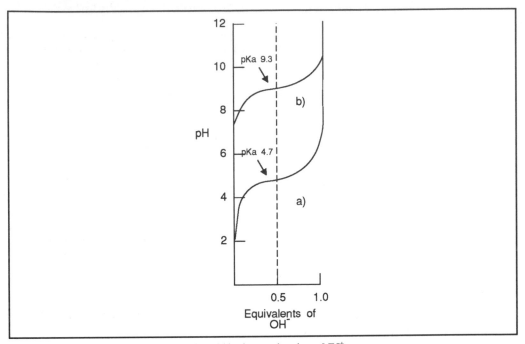

Titration curves for weak acids a) Acetic acid b) Ammonium ions, NH_4^+.

This part of the titration curve is centred on the pKa value of the ionising group and represents buffering action. Buffering action extends over the pH range pKa ±1 and is a consequence of the release of H^+ from the acid (ie HA). As OH- ions are added raising the pH, the acid partially deprotonates (because the pH has been raised). The released H^+ neutralises the added OH^- ions (ie $H^+ + OH^- \rightarrow H_2O$), thereby providing buffering action. This is the basis of all buffers. The ability to resist change in pH following addition of OH^- only extends to about 1 pH unit above the pKa value. Once this pH is reached (pH 5.7 for acetic acid, see figure), the pH rises rapidly.

buffers

Close inspection of the figure above shows that buffering ability is confined to the pH region in which a change in ionic status occurs. The ability to buffer totally depends on the deprotonation of weak acids and the ability of their conjugate bases to absorb protons

ie HA $\rightarrow$ A$^-$ + H$^+$, where H$^+$ can neutralise added OH$^-$

and A$^-$ + H$^+$ $\rightarrow$ HA, where A$^-$ can absorb added H$^+$. Thus, to act as a buffer resisting pH change caused by addition of either acid or base, both forms (HA and A$^-$) must be present.

buffering
capacity is in
the range pKa
±1

Deprotonation of ammonium ions proceeds in an exactly similar way to that of acetic acid but the titration curve is displaced on the pH axis. Essentially no change occurs in the NH_4^+ ions until pH 7-8 is reached, when deprotonation begins. Buffering (resisting change in pH) is confined to the pH range 8.3 to 10.3, which is centred around the pKa value (9.3). Ammonium ions/ammonia will not act as a buffer below about pH 8.3 or above pH 10.3. Acetic acid/acetate will not act as a buffer above pH 6 or below pH 4.

We have seen that the pKa is the pH at which a group is half deprotonated. Through the Henderson-Hasselbalch equation, we have found how the ratio between protonated and deprotonated forms is related to pH and pKa, and can be predicted from knowledge of the pKa. We have also seen that the ability to act as a buffer is related to ionisation and requires the presence of a weak acid and its conjugate base.

∏ Write down the pH range(s) over which L- alanine (pKa values 2.3 and 9.7) would act as a buffer. Why are 2 pKa values given for this amino acid?

The effective ranges for buffering by L- alanine are 1.3 - 3.3 and 8.7 - 10.7. Two pKa values are given because there are 2 groups which can ionise (the α-amino and α-carboxyl groups). Each of them deprotonates over a characteristic pH range and each has its own pKa value, which marks the pH at which dissociation is half completed. Buffering is associated with groups deprotonating/protonating; the effective buffer range of a single ionising group is pKa ±1. Within this range, the acid goes from representing 91% (at pKa -1) to 9% (at pKa +1) of the total HA + A⁻.

∏ Sketch the titration curve for a weak acid whose pKa is 7.0 onto the figure on the previous page.

The figure below shows the contents of the previous figure with the results for a weak acid of pKa 7.0 added.

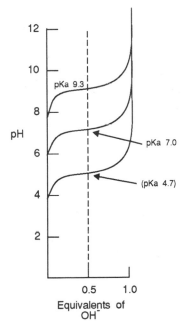

You should have a similar curve to that of acetic acid but displaced along the pH axis so that the centre of the plateau region is now at pH 7.0. Note that with the acetic acid, this was centred at pH 4.7. Remember that pKa represents the half-titration pH for any weak acid.

Ⅱ Which of the following compounds would be useful as a buffer at pH 7.0? Over what range of pH is it likely to be useful?

Compound	pKa value(s)
L-Threonine	2.1, 9.1
Imidazole	6.9
Diethanolamine	8.9

Your answer should have been imidazole and that the pH range is approximately 6-8. The key to buffering action is the ability to bind added protons (the generalised reaction is $A^- + H^+ \rightarrow HA$) and to release them so as to neutralise added hydroxyl ions (ie $HA \rightarrow A^- + H^+$; $H^+ + OH^- \rightarrow H_2O$). To do this both forms HA and A⁻ (ie acid and conjugate base) must be present simultaneously. The pKa gives the pH at which dissociation of any group is half completed, and thus [HA] = [A⁻]. Buffering is centred on the pKa and compounds only buffer in the vicinity of their pKa: in fact, as is explained in the text, buffer range is effectively pKa ±1 pH unit. Neither of the other two compounds will be any use as a buffer around pH 7.0. The fact that L- threonine has 2 pKa values is totally irrelevant.

If you are in any doubt about the argument used above, re-read the previous part of this Appendix. It is worth noting that you do not have to know what the structures of buffers are, just their pKa values, in order to make use of them! There will, of course, be circumstances when a particular compound, apparently suitable on pKa considerations, is none-the-less unsuitable (perhaps it inhibits the enzyme you are trying to study!).

Ⅱ Assume that you have prepared a 0.1 mol.l^{-1} solution of pyruvic acid ($CH_3.CO.COOH$), adjusted to pH 2.7 with NaOH. What is the precise concentration of the conjugate base of pyruvic acid at this pH, given that the pKa of pyruvic acid is 2.4.

Your calculation should have given 0.067 mol.l^{-1} as the answer. The conjugate base is the deprotonated form of pyruvic acid ie pyruvate ($CH_3.CO.COO^-$) and is formed as a result of deprotonation of pyruvic acid. The precise concentration is calculated as follows:

$$pH = pKa + \log \frac{[A^-]}{[HA]}$$

$$\therefore \log \frac{[A^-]}{[HA]} = pH - pKa = 2.7 - 2.4 = 0.3$$

$$\therefore = \text{antilog } 0.3 = 1.995$$

Thus ratio [A⁻] : [HA] is 2:1.

If total $[A^- + HA]$ is 0.1 mol.l^{-1},

then $[A^-] = \dfrac{2}{3} \times 0.1$ mol.l^{-1} = 0.067 mol.l^{-1}.

This section began with a question: 'what determines the extent to which the ionisation of a weak acid occurs?'. The preceding explanations should have shown that the extent of deprotonation is a consequence of the pH and the pKa of the group. Changes in ionic status follow a predictable relationship and are confined to pHs close to the pKa. It may also help to think of a pKa as a measure of affinity: of how tightly does a group tend to bind a proton. If you think about it, the higher the pKa, the stronger the affinity between conjugate base and proton; ie the group is still protonated despite the fact that the surrounding $[H^+]$ is low.

Index

A

absolute configuration, 18 , 134
acetylcholinesterase, 230
activation energy, 187 , 235
active site, 196 , 228
active site directed inhibitors, 235
adenine, 74
adipose tissue, 157
affinity chromatography, 68
agar, 149
agarose, 149
agarose gel, 122
alanine, 20
aldoses, 129
alginate, 149
allosteric activation, 242
allosteric enzymes, 238
allosteric properties, 61
allosteric regulation, 230 , 241
amino acid activation, 111
amino acids, 17 , 251
    chromogenic reactions, 37
    electrophoresis, 28
    identification, 37
    importance of, 35
    individual amino acids, 17
    ionisation, 24
    isoelectric point, 32
    separation, 34
    sidechains, 19
amino acyl-tRNA synthetase, 111
amino sugars, 147
aminoacylase, 251
AMP, 79
amphipathic, 159 , 173
ampicillin, 119
amylase, 145 , 147 , 251
amylopectin, 143
amylose, 143
anabolism, 7
Anfinsen, 62
anomers, 137
anti-codon, 110
anti-parallel strands, 88
antibiotic resistance, 119
antibodies, 53 , 121
apoenzyme, 233
archaebacteria, 4
arginine, 23
ascorbic acid, 139
asparagine, 21

aspartate transcarbamylase (ATCase), 243
aspartic acid, 22
assay of enzymes, 201
asymmetric carbon atom, 18 , 130
ATP, 72 , 80
autoradiography, 123

B

bacterial cell wall, 147
bacteriophage fx174, 104
ball and stick model, 132
basepairing, 88
beeswax, 166
bile acids, 162
binary fission, 7
biosynthesis, 7
blood clotting, 165
bond distortion, 237
bond lengths, 41
bond rotation, 41
bone calcification, 164
brain cells, 160

C

C-terminal amino acid, 42
calcium absorption, 164
carbohydrates, 128
carboxypeptidase A, 233
carotene, 163
carotenoids, 164
catabolism, 7
catalase, 190
catalyst, 186
cDNA, 116
CDP carriers, 81
cell - cell interaction, 171
cell theory, 2 , 4
cell walls, 11 , 147
cellobiose, 141
cells
    animal, 8
    eukaryotic, 4 , 8
    plant, 11
    prokaryotic, 4 , 5
cellulase, 147
cellulose, 146
cellulose fibres, 146
centrifugation, 66 , 99
centriole, 10
centrosome, 10
Chargaff's rules, 86 , 104
chloroplasts, 12